WITHDRAWN

P9-AZW-353

A CURE *for* DARKNESS

THE STORY OF DEPRESSION AND HOW WE TREAT IT

Alex Riley

SCRIBNER

New York London Toronto Sydney New Delhi

Scribner
An Imprint of Simon & Schuster, Inc.
1230 Avenue of the Americas
New York, NY 10020

First Scribner hardcover edition April 2021

Manufactured in the United States of America

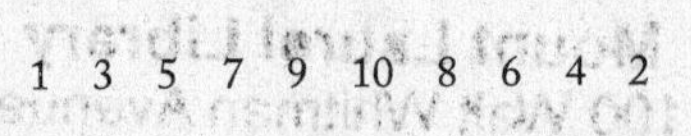

1 3 5 7 9 10 8 6 4 2

Library of Congress Cataloging-in-Publication Data has been applied for.

ISBN 978-1-5011-9877-9
ISBN 978-1-5011-9879-3 (ebook)

A portion of this book was previously published in a different form in "How a Wooden Bench in Zimbabwe Is Starting a Revolution in Mental Health" by Alex Riley in *Mosaic* in 2018, used under CC BY 4.0 (https://creativecommons.org/licenses/by/4.0/).

To Lucy

[W]hile we see sadness, unhappiness, and grief as inevitable in all societies we do not believe that this is true of clinical depression.

George Brown and Tirril Harris, *Social Origins of Depression*, 1978

Contents

Today, an estimated 322 million people around the world live with depression. It's the leading cause of disability, judged by how many years are "lost" to a disease, yet only a small percentage of people with the illness receive treatment that has been proven to help. An estimated 15 percent of people with untreated depression take their own lives.

Introduction

On a chilly December morning in 2019, I walked to my local doctor's office in South Bristol, just as gray clouds passed overhead to reveal a patchwork of clear blue sky. Unlike my previous appointments, some booked in an emergency, this one felt hopeful, like a milestone in my recovery. I told the doctor, a middle-aged woman with a kind smile who leaned forward in her chair as she listened, that I wanted to come off antidepressants. I had been taking sertraline—a selective serotonin reuptake inhibitor, or SSRI—every day for over two years and I wanted to see what life was like without it. Could I be rid of the side effects that had become so normalized that I had forgotten what life was like before? Would my energy levels be any different? My feeling of connection to others? My libido? SSRIs are known to take some of the intimacy of life away, and I wanted to be reunited.

The doctor asked me whether I was sure I was ready to come off these drugs. I told her that I was, adding that my partner, Lucy, was supportive of this decision. My bouts of depression had become so brief and infrequent that we hoped that I didn't need these drugs anymore. Although my thoughts still turned to suicide now and again, I felt confident that I could control them. The suicide plan that I had once sketched out didn't just seem like a distant memory, but the memory of a different person. In addition to antidepressants, I had been through two rounds of cognitive behavioral therapy (CBT) and seen therapists who practiced mindfulness and psychodynamic approaches. Consequently, I felt like I was better able to rationalize the extremities of negative thought that can make it feel like others would be happier, healthier, more content without me.

While therapists can come and go within a few weeks, antidepressants often need to be taken for years to keep depression at bay. Before sertraline, I had been on citalopram (another SSRI) for two years. Chemically accustomed to their effects, coming off these drugs can be a tortuous experience for many people. Withdrawal symptoms include dizziness, sweating, confusion, brain zaps, and—if recent anecdotal reports are confirmed—a heightened risk of suicide. That's why I met with my doctor and agreed to take things slowly, over months and not the two weeks that psychiatric guidelines once recommended. After four years of elevated levels of serotonin, I was introducing my brain to a new equilibrium.

At the same time, thousands of miles away in Wuhan, China, a novel coronavirus was spreading through its new home in the lungs of humans. Silent and unknown to science, this was the germination of a pandemic that would thrive on proximity, bring health care systems to their knees, and demand widespread quarantines in the general public. Had I known all this, I may have changed my decision to decrease my dose of antidepressants. After all, the consequences of a pandemic and the social triggers for depression overlap with frightening acuity. There's the death of loved ones. Unemployment and poverty. Major life transitions. Trauma. Divorce and domestic violence. All are known to precede episodes of depression. And all followed in the wake of COVID-19, a virus that could kill and debilitate, that led to some of the highest rates of unemployment since the Great Depression and forced billions of people to transition into a new and uncertain world. I reminded myself that anxiety and distress are natural responses to a global catastrophe. But if the depression returned—the crippling lack of motivation and the mental pain that can make suicide appealing—I knew I could just as easily increase my dose as decrease it.

My doctor recommended regular exercise and meditation to help with withdrawal symptoms. I had been researching this book for two and a half years by this point and knew that both can have potent antidepressant effects. Meditation is based on the positive acceptance of the present, leaving little space for the negative thoughts about the past and the future that often define depression. There is some evidence that it can even recalibrate the immune response. As low-grade chronic inflammation—the same process that underlies rheumatoid arthritis and Crohn's disease—is a common contributor to depressive symptoms, these moments of silent contemplation might be seen as a form of mental medication. As long as depressive thoughts aren't allowed to spiral and grow, meditation is an anti-inflammatory without side effects. Excepting the risk of injury and muscle tiredness, the same is true of exercise. But there is also a sense of mastery that comes with jogging, practicing yoga, or weight lifting. Both the psychological and physiological benefits of exercise are valuable parts of staying mentally healthy. The latest studies show that running three times a week, for example, is as effective in reducing depressive symptoms as first-line treatments such as SSRIs and cognitive behavioral therapy.

And so, with these studies in mind, I made sure to don my running shoes and jog to the woodlands and parks near our home in South Bristol. With our dog, Bernie, chasing squirrels through the undergrowth or quietly trundling along by my side, I felt my daily concerns start to fade away with every mile. My breathing slowed and felt effortless. My mind started to wander as my muscles flexed in rhythm. It was fluid. Meditative. I didn't carry my phone or own a smartwatch, so I don't know how far I traveled. But when I'd get home, a banana tasted like heaven and a cup of tea was pure indulgence. I felt a warm glow throughout my body that could last for the rest of the day. On other days, I made time to sit, cross my legs, breathe with my diaphragm, and let my thoughts flow through my mind as neutrinos passed imperceptibly through the earth. My favorite place to meditate is anywhere with trees. Listening to their leaves rustle, their boughs creak, reminds me of life outside of myself, wondrous products of evolution that barely move even in the fiercest winds. They even share nutrients with their neighbors through their entangled roots, just as we might reach out and offer someone a helping hand.

After three months of tapering my dose toward zero, I swallowed my last chunk of sertraline on March 6, 2020, a time when the number of deaths from the coronavirus were higher in Europe than in Asia. Based on sertraline's half-life, I knew that it might take a few days for the drug to be out of my system. I was expecting the worst. Indeed, by the second day, I felt fatigued, had the occasional cold shiver down my back, and felt so tired that I had to curl up in bed and sleep in the afternoon. When I woke up an hour or so later, I didn't feel any better. Sleep wasn't restorative. As I noted in my diary, "Today [Monday], feeling groggy, like my thoughts are flowing through treacle. Brain heaviness. Goose pimples, chills, flu-like symptoms? Confused, nauseous." I decided to go for a walk. With Bernie snuffling along next to me, I passed through city streets, boggy parks, woodlands, and across the Clifton Suspension Bridge, which was illuminated against the evening sky like a floating shelf lined with fairy lights. A fine rain filled the air. After an hour or so, I was drenched, but my head had cleared. I was returning to a life soon to be placed in lockdown.

While the world adapted to life behind a screen, I remained vigilant for any symptom or sign that I might be collapsing, just as I had done so many times before.

Like thick curly hair, mental illness runs in my family. Psychosis and mood disorders, the two major groups of psychiatric classification, can both be found in the last three generations. Rumors and realities of institutionalization have been passed down my maternal line. My mum remembers the day when my grandmother, Renee, was taken away to a mental institution in the 1970s. "That is one of my worst memories," she told me over the phone in August 2017. "We didn't know whether she was going to come back." Called High Royds, it was just one so-called asylum within a short van drive from their home in Thornton, a village just outside of Bradford in northwest England. But this memory is blurred in its details and may not have even happened. She was a child then and memories can be manufactured like nightmares. There's no doubt that Renee struggled with depression, especially after her husband, Eric, my grandfather, died of lung cancer in 1975. At a time when SSRIs like sertraline hadn't yet been put into prescriptions, she was given diazepam (Valium) to calm her anxiety and dampen her grief. When at her worst, she described her mental anguish as "tearing down the wallpaper." But no one else in the family remembers Renee being taken away for any period of time. The patient reports from High Royds, although incomplete with whole years missing, hold no trace of her.

I never met Eric or Renee. Lung cancer also claimed Renee's life, in April 1987, just a few months before my mum gave birth to her first child, my older brother. Renee had named the bump "Rupert" and, after she died, my brother couldn't be called anything else. His middle name is Eric. Her parents' passing left a dark shadow over my mum's future, one that she would never truly recover from. She pined for her parents as she became a parent herself. In her mind, it was the greatest tragedy that Renee and I were never able to meet. She always said that I was a lot like her. There was my love of nature. My studiousness and passion for science and literature. Today, I wonder whether we also suffered similarly.

Whatever our life experience, the main difference between Renee and myself is the treatments that were available to us. While she was prescribed a mild tranquilizer, I had the option of a range of SSRIs and evidence-based talk therapies such as cognitive behavioral therapy and interpersonal therapy. These two options—the biological therapies such as antidepressants and the psychology-based talk therapies—have been a central theme in the treatment of depression for decades.

With the image of Renee being taken away from her, my mum has always held a deep mistrust of psychiatric treatment. She has never reached out for professional help for her own mental struggles. Her own depression was a common feature of my childhood. There was the lack of sleep, the waking up at three in the morning and pacing downstairs. There was the drinking. The thoughts of being better off dead. Looking through old family photographs, I can see the hallmark signs that have followed me through my adulthood. In the rare photograph in which she appears—her long, dark, and curly hair like that of her girlhood crush Marc Bolan, the lead singer of the glam rock band T. Rex—she seems distant, as if there is a gulf between her and the child sitting on her lap. Her thick glasses frame a glazed expression. As a child, I often felt like there was an impenetrable gap between us. Today, I know that we share more than we care to imagine. I wish I could reach back into my childhood to tell her that it's okay to be struggling. She might not feel a connection to her children at times, but that was perfectly normal for someone who's been through the trauma that she has. After years with barely a word between us, we can now speak over the phone and I can think of all the things, despite her past, that she taught me. My cooking, my northern twang, the comfort of making do with what you have: I hold them all dear, and they all came from her.

My own experience with depression has provided a new perspective on my mum's past. Thinking of the years just before I was born, I now wonder whether her move from an apartment in the city of Bradford to an old farmhouse in the sticks of the Yorkshire Dales was her way of starting afresh, hoping to leave the past behind her. Moving with my dad—then a skinny, half-marathon-running, mullet-sporting Queen fan who wore tight white trousers—she helped transform a dilapidated house into a home. With rotten floorboards, moss growing in the dining room, and only one fire for heating and cooking, they were spartan beginnings. For two years, Dad tried his hand at raising and milking goats, ended up losing money, and returned to his former occupation in construction. Mum was a mobile hairdresser who would drive through twisting country lanes to reach her clients. (There was also a fish-and-chips van that similarly brought one of the norms of towns and cities into remote villages.) She was also skilled on the sewing machine, fashioning trousers for me and my brother out of odd bits of material and old curtains. For family weddings, she would make waistcoats and smart trousers that were strange combinations of tailored and ill-fitting.

It was a healthy upbringing, but there was a shadow following all of us during these times. Although it isn't infectious like a virus, depression thrives on proximity, traveling down familial attachments, especially from mother to child. The latest studies show that having a depressed parent increases the risk of depression in children threefold. This predisposition isn't destiny, however. Nature, as far as depression is concerned, is nearly always bound by nurture. Only in the presence of environmental triggers is this underlying risk activated.

In 2015, after years of mistrust, arguments, and silence, my parents separated after thirty years of marriage. It wasn't unexpected, but it shattered the life I once knew. The family home—the only home I knew—was put up for sale. I visited my parents separately in their unfamiliar rented accommodations. At the same time, I was transitioning into a new career. I had left my PhD at the University of Sheffield and was working as a researcher at the Natural History Museum in London. Using the latest CT-scanning technology, I studied the teeth and skeletons of sharks and rays, both living and extinct. There were aptly named cookie-cutter sharks, rays with whiplike tails, and deep-sea oddities known as chimeras or "ghost sharks" for their translucent skin. After a childhood of scrapbooks filled with dinosaurs and other megafauna from different continents and geological eras, this was a dream job. On my breaks I walked around the museum's maze of corridors and came to know each specimen like an old friend. Giant marine reptiles splayed out on a wall, stuffed lions and pandas with fur faded by time, and the skeletons of dinosaurs that I had learned the names of when I was so young that I couldn't yet tell time or ride a bike. A giant sloth stood at the entrance to our Earth Sciences Department.

After seven months at the museum, the first—and only—scientific paper I coauthored was published: "Early Development of Rostrum Saw-Teeth in a Fossil Ray Tests Classical Theories of the Evolution of Vertebrate Dentitions." Quite a mouthful. Peer-reviewed research is the bedrock of science, but I didn't think much of it. Although I had worked hard—writing sections of the study, analyzing the data, and drawing schematics of our theories—I didn't think my name should grace the paper. Any positivity in my life didn't seem to sink in. I wasn't worthy. Like the fossils that I pored over, I felt delicate and brittle, capable of breaking at the slightest drop.

A few months later, I was sitting in my supervisor's second-floor office, ancient fish fossils, cardboard boxes, and paleontological journals covering every surface. She was a prolific writer, a respected name in paleontology. I told her that I was struggling, that I couldn't continue. I began to cry. I thought I was a failure. We agreed that I needed to take some time off. I never returned to academia.

Unemployed and trying to make ends meet as a freelance writer (a hobby that I had started in 2012), my motivation waned and my interest in love and life ceased. My girlfriend told me that I was depressed and that I needed to talk to a doctor or a therapist. She couldn't support me anymore. We broke up soon afterward and, although I had seen it coming for months, I was devastated. Shortly after, the old farmhouse was sold. I felt unwanted, unloved, unmoored: drifting into a dark place that threatened to swallow me whole.

Ever since then, I have sought a deeper understanding of my own experience with depression. Where did our fixation on SSRIs and cognitive behavioral therapy come from? Are there alternatives that might be better suited to my symptoms? Rather than handing out antidepressants through trial and error, can we predict who will respond to a certain treatment before they are prescribed? Can depression be prevented before it arises? What novel treatments might be available in the future?

As I learned, my experience is not representative of depression as a whole. Just as I had once studied the diversity of the animal kingdom, I soon started to discover a similar diversity of depression, the many shapes and forms it can take, and treatments that I once thought to have been long dead and buried were transformed into modern miracles. What's more, I learned that the word *depression* is almost meaningless when seen from a global perspective. Other idioms of distress are better suited to this particular form of mental suffering. Whether it's known as heart pain or thinking too much, a person's language and culture need to be taken into account before talk therapies can be effective. While I personally sought reassurance that more effective treatments are on the horizon, I found that one of the most important missions for psychiatry today is expanding the reach of current therapies to people who have no access to mental health care. As well as invention, the treatment of depression depends on investment.

It is hardly surprising that a single diagnosis doesn't capture the reality of depression. It is a syndrome, a collection of different—but overlapping—mental states. Depression is a product of upbringing, trauma, financial uncertainty, loneliness, social bonds, diet, behavior, sedentary lifestyles, neurotransmitters, and genetics that cannot be encapsulated in a word. It can be mild or severe, recurrent or unremitting. It can emerge once and never appear again or it can cast a dark shadow throughout adulthood. Some people sleep too much, others suffer from insomnia. Some eat too much while others shrink toward starvation. Depression can emerge alongside cancer, heart disease, diabetes, and dementia and can make these diseases more lethal; it is a catalyst of mortality.

But it is also highly treatable. From antidepressants to talk therapies and electroconvulsive therapy, to some of the more novel treatments such as exercise, diet, and psychedelic substances, there are a range of options that can lead a large percentage of people back to a healthy state of mind. Current treatments are far from perfect and some remain unproven, but when faced with such a diverse disorder as depression, it's amazing that there are a range of options that are effective. "We should be aiming towards complete remission for everyone with depression," says Vikram Patel, professor of global mental health at Harvard Medical School. "The problem is that these treatments are seen as competing with each other, which is not the case." They work for different severities or symptoms of depression. Even the most debilitating forms of depression—mental disorders that come with a high risk of suicide and crippling delusions of guilt—can be treated within a few weeks. For people who have failed to respond to every type of treatment available, experimental methods such as deep brain stimulation have the potential to lift decades of suffering in seconds.

Another problem is that current treatments aren't always used correctly. Talk therapies are often restricted to four or five sessions that can be insufficient when dealing with the complexities of depression. Antidepressants aren't given at the right dose or for a long enough course. A healthy diet and exercise can be as effective as SSRIs and cognitive behavioral therapy, but aren't prescribed alongside these first-line treatments. And many psychiatrists are no longer trained to provide electroconvulsive therapy, a potentially lifesaving treatment that has been used for decades. History teaches us to learn from our mistakes, but it also reminds us not to leave our successes behind.

Although I feel comfortable talking about my experience with depression, there is still a lot of stigma and misinformation about the treatments that are available. If you see a therapist you are weak, unable to cope with stresses and strains of everyday life. If you take antidepressants you're a machine with broken parts. Even if electroconvulsive therapy saved your life you can't discuss—never mind celebrate—the reason for your recovery, for fear of being judged. All treatments have the potential to do more harm than good, but each one has to be seen as a balance between the potential benefits and the risk of side effects. Does the pain of depression or the chance of suicide make any side effects relatively unimportant? Treatments also have to be seen in the context of history; why were they initially developed, what other options were available, and how have they evolved over time? When seen through this temporal prism, stigma can vanish and treatments that were once seen as barbaric can be welcomed back into medical practice.

When I first started writing this book in the summer of 2017, I was torn about where it should start. Should I allow history to unfold from the dawn of medicine and move forward—century by century—into the modern age? While this would be a logical structure, I thought that it didn't capture the essence of the story I wanted to tell. Instead, before traveling back into ancient Greece and the booming civilizations of Mesopotamia, it seemed more appropriate to begin with two figures who have shaped modern psychiatry. Born three months apart in 1856, Emil Kraepelin and Sigmund Freud would come to embody two opposing fields in the treatment of depression. Was it a biological disease in need of physical treatments such as surgery, electrical stimulation, or drugs? Or was it an illness of environmental influences—upbringing, trauma, social bonds—that was in need of psychological therapy? From antidepressants to cognitive behavioral therapy, and from electroconvulsive therapy to psychoanalysis, the work of these two patriarchal figures in psychiatry flows through this history like two rivers feeding into the same ocean. Often kept separate, but occasionally crossing paths, the biological and the psychological themes of depression have formed the basis of my own treatment since I first walked into my doctor's office in South London in 2015, four years before I decided to withdraw from sertraline.

PART ONE

Cutting Steps into the Mountain

The day will come, where and when we know not, when every little piece of knowledge will be converted into power, and into therapeutic power.

Sigmund Freud (1915)

If a fright or despondency lasts for a long time, it is a melancholic affection.

Hippocrates

The time has gone by . . . when the unhappy insane could be cast into mismanaged Hospitals and, as too often is the case, left in jails and poor-houses, festering in heaps of filthy straw, chained to the walls of dark and dreary cells . . . Much has been done, but more, much more, remains to be accomplished.

Dorothea Dix (1848)

The Anatomists

Strolling through the port town of Trieste, nineteen-year-old Sigmund Freud passed women in elegant English dresses carrying small white dogs and smelling of patchouli. The wives of rich merchants, their fragrance and finery were in stark contrast to his study subjects inside the Trieste Zoological Station, a research institute that was "five seconds from the seashore." Wrested from the dark recesses of the Adriatic Sea—a thumb-shaped lobe of the Mediterranean between the western coast of Italy and the Balkan Peninsula—the young Freud spent his hours dissecting hundreds of slimy, stinking eels. Mouth agape, eyes glazed in the blank stare of death, and lacking the scales that define other species of fish, it was like slicing through a particularly long, and wholly unappetizing, sausage. A fresh-faced student at the University of Vienna, Freud had been given a project that had stumped many great anatomists since Aristotle: Do eels have testicles?

In 1874, a couple of years before Freud's student project, another researcher, Simon Syrski, also working in Trieste, had claimed to have found their highly sought-after testes. He called them small "lobed organs." Freud, already a confident and gifted researcher, wasn't convinced. In a letter to a childhood friend, he wrote, "Recently a Trieste zoologist claimed to have discovered the testicles, and hence the male eel, but since he apparently doesn't know what a microscope is, he failed to provide an accurate description of them." Freud certainly knew what a microscope was and, during his time on the Adriatic coast, honed his skills even further as he subjected a total of four hundred dead eels to intense scrutiny, his hands "stained from the white and red blood of the sea creatures." Later in life, Freud's intense gaze would become a trademark—along with circular spectacles, cigars, and a rug-covered chaise longue—that could penetrate deep into the soul of any interlocutor.

Sitting at his workbench with a view over the seashore, he couldn't penetrate the mystery of the eel testicles, however. The life of an eel was far more secretive and mysterious than he, or any other scientist, could have dreamed of. Although found in rivers and streams across Europe, *Anguilla anguilla* reproduce en masse somewhere in the Sargasso Sea, a three-thousand-mile journey that crosses the boundary of fresh water and salt water, the Mediterranean, and passes over the mid-Atlantic ridge. Only during the commute to this communal tryst do males develop their testes, just as deer and elk only grow antlers before their seasonal combat. For the rest of their lives, they don't need them or have them. Freud was looking for something that, at that specific time and place, didn't exist. Although he had failed in his task, Freud shouldn't have been too hard on himself. Even with modern science's ability to "see" the inner components of an atom (such as the Higgs boson) or detect the subtle vibrations from the collision of black holes billions of light-years away in space, the minutiae of eel reproduction are still yet to be uncovered. Although small larvae have been found in the Sargasso Sea, no one has observed their courtship rituals. Eel sex has held on to some of its most intimate secrets that have been millions of years in the making.

Disgruntled, disappointed, and questioning his future as an anatomist, Freud returned to the city of Vienna, the so-called Mecca for Medicine, and probably never looked at eels the same way again. They were a delicacy to some—delicious when baked in a pie. To him, however, they were nothing but a reminder of failure. Having excelled at school and university (often at the top of his class), learned seven languages in addition to his native German, and read Shakespeare from the age of eight, it must have been an unfamiliar feeling. But his time in Trieste wasn't a complete waste. Freud had honed his skills as an observer, a scientist: someone who reaches into the unknown and hopes to bring back something new.

The cosmopolitan city of Vienna was an exhilarating place to be for a young scientist. And nowhere was this more the case than in Ernst Brücke's physiology department. Before they were cleared away for lectures, tabletops were covered in devices for the measurement of physics and chemistry: "kymographs, a myograph, compasses, ophthalmometers, scales, air-pumps, induction apparatus, spirometers and gasometers formed the normal equipment of an institute in which all rooms were dominated by Galileo's command: to weigh all that was weighable and to measure all that was measurable," wrote the historian Erna Lesky in her book *The Vienna Medical School of the Nineteenth Century*. "[T]he work," she added, "even in deprived conditions, attracted students of all nationalities." People from Germany, Hungary, Russia, Romania, Slovakia, the Czech Republic, Greece, and even the United States came here to study, all speaking in different tongues and accents, but still communicating their latest discoveries in science.

Away from the lens of a microscope, however, Vienna was a relatively unpleasant place to live. For much of the nineteenth century, it was an overcrowded city with shortages in housing where sewage flowed uncovered. At high tide, the River Vienna—a tributary of the Danube that flows through the capital cities of Hungary, Serbia, and Romania—would overflow and push filthy water into the streets. Infectious diseases such as tuberculosis spread quickly and easily. Although an aqueduct had been built when Freud was a university student, the city was still in need of a good cleaning. The heavy traffic of horse-drawn carts and trams chipped away at the roads and curbs like stone mills through wheat, emitting a fine dust of granite into the air and residents' lungs. "In the streets, in the squares, in the public buildings, everywhere we can smell and breathe in quantities of dust, garbage, and discharged gas," one researcher at the University of Vienna wrote. While England had introduced a Nuisances Disposal Act as early as 1846, Vienna had no such progressive public health legislation.

Pollutants weren't just in the air and water: anti-Semitism cast a dark shadow over the crowded city. As one author later wrote, the discrimination against Jews was like "an evil-smelling vine that twined about the whole social structure of Vienna, choking so many green hopes to death." Freud had zero tolerance for anyone who might insult his Jewish heritage and it was a constant battle, one that made him want to pack up his things and leave town for good. He dreamed of England, the place where his stepbrother had settled when he was a child. He imagined that the discrimination wasn't as prevalent there as in Germany or Austria; England was a country where a Jew could walk down the street and not have his hat flung into the gutter.

Long before he created the field of psychoanalysis, Sigmund Freud's primary tool of observation was the tabletop microscope. Crafted from brass and steel with a horseshoe-shaped stand, the microscopes built by Edmund Hartnack in Paris were his particular favorite. With a mirror to reflect sunlight, a few corrugated knobs to focus the sample into view, and the all-important curved lens that looks down onto the microscope's stage like a telescope points toward the stars, anatomists like Freud could zoom into previously unseen worlds. It was a cornerstone of scientific inquiry—and of psychiatry—in the late nineteenth century. Freud spent most of his waking hours looking at slices of life blown up to three hundred or five hundred times their normal size. After his time in Trieste, he worked in Brücke's physiology laboratory on the outskirts of Vienna. It was a cramped room in a building once used as a rifle factory and, before that, a stable. It was here that Freud moved from eel testicles to the nervous system of fish, crustaceans, and humans.

It sometimes felt like he was looking back in time. The delicate nerves of these animals were like spindly road maps into our ancient ancestry. Whether he was peering down at the stained nerve cells of river crayfish, sea lampreys, or brain samples from human cadavers, they all showed a remarkable consistency in form. "[N]othing else than spinal ganglion cells," Freud laconically wrote of his fish samples. Although different shapes and sizes, they shared ancient anatomical threads, and he could see them when he looked closely enough.

Next to his bronze microscope were sheets of paper and a sharp pencil. At this time before microphotography, anatomical drawings were the only means of conveying what one saw down the microscope, providing the basis of science communication between peers and colleagues. Freud was particularly talented at translating his observations. In his study of crayfish, for example, one writer admiringly states that "the cell bodies [are] shaded so carefully that they appear three-dimensional, alive, alien eyeballs bobbing in space." Art, science, and a sense of awe merged on the page in front of him.

Keen observation, detailed translation, patience: these were qualities that would remain with Freud through the twists and turns of a long career. Another was revealing the unseen, making the invisible visible.

Nerves, the individual units of nervous systems, are delicate and often invisible to the human eye. Not only are the majority unimaginably small, but they are also transparent. Freud developed his own method of tissue staining—known as his "golden stain" for its use of gold chloride not its color—that not only kept the delicate nerves intact but made them observable under the microscope. "[B]y this method," Freud wrote, "the fibres are made to show in pink, deep purple, blue or even black, and are brought distinctly into view. . . . [I]t is believed that it will prove of great service in the study of nerve tracts." After viewing Freud's tissue slices under a microscope, Brücke, a balding man with wispy whiskers over his pale cheeks, confessed, "I see your methods alone will make you famous yet."

Despite his hopes for its dissemination and success, Freud's golden stain didn't make it far beyond the walls of the laboratory in Vienna. Ten years previous, Camillo Golgi, an Italian anatomist and future Nobel laureate, had developed a silver-based stain that was far simpler and, therefore, more popular. Freud had come so close to a scientific discovery that he could almost reach out and touch it. Unfortunately, it wouldn't be the last time. This story—with varying details—would be a leitmotif of Freud's early career. In 1884, for example, he came excruciatingly close to the theory that nerves weren't interconnected like a huge net within the body, but were separated by tiny gaps. As he wrote, "[T]he nerve cell becomes the 'beginning' of all those nerve fibres anatomically connected with it," and then "the nerve as a *unit* conducts the excitation, and so on." A few years later, Santiago Ramón y Cajal, a neuroanatomist in Barcelona, would publish his foundational theory that stated each nerve was a separate unit—a neuron—separated by a space called the synapse, thus helping to usher in a new era of neuroscience. Both Camillo Golgi and Ramón y Cajal would share a Nobel Prize for their research in 1906, "in recognition of their work on the structure of the nervous system." In his publications, Cajal cited Freud's work.

Fame often comes with fortune and Freud desperately needed some money. Ever since he was a child growing up in Leopoldstadt, the northeastern district of Vienna that attracted Jewish immigrants from across Europe, he was accustomed to an impecunious way of life. As a young adult, however, he had grown tired of not knowing whether his parents and his seven younger siblings were going to go hungry from one day to the next. He sent them some cash when he could afford to and splurged any remainder on his two luxuries in life: books and cigars. He regularly borrowed money from his colleague, friend, and member of the Viennese aristocracy, Josef Breuer, someone who was, thankfully, willing to support him when he could. With each gulden spent, Freud sunk deeper into debt. As he once wrote in his characteristically sarcastic manner, "It increases my self-respect to see how much I am worth to anyone." Even with Breuer's charity, the stains on young Freud's suit remained unchallenged and his diet of corned beef, cheese, and bread unchanged. He sought a scientific discovery not only for his personal curiosity as a scientist but for the money it might bring him.

Freud's frustrations with money reached a tipping point in 1882, the year he turned twenty-six years old and fell in love with Martha Bernays. Two months after their first meeting in his family home, Freud and Martha, a close friend of his sisters, were engaged. But he couldn't afford a wedding. He was far too deep in debt. Miserably, he couldn't even afford the two-day train journey from Vienna to Wandsbek in Hamburg, the township that Martha had recently moved to with her mother. The first four years of their relationship unfolded underneath their pens. In a world where telephones weren't yet a regular feature in the home, the young couple wrote letters to each other nearly every day. Freud used scraps of paper, pages torn from his lab notebook, and old envelopes to express his frustrations over not being with his love and having the chance to kiss. Penniless, all he had to offer was his devotion. "Do you think you can continue to love me if things go on like this for years: I buried in work and struggling for elusive success, and you lonely and far away?" Freud wrote. "I think you will have to, Marty, and in return I will love you very much."

Together with his financial struggles, his long work hours, and the regular undercurrent of anti-Semitism in Vienna, Freud's isolation from his fiancée certainly took its toll on his health. He was frequently struck with bouts of fatigue, irritability, and a crippling low mood. These spells would appear out of the blue and leave him miserable and unable to work. "I have been so caught up in myself, and then I have days on end—they invariably follow one another, it is like a recurring sickness—when my spirits decline for no apparent reason and I tend to get exasperated at the slightest provocation," he wrote to Martha in January 1884, describing a common experience of depression. "I just can't stop worrying today." Lovestruck, he thought that his only cure was to be with Martha. "I think I can feel happy nowadays only in your presence," he wrote, "and I cannot imagine that I shall ever be able to enjoy myself again without you."

In the nineteenth century, the modern view of depression was taking shape. From the 1860s onward, the idea of a "mental depression" was being forged into a stand-alone diagnosis. Before this, there was a diversity of interchangeable terms—*vapors*, *spleen*, *hypochondriasis*—to describe depressions without delusions of guilt or sin, the kinds that might be seen in the community and not in mental institutions. And then there was *melancholia*, a word that had initially been used to describe a state of prolonged fright and dejection that was centered around a single delusion. People thought they were dying of an unknown disease, had committed some unforgivable sin, or that the world was about to fall on top of them. But in the wake of the Enlightenment period in Europe, the term *melancholia* was also associated with the idea of poetic insight, of men who were gifted in the arts and sciences for their ruminative pastimes. To be melancholy was to be gifted and damned. *Melancholia* became a vague term and was almost lost from medical parlance.

By the nineteenth century, there were numerous attempts to replace the term entirely. "The word melancholia," the famed French psychiatrist Jean-Étienne-Dominique Esquirol wrote in the early nineteenth century, "consecrated in popular language to describe the habitual state of sadness affecting some individuals should be left to poets and moralists whose loose expression is not subject to the strictures of medical terminology." He proposed a new term: *lypemania*, "a disease of the brain characterised by delusions which are chronic and fixed on specific topics, absence of fever, and sadness which is often debilitating and overwhelming." While it became quite popular in some French institutions, *lypemania* didn't catch on in other European countries and *melancholia* remained as current as ever. In the United States, Benjamin Rush, the so-called father of American psychiatry and signatory of the Declaration of Independence, suggested the diagnosis of *tristimania* (meaning "sad" and "madness"), but even his impressive accolades and political clout couldn't budge *melancholia* from its perch.

While *melancholia* remained a common description of severe depression in mental institutions, the idea of "nerves" was sweeping through the Western world, blurring the lines between the body and mind. Someone might be anxious, depressed, fatigued. They might be suffering from an inexplicable paralysis of their arm or leg. Their hands might shake or their sense of smell or sight might temporarily disappear. Everything that was unusual could be explained by a fault in their nervous system, the electrical circuitry of life.

Freud's low moods and irritability were seen as a mental byproduct of his nervous system. It was constitutional. But, like any physical ailment, it could also be exacerbated by a person's environment. The fast-paced modern lifestyle, the need to compete in the growing urban centers around the world, was thought to be the main trigger for nervous disorders in men. Freud, and thousands of people like him, felt worn-down and exhausted from within, as if the rush of modernity was abrasive to the delicate innards of the body. For women, meanwhile, caring for sick relatives or educating themselves in matters for which they were supposedly unsuited—science, politics, economics—was at the core of this illness.

The tendency to link emotional distress with the nervous system was particularly common in the middle and upper classes of Europe and the United States. Since it handily replaced looser diagnoses like insanity and madness, it made a visit to mental asylums—historically, places of chains, mistreatment, and of disgraced family names—unlikely. Those were places for the melancholic, the raving, and the poor. But with a diagnosis of an unknown nervous disease, people with depression could be seen by a doctor, just like someone who had the flu or a broken leg. Noticing this, the directors and doctors of mental asylums started to rebrand their institutions to keep pace with this surge in socially acceptable patients. Places of insanity became hospitals of "nerves" and "nervous diseases." One asylum in Bendorf, western Germany, for example, changed from a "Private-Institution for the Insane and Idiots" to a "Private Institution for Brain and Nervous Disease."

Although there was no evidence that nervous diseases existed, and no microscope powerful enough to bring such maladies into view, this idea still influenced how depression was treated in the nineteenth century. Freud, for example, spent his time resting, indulged in warm baths, and played chess to give his strained nerves time to strengthen. In the United States, such a restful approach was medicalized to extremes.

The so-called rest cure was an unusual mix of spa treatment and torture. Created in Philadelphia by the neurologist and former wartime surgeon Weir Mitchell, it was as excruciating as doing nothing whatsoever. "At first, and in some cases for four or five weeks, I do not permit the patient to sit up or to sew or write or read, or to use the hands in any active way except to clean the teeth," Mitchell wrote. "I arrange to have the bowels and water passed while lying down, and the patient is lifted onto a lounge for an hour in the morning and again at bedtime, and then lifted back again into the newly-made bed." Coddled beyond compare, the patients were separated from their family and friends, and even their mail was carefully checked and censored. In this therapeutic seclusion, they were fed and cleaned as if they were infants, but they were also subject to strict punishments if they did not acquiesce. Forced feedings, rectal enemas, and even lashings were used. Meals consisted of mutton chops, pints of milk, dozens of eggs, and Mitchell's own recipe for raw beef soup laced with a few drops of hydrochloric acid (a strong digestive solution). With greater reserves of fat, fragile patients were thought to be bolstered against the damaging effects of their weak nervous systems. To reduce muscle wastage, a mild electric current was applied all over the body. Regular massages helped encourage blood flow.

In the 1880s, Charlotte Perkins Gilman, an author and social activist, wrote that her treatment made her spiral "so near the borderline of utter mental ruin that I could see over." Her experience inspired "The Yellow Wallpaper," a harrowing short story published in 1892 in which the female protagonist gradually becomes psychotic and delusional within the tight confines of her nursery room, often seeing a woman trapped behind prison bars that lie underneath the foul-smelling, off-yellow-colored wallpaper. A modern classic that was largely ignored at the time, it is a chilling reflection of Perkins's own experience, and where the rest cure nearly led her.

Virginia Woolf, the British novelist, was equally damning about her rest cure at the fashionable Harley Street doctors in Central London. "I have never spent such a wretched eight months in my life," she wrote to a friend in 1904. Her husband, Leonard Woolf, meanwhile, wrote that her "doctors had not the slightest idea of the nature of the cause of her mental state. . . . [A]ll they could say was that she was suffering from neurasthenia and that, if she could be induced or compelled to rest and eat and if she could be prevented from committing suicide, she would recover." The founder of Hull House (an educational community center for women in northern Chicago) and future Nobel laureate, Jane Addams, not only criticized the rest cure but questioned its theoretical basis. Her depression wasn't caused by a faulty nervous system, she wrote. And, even if it was a nervous disease, it certainly didn't stem from her literary interests or her pursuit of gender equality. "She credited her own depression to feeling that she lacked a goal and an appropriate vehicle with which to confront a harsh, unsympathetic—masculine—world," wrote Susan Poirier, a medical historian at the University of Illinois at Chicago. "What she needed [was] a means of humanitarian action in the political, social, and moral mainstream of the world." Exactly the opposite, that is, to the domestic "re-education" that depressed women were being subjected to.

Meanwhile, inspired by the rest cure, the "west cure" was basically a ranching holiday in the badlands of the western states of America. The artist Thomas Eakins, the poet Walt Whitman, and a future president of the United States, Theodore Roosevelt, were some of the more famous recipients of this masculine getaway. As they herded cattle, rode horses, and slept under the stars at night, they thought it was the best time of their lives.

As with Sigmund Freud in the physiology laboratory of Ernst Brücke, Emil Kraepelin received a standard introduction into anatomical research in the nineteenth century. Working under the genial director of the mental asylum in Munich, Bernhard von Gudden, he was told that every facet of insanity was rooted in some physical aberration in brain tissue. The only window into their nature and potential diversity, therefore, was through the lens of a microscope. "[Von Gudden] thought that the only entrance into the psychiatric labyrinth could be found by anatomical dissection, and by penetrating into all the fine details of the brain's constructions," Kraepelin wrote. Not only did von Gudden advocate brain anatomy over all else, he invented one of the major tools that made it possible: a coffee table–like machine that would become known as the "von Gudden microtome." After placing the brain sample in a hollow tube in the center of the table, the researcher guided a blade neatly over its upper surface, tightening a screw underneath the table to push the specimen upward before each cut. It was not too dissimilar to a butcher slicing through a chunk of cooked ham.

The ability to neatly slice through a squishy specimen such as the brain was a breakthrough. "[I]n eight days almost an entire rabbit brain can be sectioned and beautifully stained," von Gudden wrote, "something that twenty years ago I would have not considered possible; our knowledge will now be potentiated." While other students were given birds, moles, or fish to understand the basics of brain anatomy, Kraepelin was handed the brains of reptiles. Could these simplified models help reveal some of the fundamental elements of our own anatomy? Kraepelin would never know. He didn't care how quickly or finely someone could slice up a brain like salami, or how beautiful the staining techniques were; he did not believe these methods would reveal the root causes of mental disorders in his lifetime. He didn't publish any studies on neuropathology, not one paper on reptile brains.

Although he lacked enthusiasm for this particular project, Kraepelin didn't disagree with his supervisor. He also believed that mental disorders were a psychological response to changes in brain tissue. Like von Gudden and the majority of nineteenth-century psychiatrists, he was a biologist at heart. He just thought that this scientific pursuit was a little premature in an era when microscopes were still rudimentary and a microtome table was seen as a sensation. "With the means available," Kraepelin later wrote, "it was almost completely impossible to identify changes caused by disease processes." Plus, he had poor eyesight, and looking down a microscope to tease apart subtle differences in brain matter was unlikely to be fruitful.

Moving from Munich to the University of Leipzig, from the south of Germany to the north, Kraepelin discovered a different way to reveal hidden mental processes. In 1881, sitting at his wooden desk in the experimental laboratory of the famed psychologist Wilhelm Wundt, he started drinking, injecting, and inhaling different drugs, hoping that they would each provide a new view into the mind. "In spite of the Spartan simplicity of the furnishings an industrious scientific life and great enthusiasm . . . reigned in these rooms," Kraepelin wrote.

A short, stocky man with a yellowish complexion and a thick Vandyke beard, Kraepelin's primary tool was alcohol. Diluting laboratory-grade ethanol into water, he made foul-tasting cocktails of precise strengths and drank every drop. Although he added a dollop of raspberry syrup to the cocktail, watching as the clear liquid swirled into pink, there was no masking the burning sensation of pure ethanol. As bad as it tasted, Kraepelin was excited to start his latest experiment. With the help of a colleague, he tested his reaction times, his short-term memory, and the speed at which he could read basic German literature, each result recorded by a wooden, clocklike timer on the tabletop that could stop time to the millisecond.

It didn't take long for his response times to lengthen. He became more forgetful. Reading took longer and he struggled to keep still. He was getting drunk—intoxicated for science.

In total, Kraepelin managed to persuade fourteen volunteers to drink the same awful mixture, but in slightly different concentrations. Some were more alcoholic than others. With these simple variables, Kraepelin concluded that "higher doses caused an earlier onset and more extended occurrence of impairment." It was a simple equation: more booze equals more drunk, for longer. This had been known for centuries, ever since early civilizations had discovered the wonders of fermentation, but experiments like these had never been conducted before. As an attempt to isolate the effect of alcohol from the basic action of drinking, Kraepelin had also given some volunteers carbonated water as a control. He was using an early equivalent of a placebo. No one had created such a standardized method to quantify and, importantly, compare inebriation.

His work wasn't limited to alcohol. Using the same measures of response times, Kraepelin tested the caffeine in tea, tobacco in cigarettes, and more medicinal drugs such as morphine and ethyl ether, a potent tranquilizer. Without exception, he was his own first test subject. Only after his own "safety" test did he request help from volunteers. All but one drug passed the test. An injection of morphine, he found, was both unpleasant and potentially dangerous: "I personally tolerated it so badly that proper experimenting was completely impossible in my case. And the danger of developing morphinism [addiction to morphine] precluded me from persuading other people." Tea, on the other hand, was a much more pleasant experience for everyone involved. In this experiment, rather than slowing his volunteers' reactions, like alcohol, the caffeine in the tea heightened them. "There was a clear and longer lasting facilitation of sensory and intellectual processes," wrote Kraepelin. People were focused, more responsive. Plus, too much tea only led to an edgy restlessness and a headache.

Basic in formula, Kraepelin's work held enormous potential. With such systematic and accurate observations, he hoped that his research would begin to align psychology—the study of how the mind works—with the serious and respected disciplines of the natural sciences, such as chemistry and physics. They were each rooted in measurements, quantification, theories that could be tested in the laboratory with experiments. Just as chemistry was built on concentrations, catalysts, and changes over time, Kraepelin wanted to change how the mind was studied. By introducing healthy volunteers to a single variable—alcohol, for example—he sought to understand how everyday consciousness responds to a change in the environment.

Working at a time when German was the language of science, he had high hopes for success. "We felt that we were trailblazers entering virgin territory," Kraepelin wrote about himself and his supervisor, Wundt, "creators of a science with undreamt-of prospects." He named this new science "pharmacopsychology," a definition that reflected his focus on how drugs (pharmacology) could influence the mind (psychology). His work was one of the earliest roots of Big Pharma and the drugs—such as antidepressants and tranquilizers—that revolutionized mental health care in the wake of World War II. Although the methods have changed over the decades, the same principles are to be found in Kraepelin's experiments and contemporary pharmacological research: placebo controls, precise concentrations (doses) of drugs, and the repeated observations of behavior changes over time.

Although his experiments were certainly scientific, Kraepelin knew that they fell well short of a robust science. His sample sizes were too small. Just over a dozen people drinking alcoholic cocktails isn't going to be representative of everyone. Age, sex, and current drinking habits could all influence the end result. To account for such inevitable variability, large sample sizes were needed. Although Kraepelin would continue to add to his data sets for the next ten years—working from three different cities in two countries—he never felt like he achieved his initial hopes for his new science. Indeed, his supervisor, Wundt, didn't even mention his experiments in a comprehensive review of his laboratory's contributions to psychology, written in 1910. This part of his life—these primordial stages of psychopharmacology—is considered insignificant compared to his major contribution to the study of the mind and mental disorders. Kraepelin found his true calling when he moved from experimental psychology to the bedside of mentally ill patients. With the same penchant for precision and systematic observation, he would start to create a grand classification system on which the field of modern psychiatry would be built.

The origins of this work began during the Easter holidays of 1883. Over three weeks, Kraepelin sat down to compile a textbook on the current state of psychiatry. Though he was hesitant to write anything at first (it was Wundt's idea), his *Compendium of Psychiatry* provided a creative outlet for his experimental observations and, he hoped, would bring in some much-needed money for a young researcher. In its three hundred pages, Kraepelin expressed his frustration with the current state of psychiatry, how a student such as himself could become so quickly dissatisfied with the anatomical approach and the embarrassing lack of effective treatments. In Munich and elsewhere, he felt helpless when treating the "confusing mass" of mental illness that he saw daily. Liters of alcohol were used as both a stimulant and a sedative, a balm for the manic and the depressed. Over the years, Kraepelin's *Compendium* would grow in size and popularity, becoming the go-to guide of clinical psychiatry for generations to come. With flowing prose and case studies that sometimes read like science fiction, he slowly revealed the alien worlds of the mentally ill.

Über Coca

Sigmund Freud sat in an overcrowded laboratory, looking down at his latest staining experiment, and mused over his own weakened nervous system. If only he could see what a weakened nervous system looked like, understand its biology in detail. Would he then be able to find a way to make it stronger? Might he reverse his nervous spells? In 1884, he believed he had found a solution: chemistry. Freud read that a military doctor had given his troop of fatigued soldiers a new drug that quickly brought their stamina back. A white powder that dissolved easily into water, this drug was called cocaine. "A German has tested this stuff on soldiers and reported that it has really rendered them strong and capable of endurance," Freud wrote to Martha in April 1884. "Perhaps others are working at it; perhaps nothing will come of it. But I shall certainly try it, and you know that when one perseveres, sooner or later one succeeds. We do not need more than one such lucky hit to be able to think of setting up house."

As he moved from Brücke's laboratory to the adjacent medical hospital, from tissue specimens to sick patients, Freud saw cocaine as a significant side project, one that had the potential to bring him fame and fortune while also easing his mental troubles. If it could increase the endurance of soldiers, could it reverse his fatigue and low mood? If his name was associated with a medical discovery, would that mean regular royalties and scientific renown? Although it is well known to be a dangerous and addictive drug today, cocaine was a relatively new addition to the pharmaceutical arsenal of the nineteenth century. Not much was known about it other than it had some stimulating qualities. Many pharmacologists thought that it was very similar to the caffeine in tea.

Purchasing his first batch of the drug from the German pharmaceutical company Merck, Freud dissolved one part cocaine to ten parts water and gulped the viscous solution down. The first sensation was the taste: floral and not at all unpleasant. His lips then felt furry and numb. A sensation of heat spread through his skull just as a shot of whiskey warms the heart. Light-headed and dizzy, he felt a rush of energy, a feeling of activity and purpose. It was stimulating, refreshing, and restorative. "In my last severe depression I took coca again and a small dose lifted me to the heights in a wonderful fashion," Freud wrote to Martha on June 2, 1884. "I am just now busy collecting the literature for a song of praise to this magical substance." Given access to a colleague's well-stocked library in Vienna, Freud read every book he could on the topic of cocaine, from its historical significance in coca leaves to its burgeoning use in Western medicine.

A month later, rain was pouring outside Freud's bedroom window as his comprehensive review, "Über Coca," was published. Over two dozen pages, he wrote with excitement and adoration about this "treasure of the natives." The leaves of *E. coca*, Freud noted, had been chewed by Native Americans for centuries. According to Peruvian mythology, the coca leaf was a gift from the sun god to the rising Inca Empire. Generations of people in South and Central America were strengthened against long walks in the pursuit of prey, fruiting trees, or new territory. With coca in the mouth, mountains became molehills. Days without food generated no pangs of hunger. Even in death, coca leaves were placed in the mouth to help the deceased make the journey into the afterlife. "It was inevitable that a plant which had achieved such a reputation for marvellous effects in its country of origin should have been used to treat the most varied disorders and illnesses of the human body," Freud wrote.

In 1860, a young graduate student working in the university town of Göttingen isolated a single molecule from the oval-shaped leaves of *E. coca*. Cocaine was just one among many compounds that might have held some medicinal properties of the whole leaf, but it was the one with the most obvious effects. In the blink of an eye, thousands of years of cultural traditions were boiled down to a crystalline white powder.

In addition to exhaustion and fatigue, Freud noted that cocaine had been used in the treatment of digestive disorders, addiction to alcohol and morphine, asthma, diabetes, and was also used as an aphrodisiac. As the pharmaceutical company Parke, Davis & Co. reported: "An enumeration of the diseases in which coca and cocaine have been found of service would include a category of almost all the maladies the flesh is heir to." In addition to Freud's own limited experience, people with severe melancholia were shown to benefit from this new drug. Patients who had been unresponsive for weeks or months, held down in a depressive stupor, started talking again. Jerome Bauduy, a professor of diseases of the mind at Missouri Medical College, said that he "frequently witnessed the morose, silent, taciturn, patient, a prey to the most profound grief or sadness, recover his normal self, begin to talk about his case and wonder how he could have ever experienced such gloomy ideas." Four minutes after a single injection, a seventeen-year-old boy referred to as "W.H." from St. Vincent's Hospital in St. Louis, Missouri, started to openly discuss his troubles with his doctor. "Cocaine changed the scene as if by magic," one witness wrote.

Freud thought that such "magic" should be placed firmly inside pharmaceutical cabinets around the world. Nearly a year after his first taste of cocaine, he had a rare opportunity to make this into a reality. Smartly dressed, with his beard trimmed and hair slicked into a perfect parting, he stood before a meeting of the Vienna Psychiatric Society. Some of the best medical minds in the city—perhaps in the whole of Europe—were there, watching. Eloquent but exposed, Freud was like an insignificant specimen placed underneath the lens of his tabletop microscope. He told his audience about cocaine, a substance that, he thought, they urgently needed. "Psychiatry is rich in drugs that can subdue over-stimulated nervous activity," he said, "but deficient in agents that can heighten the performance of the depressed nervous system. It is natural, therefore, that we should think of making use of the effects of cocaine." His underlying message was simple, if a little muddied by medical jargon: there are sedatives to calm the manic or violent patient, but nothing for those who were depressed. Although the term wouldn't be used for another seventy years or so, Freud was offering psychiatry the first antidepressant.

A popular and oft-used portrait of Freud shows a besuited, gray-haired man with one arm held akimbo while his free hand cradles a thick brown cigar (smoking was a lifelong vice, one that would eventually kill him). His thick white beard surrounds a straight-lipped mouth, while his famously dark eyes seem to peer out of the photo, out of history, and deep into the thoughts and feelings of whoever meets his gaze. He is an unmistakably acute observer of the human condition. Although it is a static image, a single frame from one of the many photo shoots his son-in-law provided over his lifetime, it nonetheless captures the confidence and nonchalant demeanor of a paternal figure of psychiatry. Along with a pair of circular glasses and a rug-covered chaise longue, this is the image of Freud that would become instantly recognizable as the creator of psychoanalysis, a field of dreams, repressed memories, and the unconscious.

The name Sigmund Freud is so closely aligned with psychoanalysis—like Santa Claus and Christmas—that it is easy to forget that he was a biologist for much of his early adulthood. During his years at the University of Vienna and his medical training at Vienna General Hospital, there was little if any discussion of the unconscious, no interpretation of dreams, and the Oedipus complex—the theory that young boys are sexually attracted to their mothers and repress their innate desire to kill, and usurp, their fathers—would have sounded bizarre even to Freud himself. He published his first article on psychoanalysis in 1896 when he was forty years old, and only became widely known for this work twenty or so years later when he was in his sixties.

There wasn't a neat division between one career path and the other. There was no T-junction at which Freud turned away from medical science and toward psychoanalysis. Like most things in life, it was a gradual process dictated by chance encounters. His first step away from anatomy and cocaine took place outside of Vienna, a city obsessed with brain slices, in 1885, the year he was awarded a research grant to study in Paris. It was probably the most important break of his life, a trip that would open a whole new world of opportunity in the treatment of psychiatric diseases.

It was a far cry from the physical-chemical teachings of his former tutor, Ernst Brücke. In Paris, doctors studied and employed hypnosis, a mesmeric method of therapy that was closer to magic than the microscopes other psychiatrists attached themselves to. Drinking cocaine to calm his insecurities over his rudimentary French, Freud was placed under the wing of Jean-Martin Charcot, a man whom he later described as "one of the greatest physicians" he knew and a "genius." "[He] simply uproots my views and intentions," Freud wrote. "After some lectures I walk away as from Notre Dame, with a new perception of perfection." Charcot was a solitary figure in the increasingly international world of medical science, one that Freud would soon emulate. His isolation was largely due to his theories on hysteria, a mysterious disorder of convulsions, speech impediments, paralysis, euphoria, hallucinations, and a dizzying array of other symptoms that seemed to defy any logical explanation. The only thing that had been considered gospel was that it was a disorder of women. *Hysteria* comes from the Greek *hystera*, meaning "uterus," a hangover from the days when it was thought that this female organ could move around the body like a homegrown parasite, making mischief wherever it settled. First, an arm might stop working. Then speech degrades. Finally, the patient starts seeing things that aren't there.

Confident, thick-skinned, and with a certain Parisian nonchalance, Charcot had the audacity to suggest that men, absent of a uterus, could suffer from hysteria. (He wasn't the first doctor to propose this; the French physician and psychologist Paul Briquet wrote about male hysteria in 1859.) Charcot moved between physical and psychical realms of medicine—between the brain and mind—at the

Salpêtrière Hospital, causing his own mischief as he did so. For six weeks, Freud worked in his pathological laboratory, using his adept microscopy skills to study cerebral paralysis and aphasia (a loss or impairment in speech) in children. But it was his observations on hypnosis and the treatment of hysterical patients that made the biggest impression. As Charcot soothed some of his patients into a trancelike state, he could introduce an idea that would become fixed into their bodies after they came to. If he explained that their arm was paralyzed, for example, his patient would "wake" up and make this into a reality. Their otherwise healthy arm would be paralyzed. It was as if, through hypnosis, Charcot had reached into a part of their minds that was usually locked away, a mental safe house of some sort, and managed to place a rogue thought inside before the gates closed with the return of consciousness. Freud was stunned. "I had the profoundest impression of the possibility that there could be powerful mental processes which nevertheless remained hidden from the consciousness of man," he wrote. This was known as the unconscious, a place where repressed thoughts, traumatic memories, and the ancestral urges of our species were thought to play out behind bars, a psychical labyrinth locked away from everyday life. The word *unconscious* predated Freud. He didn't coin the term. But it would nonetheless make him famous.

Moving back to Vienna in February 1886, Freud didn't keep in contact with his French contemporaries. Charcot died eight years later, in 1893. He held the lessons he had learned, the things he had seen, close to his chest, shielding them from his medical supervisors and peeking at them when they weren't looking. It was for the best. Theodor Meynert, Freud's guide to psychiatry at Vienna General Hospital, thought that hypnotism "degrades a human being to a creature without will or reason and only hastens his nervous and mental degeneration. . . . It induces an artificial form of alienation." He called its growing use a "hobby of nonmedical charlatans . . . which was not even worth thinking about." To Meynert, psychiatry was still a branch of anatomy. "The more psychiatry seeks, and finds, its scientific basis in a deep and finely grained understanding of anatomical structure [of the brain], the more it elevates itself to the status of a science that deals with causes," he wrote.

Freud would soon be creating his own science, his own hypotheses on the causes of mental disorders.

By the late 1880s an epidemic of drug addiction was spreading from city to city in Europe, and many doctors blamed Sigmund Freud, the passionate and persuasive author of "Über Coca," for the growing public health crisis. According to Louis Lewin, a pharmacologist from Berlin, Germany, and one of Freud's most vocal critics, cocaine dens of "unimaginable depravity and filth" had started to coagulate in the recesses of European cities, including his own. "These unfortunate beings lead a miserable life whose hours are measured by the imperative necessity for a new dose of the drug," Lewin wrote. "Those who believe they can enter the temple of happiness through this gate of pleasure purchase their momentary delights at the cost of body and soul. They speedily pass through the gate of unhappiness into the night of the abyss." Albrecht Erlenmeyer, another leading doctor from Germany, blamed Freud directly for introducing the "third scourge of humanity," a drug just as damaging as morphine and alcohol.

There had been warning signs from the beginning. After buying his first batch of the drug in April 1884, Freud had sent cocaine packages to his favorite sister, Rosa, and to Martha, and provided the first few grams that his good friend Ernst von Fleischl-Marxow used in order to wean himself off morphine (a drug he had become dependent on after a painful surgical operation to his thumb). Long into the night, Freud would watch his friend sit in the bathtub as he injected more and more cocaine into his bloodstream. Freud held firm that he wasn't just substituting one addiction for another, but that the cocaine infusions acted like nicotine strips for people trying to quit smoking today. Soon, Fleischl-Marxow was injecting over a gram of cocaine into his body every day, a tremendous amount compared to the twentieth of a gram that Freud occasionally swallowed during his episodes of irritability, fatigue, and low mood. Some nights the two doctors were able to discuss exhilarating topics of philosophy and science together. On others, Fleischl-Marxow became delirious in the bathtub and saw white snakes and insects crawling over his skin. Freud had once described his friend as "a thoroughly excellent person in whom nature and education have combined to do their best." But this once-bright light was starting to fade.

Fleischl-Marxow's hallucinations didn't dent Freud's hopes for cocaine's future in medicine. Still in debt and struggling to make ends meet, he was determined to show that it had a golden future. Stubborn, emotionally explosive, and unmoved by the evidence in front of him, he was beginning to isolate himself from the field that had made him into a promising young anatomist. He was turning his back on science and medicine. In response to his critics in Europe, Freud wrote the short rebuttal "Craving for and Fear of Cocaine," a title that sounds like an unpublished Hunter S. Thompson novel. It was similarly outlandish. Published in July 1887, Freud stated that only morphine addicts became addicted to cocaine, as a result of an already "weakened will power." "Cocaine has taken no other victims from us, no victims of its own," he wrote. He decried Albrecht Erlenmeyer's statement that cocaine was the "third scourge of humanity," while insisting that, for him, after years of drinking his cocaine solutions, the drug still showed no sign of having addictive qualities. "On the contrary," Freud wrote, "more often than I should have liked, an aversion to the drug took place, which caused the discontinuation of its use." He accepted, however, that not everyone had his agreeable disposition and that the risk of misuse—especially among addicts and those who injected the drug—was greater than that of morphine.

This paper was largely ignored. "It made no impression on his contemporaries," wrote Siegfried Bernfeld, an early biographer of Freud. "The paper was probably judged and discarded as the stubborn reiteration of one who could not admit error frankly." To everyone but Freud the conclusions were clear: cocaine was no wonder drug for psychiatry, no cure for weak nerves or its concomitant bouts of irritability and depression.

Criticized and pushed to the fringes of academia, Freud nevertheless achieved one of his goals in life. On June 17, 1886, after four long years of engagement, he and Martha were married in Wandsbek. "Freud had at last reached the Haven of happiness that he had yearned for," wrote his colleague and biographer Ernest Jones. In more debt than ever before (largely due to his cocaine habit), the majority of the money for the ceremony came from Martha's wealthy aunt and uncle. In 1890, the young couple moved into Berggasse 19, a second-floor apartment in the northern district of Vienna, where Freud would live and work for the next forty-eight years. He hung reproductions of Rembrandt's *Anatomy Lesson* and Fuseli's *Nightmare*, and covered every surface with ancient artifacts from Greece and Egypt. As a housewarming gift, one of his patients, Madame Benvenisti, bought him a chaise longue, a piece of furniture that would become a centerpiece of his future.

Although he didn't publish any more papers on the drug, Freud continued to drink his cocaine solutions up until the mid-1890s, over a decade of use. He no longer sent packages to friends and family and he certainly didn't prescribe cocaine injections to his patients. Nevertheless, his interest in this drug still led to one of his most prescient insights into the potential cause of depressive disorders. If a drug could lift a person back into well-being, he reasoned, then the underlying cause of depression must be rooted in some "unknown central agent, which can be removed chemically." In the late nineteenth century, before anyone knew about the existence of neurotransmitters in the brain, before any antidepressant had been put into prescription, Freud was flirting with a theory of depression that wouldn't gain traction for another eighty years or so. As he dove deeper into more psychological matters—away from the fine neurons of the brain and into the strata of the mind—Freud never truly lost his faith in drug therapy, even as Emil Kraepelin became the figurehead for biological psychiatry, the field that saw depression as a bodily disease in need of physical treatments.

"Psychiatry's Linnaeus"

Kraepelin had dreamed of becoming a professor of psychiatry by the time he was thirty years old. And he achieved it—just. Thanks to his psychological experiments and the unexpected popularity of his *Compendium of Psychiatry*, he had moved from Leipzig to become the professor of psychiatry at the University of Dorpat (also known as the University of Tartu) in Estonia in 1886, continuing his experiments into the effects of drugs on the mind. He was greeted at the railway station by a fellow professor from the university and, in a cart drawn by two brown horses, he was taken through town, over the River Embach, to his new home. In contrast to this warm reception, Dorpat was a miserable, isolating place with bitingly cold winters and heavy snowfalls. Kraepelin would later say that it was five years in a "kind of exile." "Going for walks offered little attraction, because the surroundings were monotonous," he wrote. "It was not worth going for trips by rail [either], because the trains came so rarely and were slow." The deaths of his children compounded his misery. In 1890, his third daughter, Vera, died of diphtheria at eighteen months. A year later, Hans, his first son, died of sepsis. Out of four births in five years, only Antonise, his second daughter, survived.

To distance himself from grief, Kraepelin buried himself in his work, adding to the textbook he had started in 1883. From its 380 pages, his "little book," as he called it, would grow into a 3,000-page, four-volume tome by 1915, each volume dedicated to the memory of his former teacher Bernhard von Gudden, who had mysteriously died—along with the so-called Dream King of Bavaria, Ludwig II—in Lake Starnberg, in 1886.

On November 9, 1890, just as winter started to grip the city of Dorpat in its vise, Kraepelin received an offer of employment from Heidelberg University, an academic hub back in Germany. He was to be the new chair of psychiatry. Kraepelin gratefully accepted and moved in March 1891 with his wife and daughter. It was a huge step forward in his career and provided a healthier lifestyle for his family. "My wife and I are fine now," Kraepelin wrote in the autumn, "the marvelous landscape and the joy of being in Germany again as well as a long vacation have gradually soothed the pain and made us more and more enjoy the great luck of our return from grim fights." Compared to Dorpat, Heidelberg was more clement in its weather and offered tree-lined walks around an ancient castle and hiking routes over two neighboring mountains, each as smooth and green as two heads of broccoli.

Heidelberg is a city steeped in philosophy and academia. For much of the eighteenth and nineteenth centuries, famed poets and physicists walked up and down its streets and, particularly, the wooded path that ran behind Kraepelin's house. When not working at the hospital clinic, he would follow in their footsteps with his Great Dane in tow. The two of them, a short and stocky man and his giant dog, must have made quite the sight. "As long as my professional life allowed, I fled into the fresh air," he later wrote. In a city known for its unusually tropical climate, he was surrounded by an ordered display of sun-loving plants such as the bulbous-leaved stonecrop, purple-flowered ivy-leaved toadflax, and the glossy teardrop-shaped leaves of spreading pellitory. Among this botanical backdrop were ruby-tailed wasps, iridescent wild bees, and wall lizards.

As a boy growing up in northern Germany, Kraepelin had walked through the countryside with his older brother, Karl, analyzing plants according to the classification system of the Swedish botanist Carl Linnaeus. Different leaf shapes, colors, and habitats could identify a single species, each one formalized with two Latin names. A bright explosion of azure petals, for instance, often defined cornflowers (*Centaurea cyanus*), the national flower of Germany. From his earliest days, Kraepelin learned that even the endless forms of evolution could be put into neat groups. In his thirties, he attempted to furnish psychiatry with a similar system. For his efforts, he would later become known as "Psychiatry's Linnaeus."

Kraepelin's days at the university clinic were filled with a diversity of mental diseases, a veritable cocktail of disorder. Slowly, judiciously, and with the same respect for accuracy that he had shown in his laboratory experiments, he started to document what he saw on a standardized set of counting cards, or "*Zählkarte*." Although similar *Zählkarte* had been used in German hospitals for decades, they had never been so treasured by one man before. Rather than taking a snapshot of a patient's symptoms and jumping to conclusions, Kraepelin monitored their progress over months and years. If they were discharged from the hospital, he tried to keep in touch through their doctor or local psychiatric clinic. If they were transferred to another mental asylum, he would take note of where they were going and how he might be able to receive updates of their health. Collectively, his counting cards created a timeline. As new details were added to the original notes, they became an invaluable insight into a patient's unique mental state and how they changed—for better or worse—over time. From his millisecond experiments in the laboratory to the months of his patients' mental illness in the clinic, Kraepelin turned to time as a valuable resource.

After amassing thousands of such cards, Kraepelin would pack up his boxes and travel to his "place in the country," a villa on the tranquil shores of Lake Maggiore, Italy. Like working on a huge jigsaw puzzle, he was constantly trying new combinations until a pattern seemed to click into place. As one of Kraepelin's colleagues recalled, "[A]gain and again [he] worked through thousands and thousands of his patients' files in order to group and re-group them." Were there a few core similarities that linked them? Was there a final stage, a terminus, for a common set of symptoms? In short, did they fall into discrete categories based on similarities and differences, just as plants and animals can be sorted into different kingdoms, families, genera, and species?

Meticulous, obsessive, and often frenetic in his desire for new information, Kraepelin soon became convinced that there were two main groups of mental disorder. One group he deemed incurable. He called this disorder "dementia praecox," a disease of hallucinations and paranoia that would be renamed schizophrenia in 1909. Once diagnosed, usually in one's adolescence, Kraepelin was sure that it was a downward slope toward dementia (this is now known not to be the case). The disorder that he frequently saw patients recover from he called "manic-depressive insanity." He noted that the first episode arose in young adulthood (before the age of twenty-five), it was more common in women (especially in those that were pregnant or recent mothers), and each bout ran a finite course, often lasting a few months to a year. Although commonly mistaken for the modern diagnosis of bipolar disorder, Kraepelin's term actually represented a collection of states that included depression, mania, and the cyclic disorder that flipped between both these extreme lows and highs of mood. (Bipolar wouldn't be classified as a distinct mood disorder until 1957.) "During the course of years," Kraepelin wrote of manic-depressive insanity, "I have become more and more convinced that all the pictures mentioned are but manifestations of one disease process."

Manic-depressive insanity was, Kraepelin thought, a disorder primarily of the body. Of all the potential causes, he wrote, "defective heredity is the most important, occurring in seventy to eighty percent of cases." In a time before the discovery of DNA, Kraepelin believed that any traits—even those acquired during a person's lifetime—were passed on from parent to offspring. "[S]o-called psychic causes—unhappy love, business failure, overwork—are the product rather than the cause of the disease," Kraepelin added. "[T]hey are merely the outward manifestation of a pre-existing condition; their effects depend for the most part on the subject's Anlage," their cellular development and constitution.

As he embraced the biological theory of depression, Kraepelin also practiced elements of counseling. Chatting with his patients at the university clinic in Heidelberg, he learned about their previous life experiences, when their symptoms first arose, and their problems back home. He advised his depressed patients to live a quiet life, avoid the stress of marriage, and, for those women who were suffering from postnatal depression (a sizable chunk of his patients), not to have any more children. Since he thought that the majority of psychiatric disorders were the expression of "defective heredity," he believed that all "sources of emotional disturbance should be avoided, such as the visits of relatives, long conversations, [or the sending or receiving of] letters," he wrote. "Attempts to comfort the patient in the height of the disease seem to be useless."

Despite his influential contributions to psychiatric theory and classification, Emil Kraepelin isn't a household name like many other historical figures of his field. Unlike Alois Alzheimer or Hans Asperger, there are no diseases that bear his name. Unlike Elisabeth Kübler-Ross (the creator of the five stages of grief), he has no eponymous model of psychological behavior. He doesn't have anywhere near the global fame of Sigmund Freud. But he was no less influential. Many would argue that he is the most influential of them all. "Modern psychiatry begins with Kraepelin," one textbook from 1969 states. As two historians wrote in 2015, Kraepelin was "an icon who helped guide us to the now dominant view of psychiatry that is medical in orientation, diagnostic in focus, and predominantly based on the brain." His great divide between psychotic disorders and mood disorders still exists today, both in how we diagnose mental disorders and how we treat them. Schizophrenia is often kept separate from depression, for instance, one being treated with antipsychotics, the other antidepressants. As the Swiss psychiatrist Eugen Bleuler wrote in 1917, by "cutting steps in the mountain," Kraepelin created a path for other psychiatrists to climb into the future.

In 1899, however, his importance wasn't in how his work would shape the future but how it had provided a break from the past. First published in the sixth edition of his popular *Psychiatry* textbook, Kraepelin's manic-depressive insanity had succeeded where other terms had failed: he added a vital update to nearly two thousand years of melancholia.

A Melancholic Humor

Just as the natural was made up of the four elements earth, fire, air, and water, Hippocrates, the famed philosopher of ancient Greece, thought our bodies were composed of just four basic components. "The Human body contains blood, phlegm, yellow bile, and black bile," he wrote. "Health is primarily that state in which these constituent substances are in the correct proportion to each other, both in strength and quantity, and are well mixed." Hippocrates viewed the body as a system of fluids, pumps, and valves, like a well-oiled machine, that needed to be kept in check. Too much or too little of one particular fluid—or "humor"—risked damaging the cogs that maintained this harmony, ratcheting a person away from the smooth ride of health and into a breakdown of disease and disorder. Influenced by the three "doshas" of ancient India, the underlying theory was rooted in balance and imbalance, health and disease.

People with too much phlegm, for example, were resigned to life, or "phlegmatic." Too much blood imbued a sanguine character. Yellow bile was associated with aggression. Finally, an excess of black bile led to two of the most feared diseases. If it became trapped in the body, clogged, it formed cancerous tumors that sent thin tendrils through the body like the spindly legs of a crab. In an unhealthy abundance, meanwhile, black bile led to the desolate and hopeless state known as melancholia.

In ancient Greek, *black bile* literally translates as *melas khōle*, and there are significant overlaps between this state and the more severe forms of depression known today. In fact, *melancholia* is still used to describe a state of depression that is generally defined by a key set of symptoms: insomnia, early-morning waking, severe weight loss, a sluggishness that borders on physical disability, and delusions of wrongdoing or guilt that are often repeated like a daily confession of sin. This isn't your everyday case of the blues. It is a life-threatening disorder that responds remarkably well to biological treatments such as antidepressants and electroconvulsive therapy (ECT) and has historical roots that reach back to the dawn of medicine.

Aristotle, Plato, Socrates, and Pythagoras: these household names all had an influence on generations of philosophers who tried to understand how the mind worked in both health and disease. But the most important writer on the topic of melancholia lived in Rome in the first century AD, four centuries after Hippocrates. Claudius Galen of Pergamum started his career as a physician to gladiators. As he patched up their wounds, he also provided individual dietary regimens to keep them at their fighting best. He then moved to Rome and became the personal physician to the emperor Marcus Aurelius and his family. No matter where he settled, he was a prolific writer, reportedly publishing four hundred books before he died in 210 AD when he was in his seventies. Most of the ideas weren't his own, however. "He was a great borrower," Franz Alexander and Sheldon Selesnick write in their 1966 book, *The History of Psychiatry*, "he plagiarized, synthesized, embellished, and copied. The perpetuation of his influence was due to his staunch advocacy of Hippocrates and his acceptance of a Creator to whom he frequently dedicated his works. Christianity in the Dark Ages could find no exception to Galenic monotheism." As religion became the ultimate truth in place of medical science in the medieval ages, Galen's works could still be seen as gospel for fifteen hundred years of Western thought.

What is left of his medical writings (or their translations) shows that Galen thought that melancholia wasn't a single disease. There were at least three types of melancholia that could be classified according to their symptoms and presumed location in the body. First, there was the imbalance of black bile within the blood. Second, a condition in which the black bile had poisoned the brain. And third, a gut-based disease that he called "melancholia hypochondriaca" that was associated with flatulence and stomach pains, constipation and indigestion. This was also called the "gassy disease." As future scientists would discover for themselves, the link between the brain and the gut, particularly when discussing depression, is a vital one.

Galen knew that different diseases require different treatments. For his blood-based melancholia, he aimed to reduce the amount of black bile in a person's body by draining them of some blood. By cutting a vein on the inside of the elbow joint, Galen hoped that such bloodletting would bring his patients back to reason, even in those cases in which the black bile had poisoned their brain. In a world without antibiotics, sterile practices, or any knowledge of bacteria or viruses, this was a very dangerous approach. Any cut could be fatal, every blade a potential vehicle for infection.

In cases of the gassy disease, however, bloodletting wasn't suitable. In fact, Galen actively argued against its use. Instead, he recommended herbal remedies, such as black hellebore, that induce vomiting, or emetics that helped flush the disease from both orifices. As he purged his patients, Galen also advised them on what they should put back into their bodies. As in Ayurvedic medicine, diet was an integral part of every treatment. In particular, the avoidance of foodstuffs that were thought to increase the black bile, or, in Galen's terminology, "atrabilious blood," in the body. "I can tell you," Galen wrote, "that atrabilious blood is generated by eating the meat of goats and oxen, and still more of he-goats and bulls; even worse is the meat of asses and camels . . . as well as the meat of foxes and dogs. Above all, consumption of hares creates the same kind of blood, and more so the flesh of wild boars . . . so do all kinds of pickled meat of terrestrial animals, and of the beasts living in the water, it is the tuna fish, the whale, the seal, dolphins, [and] the dog shark." It wasn't just a meat-based diet that he thought problematic, but certain vegetables including cabbage, sprouts, and lentils. Red wine was a luxury best avoided for those at risk of melancholia.

Galen much preferred dietary changes to bloodletting. But he often found himself torn between the two. Without any reliable markers of the different types of melancholia—blood, brain, or gut—he sometimes wasn't sure which line of treatment would be most suitable. In such cases, he undertook a tentative course of bloodletting at the elbow. "If the outflowing blood does not appear to be atrabilious, then stop at once," he wrote in his book *On the Affected Parts*. "But if the blood seems to be of this kind, then withdraw as much as you deem sufficient in view of the constitution of the patient." It was as if he could see the dark and gloopy humor drain out of his patients' arms, each atrabilious drop bringing them closer to health, harmony, and away from the dreaded melancholia.

Before the texts of Hippocrates and Galen crumbled into dust like the columns of the Greek and Roman Empires, a few Christian missionaries made copies of their rolls of parchment and fled to the booming civilizations of Mesopotamia. Located in the middle of the Silk Road that connected East to West, countries that are now called Iran, Iraq, Saudi Arabia, Afghanistan, Syria, and Turkey became vital lifelines for modern medicine. The Dark Ages—the medieval period between the fall of Rome in the fifth century and the Enlightenment of the seventeenth—were only dark in western Europe. Although this was an era defined by Islam, it was a bustling fusion of cultures from around the world. Christians and Jews learned and wrote in Arabic. Cities such as Damascus, Cairo, and Baghdad attracted immigrants from all over the world, if not to stay then to visit and learn from some of the great polymaths that resided in these centers of civilization.

In 849 AD, Abu Zayd al-Balkhi was born in a small village in what is today Afghanistan. Homeschooled by his father, he was a precocious child with interests as varied as the culture that surrounded him. As a young adult, he moved to Baghdad and spent eight years researching any topic that was worthy of his attention. During the so-called Golden Age of Islam, there was a lot to take in. Shy, introverted, and with a preference for isolation over any form of socializing, Abu Zayd wrote at least sixty books in his lifetime, one of which merged the bodily theories of ancient Greece with psychological treatment. Although he didn't doubt that some depressions came from within, from an imbalance of black bile, say, he also put forward the view that this illness could arise from purely mental means. Negative thinking was at the heart of this condition. Abu Zayd recognized that how we view and interpret the world can be far more important than the world itself. Our reality can be shaped by our own perception, preconceived expectations, and thought patterns. Something that might be neutral or even slightly positive can be twisted into an experience that was negative and personally detrimental. This "cognitive theory" of depression would be made famous by Aaron Beck, a psychiatrist working in Philadelphia in the 1960s. Although each came to their theories independently, we can say that Abu Zayd predates this modern treatment by ten centuries. "No scholar before him had written a medical treatise of this kind," Malik Badri, a historian and psychologist from Sudan, writes.

In his book *Sustenance of the Soul*, Abu Zayd separated depression into three types. First, there was sadness, a normal reaction to life. Second, there was depression triggered by stressful life events. And third, a form of the illness that was heritable and seemed to be unrelated to external circumstance. In short, Abu Zayd used a classification system that would only become popular during the twentieth century: he saw depression not only as a spectrum from sadness to disease, but that there was a "reactive" type and an "endogenous" type. The former was treatable with psychotherapy and guidance to a more positive outlook. The latter was not. Abu Zayd thought that bloodletting was a barbaric treatment and vaguely described a form of "blood purifying" that may have included dietary changes or herbal remedies.

Although progressive in his thoughts on depression, Abu Zayd was largely ignored in subsequent centuries. His texts were only recently rediscovered and translated. In the 1970s, the texts were translated into German and, in 2014, into English. Today, you can read a section of *Sustenance of the Soul* on your e-reader.

This could have been a very different history if Abu Zayd's works were better known. Bloodletting might have stopped here. Instead, the most influential philosopher working in the city of Baghdad was Ibn Sina, or "Avicenna." It was his *Canon of Medicine* that cemented the humoral theories of Galen into the Middle East and then transported them back into western Europe. In the eleventh and twelfth centuries, two missionaries, Constantinus Africanus and Gerard of Cremona, translated this Arabic text into Latin, the language of the Western world.

The works of Avicenna were used as educational tools in medicine for centuries. As his words were passed on and reprinted, they were also merged into the prevailing Western worldview of the time: Christianity. In her twelfth-century *Book of Holistic Healing*, for instance, Hildegard of Bingen wrote that black bile was a marker of the original sin, one that stretched right back to Adam in the Garden of Eden. Black bile, she wrote, "came into being at the very beginning out of Adam's semen through the breath of the serpent since Adam followed his advice about eating the apple." Melancholia, she added, "is due to the first attack by the devil on the nature of man since man disobeyed God's command by eating the apple. . . . [I]t causes depression and doubt in every consolation so that the person can find no joy in heavenly life and no consolation in his earthly existence."

This mix of theology and medical theory would remain little changed up until the seventeenth century. In his epic treatise, *The Anatomy of Melancholy*, Robert Burton added that "spirit" was also an important determinant of disease. "Spirit is a most subtle vapor," he wrote, "which is expressed from the blood, and the instrument of the soul." To him, melancholy was a habitual state of sadness that arose from astrological signs, curses from the Gods, differences in air, changes in the weather, and could only be cured by a simple diet, regular exercise, and mental occupation. "There can be no better cure than continual business," Burton wrote, a statement that would soon be brought into a therapeutic reality in mental asylums.

At around this time, in the era of Enlightenment, skilled anatomists had sifted through the fluids and fibers of the human body—sometimes stealing dead criminals from the gallows to use as cadavers—and found no evidence of a gloopy black substance anywhere. Blood, phlegm, and yellow bile, yes. But nothing of *melan kholē*. Black bile, a focal point of medical science for so long, didn't exist. As one author wrote in 1643, "[W]e cannot here yield to what some Physicians affirm, that Melancholy doth arise from a Melancholick humor."

How did this theory of melancholia endure for so long? For one, the understanding of how the body works was rudimentary in the West. It was only in the seventeenth century that the circulation of blood was starting to be understood, for example. The notion that humans are made up of microscopic cells hadn't yet been proposed. In fact, the first microscopes were only just being invented. Finally, and perhaps most important, the human body is infinitely complex and it demands centuries and millennia to understand in detail. Nowhere is this more obvious than depression, a disorder that has an almost comical talent of evading biological definition. Whether it's caused by an imbalance of bile or chemicals in the brain, it has a history of duping those who try and pin it to a unifying theory.

Instruments of Cure

The eighteenth century was the beginning of a seismic social change in the Western world. From 1760 onward, the Industrial Revolution fed the sprawl of factories and agriculture, fueling a population boom. Between 1800 and 1850, for example, the number of people living in Britain doubled, from ten and a half million to twenty-one million. The country was becoming an empire, one forged in the furnace and fed with coal wrested from deep below the earth. Especially in the north of the country, chimney smoke covered the landscape in soot. The black veneer on buildings and trees was so ubiquitous that a species of moth became dominated by its darker form, a "melanic" mutant that was better camouflaged in this new human-made environment. Known as the peppered moth (*Biston betularia*), those moths that were paler in color were easily picked off by hungry birds.

With any revolution there are winners and losers; inequality is the norm. While a select few got rich from coal and oil, millions of workers stayed poor. Anyone strolling through the British cities of the eighteenth century would soon stumble upon the disease, dirt, and depravity that thrived in areas where humans congregated. Poorhouses, prisons, and workhouses were places for those who failed to make a living. And with financial troubles, the death of loved ones (either from disease or accidents at work), and a very raw feeling of helplessness, insanity thrived. Mania and melancholia, the violent and the desolate, were the two diagnoses used, as different as chalk and cheese. But, in the growing number of so-called mental asylums, these inpatients were similarly chained and bled.

The nonexistence of black bile barely changed the view of melancholia and its treatment. After all, there was nothing better, no alternative explanation. The suggestion of "animal spirits" or the soul didn't offer anything concrete about the origin or treatment of depression. The terms *vapors*, *hypochondriasis*, and *spleen* were used for states of mental agitation, all still rooted in the notion of black bile being burned in the spleen and infusing the brain with darkness. Bloodletting and purging were still part and parcel of many therapeutic regimens. As the superintendent of Bethlem Royal Hospital in London wrote in the eighteenth century, "[T]hat has been the practice invariably for years, long before my time, it was handed down to me by my father, and I do not know any better practice." People with mental disorders were often chained—sometimes for their own protection—as they bled. Seen as lower animals rather than fellow humans, punishment was viewed as a suitable form of therapy. "Old and young, men and women, the frantic and the melancholy, were treated worse, and more neglected, than the beasts of the field," one asylum director wrote.

This callous approach to insanity was slowly changing, however; shifting toward methods of psychotherapy, individualized care, leisurely activity, and a healthy diet.

In the walled city of York in the north of England, William Tuke built a cottage-style "Retreat" that banned physical restraint and replaced it with entertainment, education, and empathetic conversation. A sixty-year-old tea merchant, Tuke felt an obligation to create the Retreat after Hannah Mills, a local widow and Quaker, died in suspicious circumstances shortly after being admitted to the mental asylum in York. A Quaker himself, he believed that God spoke directly through him and thought that news of Mills's death reached his ears for a reason. He began designing a retreat in York that could live up to the word *asylum*.

One of the first patients to walk over the Retreat's wooden floorboards and gaze upon its flower-filled gardens was Mary Holt. Described as "low and melancholy," she wasn't chained or bound in a straitjacket. Instead, she was fed a healthy meal three times a day, drank fresh milk from the asylum's cows, and was offered porter or wine with lunch. Her depressed state was met with compassion, and any morbid ideas or delusions were guided, through conversation with her carers, toward more peaceful thoughts. Sewing, knitting, and reading were thought to be suitable occupations for people like Mary. Mathematics and the classics, meanwhile, were hailed as the most beneficial literature. As evidenced by the Retreat's bills, opulent delicacies such as oranges and figs were a part of a therapeutic environment that would have seemed completely alien to people who would have once been detained in the workhouses, poorhouses, or prisons that dotted the English countryside, sweeping up any vagrants that were sleeping rough on the soot-covered city streets.

A cottage in the middle of nowhere is a strange place for a revolution, but that was indeed the case here. The Retreat was the physical manifestation of what became known as "moral treatment," a form of care that would spread from northern England and alter centuries of abuse masquerading as medical science. Once seen as beasts only responsive to punishment, in these institutions the manic and the melancholic were treated as humans that were unwell and in need of protection and support.

Born in 1732, William Tuke was a pertinacious man whose life was defined by loss. When he was a child, he fell out of a tree and was trephined—a procedure in which a small disk of bone is cut from the skull—to ease any swelling of his brain. Throughout his life, people would touch this boneless part of his head and feel the soft spring of his scar tissue. Both his parents died when he was an adolescent and William was raised by his aunt and uncle, Mary and Henry Frankland, who owned a grocery shop. Working as their apprentice, he would soon be given the business after Henry died and Mary, a widow without any children of her own, could no longer look after it. In 1754, when he was in his early twenties, William married Elizabeth Hoyland from Sheffield, a nearby city in northern England. She died giving birth to their fifth child in six years. "I was on the brink of destruction when my dear wife was taken from me," Tuke wrote.

Tuke turned to his religion for comfort, always attending the annual meeting of the Society of Friends (the traditional term for Quakers) held in London and even serving as their treasurer for twenty years. Running a tea business while the East India Company had a monopoly on the import and distribution of most goods from the British colonies, he was business savvy and money-minded. Thankfully, in the late 1800s, tea drinking was growing in popularity. No longer a pastime of the wealthy, it was quickly becoming an everyday beverage for an entire nation. In 1784, the British government reduced the amount of tax on tea from 100 percent to 12 percent. Tuke profited hugely, expanding his business into coffee from Turkey and chocolate from the West Indies.

Dogged in everything he did, William Tuke was also hugely empathetic and civil to anyone he met. As one of his patients' poems states:

> What tenderness with strength combined,
> Dwelt in his energetic mind.

Initially limited to local "Friends," the Retreat at York was soon admitting new residents from as far away as Bristol and Penzance in the southwest of England, the opposite side of the country from this new asylum. Traveling by horse and cart, the journey was long and arduous. But upon entering the enclosed garden through a row of beech trees and stepping toward a farmhouse in full bloom, it was worth it. For the family member who had accompanied their manic or melancholic kin, it might have seemed like they were dropping them off at a luxury holiday destination with free booze, an island of peace and pleasure where there were once only shackles, seclusion, bloodletting, brutality, and despair.

The Retreat at York was used as a model for dozens of new asylums being planned across the country. Physical restraints were thrown out and mental asylums in England and across Europe were built with moral treatment in mind. Hanwell Asylum in North London, for example, was filled with the fresh air of optimism. When John Conolly became the medical superintendent in 1839, there were six hundred instruments of restraint. Iron manacles, leather straps, chains and locks: he threw them all out. "Darkness and bonds are, even in [the most] difficult cases, never applied," he wrote in 1856, "hasty and reproachful words are forbidden; punishment is a word unknown; and as far as can be done in intervals of calmness, the patient is the object of soothing attentions; whilst during the worst paroxysms of his disorder, he is never abandoned and left to his wretchedness." To replace the methods of restraint, Conolly hired a larger troupe of trained nurses and medical attendants, and made sure that isolation was only used for the most dangerous or violent patients. Those with melancholia, now freed from their shackles, were supervised and supported, even through the long dark night, if necessary. "The old prison-air has departed, and asylums have really become . . . instruments of cure," Conolly wrote. "Not only to the maniacal patient, who comes tied up because he is violent, [but] to the melancholic creature who is also tied and bound, because he desires to end his existence, does the non-restraint system bring deliverance and comfort."

Once freed of physical restraint, patients were surrounded by activities and entertainment. "Recreation, when it is essentially innocent in its tendency, is so necessary to the preservation of mental health, that it has always formed a part in the treatment pursued in all well conducted asylums," one medical director wrote in 1844. Backgammon, billiards, and board games were available indoors. Outside, there was cricket, kite flying, and other leisurely activities on the garden's lawns. Musical instruments, literature, and art classes provided a creative outlet for people whose every moment had once been censored and subjugated. The transformation was striking. "The convalescent wear an expression of the most serene satisfaction," Conolly wrote, "and a smile, seldom seen before, plays on the countenance of the melancholic." Anything to distract these patients from their morbid thoughts was seen as therapeutic.

As with the practices of Galen in Rome, diet was a key ingredient of moral treatment. Those melancholics who were admitted into asylums half starved were initially fed a diet of beef tea and arrowroot. When they could stomach it, they were given three meals a day, the largest for lunch, including healthy portions of meat and vegetables and a few glasses of booze to wash it down. Sherry, brandy, and wine were all seen as suitable tonics. The Stanley Royd Hospital in northwest England brewed its own beer on-site. Instead of emetics or bleedings, rhubarb was often given as a natural laxative that helped ease the constipation often associated with this mental disorder. As with Galen's melancholia hypochondriaca, the digestive system was seen as a circuitous path toward a cure.

By the early nineteenth century, moral treatment had crossed the Atlantic and institutions in the United States were designed according to the religious and moral code of the Retreat at York. In 1817, the Friends Asylum was built according to the beliefs of two leading figures of the moral treatment movement: Thomas Scattergood, a "mournful prophet" who suffered from frequent bouts of melancholia and visited William Tuke in England, and Benjamin Rush who sought freedom for his mentally ill patients. Inspired by the work of Rush and Scattergood, the social reformer Dorothea Dix helped to spread the values of moral treatment throughout the mental asylums of the United States. After a nervous breakdown in 1836, Dix was treated at the Retreat at York, and became a living example of the benefits of respect, leisure, light occupation, and a gradual realignment with reason and reality.

Did moral treatment actually work? Were people more likely to recover from melancholia—or mania—if they were treated with generosity and fed a healthy diet? Without chains and enforced punishments, it certainly wouldn't have been detrimental to people's mental health. But it's difficult to know what was responsible for any recovery, especially those that occurred two hundred years ago. A healthy diet and entertainment might have helped. Or, alternatively, time spent away from the outside world might have been just what a person needed to break their cycle of depression. Precise conclusions are an impossibility. The one thing that is clear is that depression can disappear on its own given time, a fact that can make any treatment appear to be curative.

As with mania, the depressed and melancholic have long been known to spontaneously recover. (This might be the only time a depressed person can be called spontaneous.) No one knows why or how this happens, just that it does—often. It might take months or years, but a large majority of patients will get better if left alone, even if they spend a long time getting worse beforehand. Like a bout of flu or shifting seasons, depression usually has a finite and cyclic course. This might seem like a good thing, and in many ways it is. But this phenomenon has also provided an accidental platform for abuse. Without controlling for these spontaneous recoveries, many doctors and psychiatrists of yore have been tricked into thinking that their methods were effective. Whether they were given the rest cure, bloodletting, purging, or a change in diet, a patient's recovery might have happened anyway, without any intervention whatsoever. Without taking spontaneous recoveries into account, anything could be deemed to have a positive effect.

This does not mean that the treatments used today are a waste of time—far from it. Those months or years with depression can be so inexplicably painful that anything that can reduce their duration is to be lauded. Between 10 to 15 percent of people with depression die by suicide, a risk that is seven times higher for men than women. And not everyone spontaneously recovers. Chronic depression can be an endless state of despair, especially for the elderly, who may be suffering from other diseases such as cancer, diabetes, or dementia. An effective treatment for depression, therefore, can be seen as anything that can shorten the time to recovery. A remission in a few days or a few weeks, for example. Not in a few months or years. This pursuit of early remission is captured by the term *efficacy*, once defined as "the capability of an agent, demonstrably and measurably, to alter the statistically predictable natural history of the disease." But in the asylums of Britain and elsewhere, there were few options but to wait and hope that the patient recovered on their own.

• • •

In the nineteenth century, the size of mental institutions exploded and left the initial hopes of moral therapy in their ashes. While the average asylum housed just over one hundred patients in 1827, this number had risen to over one thousand by 1910, a tenfold change in magnitude. "Our numbers have increased out of all proportion to our accommodation," one worker at a county asylum in Staffordshire wrote. "We have 36 males sleeping without beds." Similar patterns were seen across Europe, in particular in France and Germany. In 1904, a total of 150,000 people were living in mental institutions in the United States. The ratio of medical attendants to patients became so small that neglect and isolation were commonplace. "The treatment is humane but it necessarily lacks individuality," one doctor wrote in the medical journal *The Lancet*, "and that special character which arises from dealing with a limited number of cases directly." Even the Retreat at York, a house originally designed for a maximum of thirty patients, was eventually home to over one hundred. As one patient said, "[T]o one that has always been used to a small family, this is just like being in a show."

By the end of the nineteenth century, well-kept gardens and cottage-like buildings were replaced with high fences, long, tree-lined driveways, and a brutal architectural design that was made for easy surveillance. A central bell tower provided a 360-degree view of the grounds, while long corridors allowed hundreds of patients to be monitored at a glance. A mental institution in Liverpool was famous for a corridor that was over half a kilometer long. Homely asylums with specific diets and entertainment were replaced by cold and monotonous institutions. The food was the same no matter your condition. Medical attendants wore matching uniforms and a jangle of keys hung from their belts as if they were prison guards.

The future seemed bleak for melancholics and their mentally ill neighbors. With overcrowding only getting worse, the hopes of moral treatment were fading. Plus, there was still no insight from anatomists, either. In the 1840s, the German psychiatrist Wilhelm Griesinger famously proposed that "patients with so-called 'mental illnesses' are really individuals with illnesses of the nerves and the brain," but there had been no breakthroughs from neuroanatomists. The autopsies of deceased patients found no brain abnormalities consistent with mania or melancholia. There was no black bile. But there was still nothing to replace it with.

At a time when the invisible and microscopic were increasingly important, thanks to new technologies, it made sense that mental disorders, too, could be the result of some previously unseen—and yet discoverable—phenomena. Bacteria and viruses were discovered in the second half of the nineteenth century, replacing the old notions that "bad air" (or "miasma") caused infections such as tuberculosis, typhoid, and cholera. And, indeed, there was one discovery that suggested that this faith in neuroanatomy wasn't misplaced. General paresis, a fatal blend of depression, mania, delusions, and dementia that had been dubbed "one of the most terrible maladies that can afflict a human" was associated with clear lesions in the brain, small pockmarks where neurons had died. It was as if a tiny but voracious organism had been munching on a patient's soft brain tissue, slowly sending them into madness. The modern name for general paresis is *neurosyphilis*, a bacterial infection that has reached into the nervous system. (This disease would help to revolutionize biological psychiatry in the early twentieth century, leading to the first Nobel Prizes awarded to a psychiatrist.)

Melancholia, meanwhile, was still no clearer than it had been in Galen's day. "Without wishing to discourage younger men from following up this line of work," George Savage, the medical superintendent from Bethlem Royal Hospital in London and author of the popular 1884 textbook *Insanity and Allied Neuroses*, wrote, "[w]e are groping in the dark for what we do not yet know. There is at present a deep ditch of ignorance between us and the true pathology, and for that matter, physiology of the mind."

Ten years later, Sigmund Freud started to craft his own physiology of the mind, a science rooted in strata of the psyche rather than anatomical dissection. At the same time, in Heidelberg, Emil Kraepelin was similarly discouraged by anatomy and would later come to doubt the psychiatric classification that he had created.

The Talking Cure

The field of psychoanalysis wasn't the sole creation of one man. Biographies and historical accounts have long painted Sigmund Freud as a maverick theoretician who dared to explore topics that others shied away from. That he was a depressed loner in an uncharted world of the human psyche was taken as fact, a testament of his courage and determination in the face of impossible odds. But the first biographers of Freud, Ernest Jones and Siegfried Bernfeld, were also his colleagues and close friends. The person in charge of his legacy, his private letters, and his unpublished documents was his daughter, Anna Freud, one of the most faithful Freudian psychoanalysts of the twentieth century. In short, the story of psychoanalysis and how it became one of the most popular fields of psychology has long been biased and airbrushed.

As new documents and details have emerged, historians of the twenty-first century have drastically changed the story of how psychoanalysis came to be. "It can be said that Sigmund Freud did not so much create a revolution in the way men and women understood their inner lives," George Makari, a historian and professor of psychiatry at Weill Cornell Medical College, wrote in his 2008 book, *Revolution in Mind*. "Rather, he took command of revolutions that were already in progress." As Freud himself wrote to Sándor Ferenczi, his colleague in Budapest, "I have a decidedly obliging intellect and am very much inclined toward plagiarism." Evolutionary theory, Newtonian physics, sexology, hypnotherapy, experimental psychology: Freud utilized them all. Theories of inhibition, the unconscious, innate sex drives, and even the idea that dreams might offer an insight into insanity were being debated long before Freud began his move from brain anatomy to the strata of the psyche. Freud's true gift, Makari concludes, was his ability to attract those ideas that were flourishing around him and condense them into his own towering edifice of thought.

This is certainly the case for his development of psychoanalysis as a form of talk therapy. Inspired by his trip to Paris and the power of hypnosis, Freud coauthored *Studies in Hysteria* in 1896 with Josef Breuer, his mentor, financial benefactor, and close colleague who had provided an insight into how mental symptoms can be removed with words. Based on a now famous case study of Bertha Pappenheim (referred to as "Anna O." in *Studies in Hysteria*), Breuer set the foundations on which psychoanalysis was later built.

Raised in a wealthy traditional Jewish family in Vienna, twenty-one-year-old Bertha Pappenheim first met Breuer in December 1880, a time when Freud was still a budding anatomist who hadn't yet met Charcot and his hysterical patients in Paris. After nursing her dying father, Pappenheim was exhibiting some highly unusual symptoms. "She spoke an agrammatical jargon composed of a mixture of four or five languages," the psychiatrist and historian Henri Ellenberger wrote. "Her personality, now, was split into one 'normal,' conscious, sad person, and one morbid, uncouth, agitated person who had hallucinations of black snakes." Pappenheim described her two personalities as "a real one and an evil one." She remained in bed for nearly four months between the end of 1880 and the spring of 1881. Breuer, a bearded man with a domed forehead and deep-set eyes that curved downward like the dark facial markings of a sloth, was the only doctor she would speak to.

Her father's death in April 1881 made Bertha spiral further into idiosyncrasy. She only spoke in English and didn't understand German. She became deeply suicidal and, for her own safety, was transferred to a country house—Breuer initially called it a "villa"—

that was within walking distance of the Inzersdorf mental asylum on the outskirts of Vienna. There, she soon felt calmer, her agitation subsided, and she enjoyed playing with a Newfoundland dog and meeting the local residents. Breuer traveled from Vienna every few days and would spend hours talking to Pappenheim. He noticed that when she explained her symptoms, the contents of her hallucinations, for example, they seemed to dissipate, just as discussing problems at home with a close friend might help relieve some of the stress they have caused. It was cathartic.

In December 1881, a full year after her symptoms first started, Pappenheim's illness took an unexpected turn. She reverted to those first days and relived them, hour by hour, all over again. It was as if her body clock had been set back by 365 days. Every tomorrow was a repeat, as if Breuer's medical notes were now being played out like a dramatist's script. "Breuer was able to check that the events she hallucinated had occurred, day by day, exactly one year earlier," wrote Ellenberger. Breuer, a valued member of the aristocracy and one of the most respected doctors in Vienna, noticed that among this repetition was another opportunity for catharsis. When he asked Pappenheim to explain when and where her symptoms had taken hold of her, he could slowly uncover their original source. Like a detective pushing pins into a city map to determine the location of a wanted criminal, Breuer could track down the offending moment and, he hoped, prevent them from doing further harm.

It was a long and tiring process, one that was made even more drawn out by Pappenheim's mercurial mental state and Breuer's busy schedule back in Vienna. It could take hours of discussion, often through a distressing chain of memories, before the original experience was brought out into the open. But the rewards were clear to see. Her recollections were restorative. By reliving the memory that inspired the symptom, Pappenheim was able to defuse each situation and the damage it was doing to her body. While nursing her father, for example, she had a vision of a black snake slithering from the walls of the room and killing him. She felt an overwhelming urge to beat it away with her right hand, but it would not move—her arm was stiff, stationary, and, to add insult to injury, her fingertips had transformed into little snakes, a terrifying hand puppet of the mythical creature Medusa. At that moment, she prayed to herself in English. After discussing this memory with Breuer, her repetitive spells were broken. The black snakes vanished. She could move her arm. Her English was replaced by her native German.

The healing power of words and recollection was remarkable. Over hours of intimate discussion (sometimes aided by hypnosis), Pappenheim's symptoms gradually lessened and sometimes disappeared altogether. It was as if, by returning to these traumatic emotions that she had feared and suppressed, she was able to relieve some pent-up energy that was powering her unusual disabilities. While Breuer called this technique the "cathartic method," Pappenheim said it was a "talking cure."

But it was no cure. Bertha Pappenheim didn't recover from her illness. Although it might have been true that she felt better after discussing her problems with a trusted doctor, Pappenheim would spend the next few years institutionalized and dependent on regular injections of morphine. Only after she found her true calling in campaigning for social equality—becoming one of the most famous feminists of the late nineteenth century—was she able to live a life outside of the institution. Although she was diagnosed as hysterical, and included as one case study (the previously mentioned "Anna O.") in Freud and Breuer's *Studies in Hysteria*, her illness has more recently been tied to her pious and sexually restrictive upbringing, her later subjugation by a patriarchal society, and her unconscious striving for meaning and purpose. "[Her] illness was the desperate struggle of an unsatisfied young woman who found no outlets for her physical and mental energies, nor for her idealistic strivings," wrote Ellenberger. As a woman raised in a traditional Jewish family, she wasn't allowed to even apply to a university. As with women subjected to the rest cure, her social niche was set within the household. Perhaps caring for her dying father had brought this expectation into an uncomfortable reality, a glimpse into a future that she feared more than any number of snakes.

Bertha Pappenheim's case reports added a fertile seed to Freud's growing interest in the psyche and, in particular, the unconscious. In Paris, Jean-Martin Charcot used hypnosis to paralyze someone's arm with words. With his cathartic method, Breuer showed that extracting painful memories could make a paralyzed arm move again. Whether it was a limb, speech, hallucinations, or morbid fears, the mind seemed to be in ultimate control of the body. Only by making the patient aware of these connections could they be guided toward the relief of their symptoms. "[Breuer] found a method of bringing into her consciousness the unconscious processes which contained the meaning of her symptoms and the symptoms vanished," Freud later said.

After publishing *Studies in Hysteria*, Freud started to separate himself from Breuer. He thought that hypnosis was largely ineffective and only worked on a select few patients. He no longer relied on catharsis—the retelling of a story. Freud became an interpreter, suggesting causal connections between memories and a person's mental state. As with his golden stain that made the invisible visible, it was Freud's task to find what his patients were hiding away, to fill it with color and meaning.

In 1899, the same year that Kraepelin first presented his all-encompassing "manic-depressive insanity," Freud published *The Interpretation of Dreams*, a book that, he thought, was his most original and important work. "Insight such as this falls to one's lot but once in a lifetime," he wrote. While Charcot had reached into the unconscious of his patients in Paris with hypnosis, Freud argued that a person's dreams provided a "royal road" into this otherwise locked cargo of the mind. By interpreting the content of a patient's dreams, Freud sought to reveal his patients' repressed memories, wishes, and desires, allowing them to relive an appropriate emotional response. No longer restricted to the idiosyncrasies of hysterical patients, dream interpretation was a means of alleviating those conflicts, perversions, and traumas that everyone hides away from their family, from their doctor, and even from themselves. Crucially, everyone dreamed at night, even if they didn't think twice about what their dreams meant.

Dream interpretation was just one component of the nascent field of psychoanalysis. The other was free association. Inspired by a favorite book from his childhood that advised writing "without falsification or hypocrisy, everything that comes into your head," it was a slow and protracted process in which Freud had his patients speak, uninterrupted, about any memory, thought, or feeling as they lay on his couch in the consulting room at Berggasse 19. Famously attentive and constantly looking for hidden meanings, Freud thought he could hear the subtle whispers of the unconscious as his patients spoke. It was as if the mind's cargo was shaking its cage and screaming to be set free. In isolation, these mutterings might be considered meaningless. But through years of analysis and interpretation, Freud thought he could connect them into a neat line of associations that might bring understanding—and relief—to his patients' suffering. He called this "a resolute pursuit of the historical meaning of a symptom."

"In every one of our patients we learn through analysis that the symptoms and their effects have set the sufferer back to some past period of his life," Freud later said. "In the majority of cases it is actually a very early phase of the life-history which has been thus selected, a period in childhood, even, absurd as it may sound, the period of existence as a suckling infant."

His patients complained of aches and pains, loss of senses, obsessional thoughts, strict routines, and a list of unusual symptoms that seemed to have no physical basis. They were known as "neurotics." In its earliest years, psychoanalysis was focused on three main forms of neurosis: anxiety-hysteria, conversion-hysteria, and the obsessional neurosis. "These three disorders, which we are accustomed to combine together in a group as the transference neuroses, constitute the field open to psycho-analytic therapy," Freud said. He met with few, if any, depressed patients. To him, they were boring. As one psychotherapist later wrote, a depressive was someone "who interacts sparsely with others, is dull and unproductive, sees the world in an impoverished and stereotyped way, and really wants to be left alone."

It was uninteresting. It didn't have the idiosyncrasies of hysteria or the psychological whodunits of the "Rat Man" (a case in which the memory of being told about a form of torture involving rats was associated with obsessional neurosis) or the "Wolf Man" (a famous dream of white wolves in a tree that Freud interpreted as the early moment of anxiety when boys and girls become aware of the other sex). By this point, the diagnosis of "weak nerves" had been separated from its psychological symptoms of anxiety and depression and was largely seen as a case of chronic fatigue, a physical state that Freud pinned to a depletion of sexual energy from too much masturbating.

Freud was sailing into uncharted seas and felt like he was sometimes speaking to the waves. His once-in-a-lifetime insight, *The Interpretation of Dreams*, sold terribly. Over six years, only 351 copies were sold, many to people who would provide scathing reviews of its contents. "The attitude adopted by reviewers," Freud wrote, "could only lead one to suppose that my work was doomed to be sunk into complete silence."

Early in November 2015, I cycled through the growing chill of winter to a psychiatric hospital near Brixton in South London. My bike's gears didn't work and the chain fit loosely around its rings, clunking forward with each pedal. It wasn't a smooth journey and what should have taken fifteen minutes took around half an hour. I got lost. When I arrived at the gates to the hospital—a low, redbrick building with colorful signs detailing the different psychiatric departments—there were a few speed bumps, a final test of the assault course. But I had made it. I locked my bike up in the parking lot, put my elasticated lights into my bag, and walked into the building. I was ready for my first session of group therapy.

My doctor had put me forward for this four-week course of cognitive behavioral therapy (CBT) a few months earlier. After leaving academia and focusing on my writing, I was still having thoughts of suicide and had a crippling lack of interest in once-enjoyable activities. He initially offered me an antidepressant, an SSRI called citalopram. I declined. I worried that such a drug would make my mind blur. My writing, too. I was just beginning to be published in magazines and didn't want to sacrifice the progress I had made. Only later would I realize that an antidepressant could be a necessary tool, especially when the alternative is depression. Instead, we agreed that a group therapy "workshop" might be a better place to start. And so, every Wednesday, I would force myself out of the apartment, cycle to the hospital, and join twenty or so other people who were also struggling with their mental health.

We were a diverse bunch. Some people sat in their comfy trousers and sneakers, while others were still in their smart-casual work attire. One guy turned up still in his flashy cycling gear (clip-in shoes included), and stored one of his bike wheels behind his seat. Some of us were young, in our early twenties, while others were middle-aged and graying. There were men and women, wealthy and poor, and a range of ethnicities. A cursory glance around the room provided a quick insight into the nondiscriminatory nature of depression: it affects us all.

I didn't know it then, but these four sessions of psychotherapy were built on the work of Sigmund Freud and Emil Kraepelin. Although it wasn't confidential or one-to-one, and there was no chaise longue in sight, it was still a platform in which an insight into mental processes—thoughts instead of memories—is used to treat depression. Plus, it was a psychoanalyst, a disciple of Freud, who created the core elements of this psychotherapy in the 1960s. Then there was the health questionnaire on my lap, a systematic method of tracking how my symptoms changed over the course of the four-week treatment. A form of psychotherapy with a set of counting cards, this was a convergence of history, a modern meeting of the two forefathers of psychiatry.

Sitting down on a slightly padded chair, I filled out the Patient Health Questionnaire (or PHQ-9) before the workshop began. I have filled out many such questionnaires since then—both over the phone and in person—and it isn't a difficult or time-consuming task. There is a list of nine common symptoms of depression and, for each one, a question: How often in the past two weeks have you felt like this? Not at all? Several days? More than half the days? Or nearly every day? Each answer is given a number from zero to three so that the number of symptoms is easily added up and the duration of life that they encroach upon can be quantified. It's a quiz you ideally don't want to score high on. For me, as I looked down at the questionnaire, four symptoms really stood out as descriptions of my recent self.

1. Little interest or pleasure in doing things. (Nearly every day, three.)
2. Feeling down, depressed, or hopeless. (Three.)
6. Feeling bad about yourself—or that you are a failure or have let yourself or your family down. (Two.)
9. Thoughts that you would be better off dead or of hurting yourself in some way. (Two.)

I don't know why, but I tried to shield my paper as I wrote down my scores. Perhaps I felt a deep shame in admitting to my problems, even if it was just a number on a piece of paper while I was sitting next to a complete stranger. Perhaps I didn't want anyone to have a quick insight into this part of my identity. After all, I was still coming to terms with it myself.

The aim of any treatment, whether it is talk therapy or drugs, is to reduce the severity of depression. When using the PHQ-9, this is gauged by the sum total of your nine scores. Score below five and you aren't considered to be depressed. Score above six, however, and you creep up through mild, moderate, moderately severe, and severe depression. (I was in the moderate to moderately severe range.) It's far from a perfect system. Everyone is likely to be biased by some degree and thus skew the results. In my case, I felt an urge to exaggerate my symptoms as a way to ensure further treatment. I didn't want to score below a threshold and be turfed out of the therapy I had waited months for. Other people, meanwhile, may trivialize their problems, as they don't think they have depression or need therapy. Either way, the PHQ-9—and other tools like it—is a quick and handy insight into emotional suffering.

With a dark November night at the window, our therapist—a broad-shouldered, smartly dressed man with a short black beard—wrote the word *RUMINATION* on the clipboard in large red letters. It was a word I knew—"to think about something deeply." Charles Darwin, for example, ruminated on the topic of evolution for a long time before writing his magnum opus, *On the Origin of Species*, in 1859. I have ruminated on the title of this book. But in CBT this word takes on a slightly different meaning: It is to cycle through negative thought patterns and become so consumed by them that you can't escape. You become trapped in your own mind, shackled by a world that you have created inside. Like training a muscle, the more you think that way, the stronger the cycle is, and the harder it is to break free.

The therapist then drew the face of a cartoon clown on the clipboard. It had a big looped smile, a red nose, and untamed hair. He asked everyone around the room to think of a word to describe clowns. They are funny. Happy. Colorful. But they are also terrifying, sad, and dark, the stuff of nightmares. All were correct, he said, it just depended on the context: the person who was answering the question. So, the field of CBT asks, why not choose the more uplifting options? If clowns are funny and happy, then you'll have a better time when you see one. Such a simple thought experiment reveals the basis of CBT. If you can choose how you think about objects and experiences in life, you can change the way you feel about yourself and the world around you.

I found these lessons to be helpful and interesting, but they made little impact on my mental health. I was silent throughout the group exercises and didn't want to share those problems that were most troubling to me. As in school classrooms or university lecture halls, there were a few outspoken people who seemed to speak for the rest of us, even though our problems might be very different. I wanted to talk about my self-doubt and feelings of guilt. I wanted to understand my suicidal thoughts, my separation from friends and family. I had turned twenty-five years old the previous month and I remember sitting at my birthday drinks feeling completely alone. I felt hollow, halfway to death on a day that celebrates birth. I didn't know then that these thoughts would later lead me so close to suicide that I shiver at the thought of not being here now.

After the four sessions of group CBT, I returned to my doctor and explained that I was still feeling depressed. We spoke for five or ten minutes; the briefest of insights into my symptoms, the course of my illness, and its potential triggers. We decided that I should take the antidepressant originally offered: citalopram. This drug would tinker with the neurotransmitters in my brain and, I hoped, make me stable again. I wasn't told of any side effects. I wasn't told that coming off these drugs could be difficult and potentially dangerous for some people. All I knew was that I was desperate for change, willing to do anything to recover. For the next four years of my life—as I started dating my partner, Lucy, moved from London to Berlin to Bristol, continued to write science articles for various magazines, signed a book deal, became an uncle, and adopted a dog—I would swallow an SSRI every day before bed.

Love and Hate

The first significant psychoanalytic theory of depression was proposed by Karl Abraham, a promising psychoanalyst from Berlin whom Freud initially kept at arm's length while trying to pull his protégé, Carl Jung, closer. A smooth-faced, stoic, and often sardonic man who, like Freud, was born to a Jewish family, Abraham felt a duty to understand and help his oft-ignored depressed patients. "While the conditions of nervous anxiety have been treated in detail in the psychoanalytic literature, the depressive conditions have not found similar consideration," he told an audience at the third International Psychoanalytic Congress in Weimar, Germany, in 1911. "And yet depressive affect is just as widespread." His first insight into depression came from Giovanni Segantini, a famous Swiss painter whose oeuvre he much admired. (During their correspondences and publications, it was not uncommon for Freud's followers to use pieces of art, literature, or mythology to illuminate their theories of human psychology.) One of Segantini's compositions stood out from the rest. Known as *The Wicked Mothers*, it is a chilling backdrop of ice-capped mountains, frozen snow, and, in the foreground, a pallid mother entangled in the trunk of a gnarled tree. A fierce, red-haired baby feeds at her breast. Abraham looked at this painting and saw a window into the roots of depression.

As he learned more about Segantini, Abraham's two-dimensional ideas of a "Wicked Mother" figure were given shape and substance. When Segantini was a baby, his mother was depressed after grieving the death of her first child (Segantini's older brother). Then, when he was five years old, his mother died. Raised by his grandparents, he suffered from severe bouts of depression and was often institutionalized or living on the streets. Abraham interpreted Segantini's story—and his painting—as a universal metaphor for depression. Although Segantini loved his mother and longed for her affection and attention, she was unable to satisfy his emotional needs. When she died, Abraham believed that Segantini's infant mind interpreted this as a form of abandonment. It was as if she had simply packed up her bags and run out on him. Hatred welled up inside, pushed love aside, and clouded his world as an adult. This unconscious hatred tainted everyone and everything in his surroundings. Not only did he himself hate, but it felt like everyone else hated him back. Segantini felt alone, castigated, and judged from every angle for his incapacity to love, the core features of many types of depression.

Love and hate are opposing forces in psychoanalysis. Eugen Bleuler, an early advocate of psychoanalysis working in Zurich, called the continual battle between these two emotions "ambivalence." Adopting this phenomenon into his own practice, Abraham proposed that his depressed patients didn't love their mothers when they were children and that hatred had been victorious in their emotional war of ambivalence. Absent of affection, feelings of hostility found an empty space in which to thrive. Abraham summarized his patients' experiences with the maxim "I cannot love people; I have to hate them." In his private practice, he noted how his depressed patients' dreams were often full of masochistic desires and seething with anger and frustration.

Abraham saw his own life reflected in his analysis of Segantini. "A recurring theme in Abraham's work is traumatic early abandonment by the mother, of the kind he experienced when he was two," the psychoanalyst and historian Anna Bentinck van Schoonheten writes in her 2015 biography of Karl Abraham. After a tragic fall down a flight of stairs, his mother miscarried her second child, a daughter, and a moment of intense grief turned into unshakable despair. "His mother was still physically present, but unreachable because of her depression," wrote Bentinck van Schoonheten. Too young to empathize with her loss, little Karl felt like his mother had abandoned him, sapping his childhood of the warmth and comfort that he most desired. Although he kept much of his personal life private, Bentinck van Schoonheten concludes that he suffered from severe depressive episodes for much of his adult life, periods when his letters become infrequent, impersonal, and nihilistic.

Juggling his growing private practice and the thrice-weekly meetings of the Berlin Psychoanalytic Institute, Abraham studied Segantini for two years between 1908 and 1910. As he learned of their similar childhoods and experiences with depression, he was unaware that this project was also a deathly view into his own future. Segantini had died from an infection in 1899. He was forty-one years old. Likewise, Abraham would die from septicemia and wouldn't live to see his fiftieth birthday.

Just as the gravitational forces of planets can slingshot passing asteroids into the icy hinterlands of space, the people whom Freud brought closest to himself were often destined to be pushed furthest away. Alfred Adler, Wilhelm Stekel, and even his "crown prince," Carl Jung, were all shown the door after they contradicted his dogmatic theories of an innate sexuality. His closest friends, Josef Breuer and Wilhelm Fliess, became his most hated former acquaintances. Abraham, on the other hand, was a harmonious force in the history of psychoanalysis. Where his leader was mercurial, he was levelheaded. Instead of disparaging any diversion from Freudian doctrines, he welcomed a broad range of views and insights from his peers. When mentally well, he was also a model psychotherapist. "He spoke little, but his silence was justified and in an extraordinary way urgent and encouraging," one patient recalled. "[H]is voice, with its dark timbre, was calm and calming. Cool and detached, but when necessary humanely intimate, he was certain of the trust of his students and patients."

After their first meeting in December 1907, Karl Abraham and Sigmund Freud wrote nearly five hundred letters to each other, a slightly greater number coming from Berlin than Vienna. Reading through their correspondence, it is clear that Abraham deeply admired his commander in chief and often expressed his desire to meet face-to-face. Whenever Freud was traveling away from Vienna for work or vacation, Abraham never failed to remind him that he would be thrilled to host him at his Berlin home, even if it was just for a few hours. Freud, meanwhile, was awkward and standoffish in his responses. Even when he did pass through Berlin on his way back from England in September 1908, he informed Abraham that he did not find the time to pop in to see him and his wife. "I was actually in Berlin for 24 hours and did not call on you," he wrote. "I could not, because I crossed from England with my aged brother and visited my sister who lives in Berlin, and between the two camps of fond relatives I saw as little of Berlin as I did of you." A few days later, Abraham replied, "Now, I must beg you, dear Professor, not to imagine that I took offense at your passing through Berlin. If you reproach yourself for this reason there is a simple way to make amends. When next you come to Berlin, please give my wife and myself a good deal of your time." When Freud was invited to Clark University in the United States to present his introductory courses to psychoanalysis, Abraham only asked whether he was able to pay him a visit on his way back. "That is what interests me particularly in this matter," he wrote. After Freud mentioned that he might be able to provide a "brief reunion," Abraham couldn't contain his excitement. "What gave me most pleasure in your letter was the hope of seeing you here in the autumn. Perhaps I may ask you already now to be generous in apportioning your time! I am so looking forward to it, and in the course of time such a great deal has accumulated that I want to discuss with you!"

Like two courting teenagers destined for a breakup, there was a disparity in affection between Abraham and Freud. Over the years, however, their relationship would become one of the most productive in the early history of psychoanalysis. "This was a partnership that paid rich scientific dividends," the psychoanalyst Edward Glover wrote. "Freud writing with magisterial clarity on the essentials of the science he had founded [while] Abraham, optimistic and sanguine as ever, reaching boldly into new territories."

His boldest territory was depression, a black hole of human psychology that few of his peers dared to contemplate. If Freud's friends, family, or colleagues suffered from depression, they were advised to travel to see Abraham in Berlin. His home—a spacious apartment opposite a fire station with a sandpit and playground in the back for the children—became an international hub for the psychotherapy of depression. For people who were often unwilling or unable to speak, never mind waxing smoothly through the practice of free association, psychoanalysis was often a painful experience for both the analyst and the patient. But Abraham found a way past these problems. As with Emil Kraepelin, he noticed that depressions came and went, that they were often separated by lucid periods in which the patient was more open to discussion, more motivated to dig into their dreams, thoughts, and painful memories. Just as the worst of the depression was subsiding, Abraham explained and interpreted his patients' symptoms, how they were the product of their ambivalence between love and hate, their relationship to their mother, and their inability to love.

He observed some significant improvements, even claiming that he had cured someone with melancholia—sometimes seen as a more chronic, less cyclical condition than depressive insanity—before it became permanent, "something which had never happened before." "By the help of psychoanalytical interpretation of certain facts and connections," Abraham added, "I succeeded in attaining a greater psychic rapport with the patients than I had ever previously achieved. . . . Psychoanalysis [is] the only rational therapy to apply to the manic-depressive psychoses." There was one patient, his sixth case of depression, that he thought exemplified the power of this talk therapy. Before entering Abraham's practice in Berlin, the patient had spent time in mental asylums and only found partial relief from his depression through such palliative care. But Abraham was able to analyze and interpret his depression, uncovering long-lost memories and linking them with his present state. Slowly, step by gradual step, the symptoms were removed as if this "talking cure" were slicing through the psychological shackles that had long held his patient down. "His severe depression began to subside after four weeks," Abraham wrote. "[H]e had a feeling of hope that he would once again be capable of work." After six months, the patient's analysis came to an end. His friends noticed a positive change in his character and, twelve months later, the depression hadn't returned. "I am happier and more care-free than I have ever been before," he reported.

Abraham knew that this was a promising start and nothing more. Six patients were hardly a representative sample of all depressed people. Plus, perhaps these patients would have recovered anyway, without psychotherapy of any kind. "Naturally," Abraham wrote, "a definite opinion as regards a cure cannot yet be given for after twenty years of illness, interrupted by free intervals of varying length, an improvement of [a few] months' duration signifies very little." Still, he believed that the improvement was in step with the success or failure of his analysis. With every memory he analyzed, every dream he interpreted, he believed that his patients looked slightly more stable, happier, and less governed by feelings of shame and hatred. As with his patients, the future for psychoanalysis in the treatment of depression was looking much brighter in Berlin. "Although our results at present are incomplete," Abraham wrote, "it is only psychoanalysis that will reveal the hidden structure of this large group of mental diseases."

Although he rarely saw it for himself, Freud was pleased with his colleague's efforts. "Berlin is a difficult but important soil," he wrote, "and your efforts to make it cultivable for our purposes deserve every praise." And by setting up shop in the booming capital of Germany, Abraham was infiltrating a country dominated by biological psychiatrists led by Emil Kraepelin.

Freud and Kraepelin never met. Like particles of matter and antimatter, perhaps the whole world of psychiatry would have collapsed if they had. Although the figureheads of two rival factions—the psychological versus the biological—they may actually have found many similarities to discuss. Born in the same year, both Freud and Kraepelin were trained by leading neuroanatomists of the nineteenth century. Their first scientific pursuits were influenced more by money and the opportunity of marriage than basic curiosity. And they each were interested in drug therapies and forms of counseling. History is never as simple as two opposing silos.

It is clear from their writings that they weren't overly fond of each other. In the eighth edition of his psychiatry textbook, published in 1909, Kraepelin wrote that "Freud and [Carl] Jung are nonsensical," adding that "the fundamental features of the Freudian trend of investigation [are] the representation of arbitrary assumptions and conjectures as scientific facts . . . and the tendency to generalization beyond measure from single observations. As I am accustomed to walk on the sure foundation of direct experience, my Philistine conscience of natural science stumbles at every step on objections, considerations, and doubts, over which the likely soaring tower of imagination of Freud's disciples carries them without difficulty." In short, this so-called science of psychoanalysis ran counter to the fundamental components of what Kraepelin would consider scientific.

In response, Freud wrote that Kraepelin was the leader of the "enemy camp" in the "blackest clique of Munich." In his famous *Introductory Lectures on Psycho-Analysis*, he added several scathing comments on the lack of understanding of such "scientific psychiatry," tying the biological approach to routine family histories and vague notions of heredity. "Psychiatry has given names to various compulsions; and has nothing more to say about them," he told his audience. "He has to content himself with a diagnosis and, in spite of wide experience, with a very uncertain prognosis of its future course. . . . [I]t is but a miserable contribution . . . a condemnation instead of an explanation." Defective heredity, strict diagnosis, long-term prognosis without treatment: although he didn't mention Kraepelin directly, he was criticizing the very core of his scientific contributions.

Psychoanalysis offered something that biological psychiatry badly needed: an understanding of a person's illness. Whether theories of love, hate, and sexual longing were correct hardly mattered. If someone believed that they were being heard, that their unique personal history was the reason for their tumultuous present, then the therapeutic gains could be wondrous. It wasn't an easy skill to learn, and psychoanalysts had to tailor their approach for different situations. Psychotherapy with a depressed person, for example, was a delicate balancing act. "There must be a continuous, subtle, empathetic tie between the analyst and his depressive patients," Edith Jacobson, a German émigré and disciple of Karl Abraham, wrote. "[W]e must be very careful not to let empty silences grow or not to talk too long, too rapidly, and too emphatically; that is, never to give too much or too little." Such patients are emotionally unstable, so prone to swing back into a depression that any disappointment or misconstrued comment can undo any progress made in the analysis. "For this reason they usually need many years of analysis with slow, patient, consistent work," Jacobson added. "With these patients, we are always between the devil and the deep blue sea."

The most basic elements of psychoanalysis—sitting down and talking with another human—were revolutionary, a model that subsequent psychotherapies would replicate. "Words and magic were in the beginning one and the same thing, and even to-day words retain much of their magical power," Freud explained in one of his lectures in 1915. "By words one of us can give to another the greatest happiness or bring utter despair; by words the teacher imparts his knowledge to the student; by words the orator sweeps his audience with him and determines its judgments and decisions. Words call forth emotions and are universally the means by which we influence our fellow creatures. Therefore let us not despise the use of words in psychotherapy."

A First Sketch

In addition to his *Psychiatry* textbooks and the neat classifications they contained, Kraepelin had guided the construction of a new scientific hub of biological psychiatry at the University of Munich. Built in 1904, the facility—a whitewashed, horseshoe-shaped building on a tree-lined street in central Munich—was replete with laboratories dedicated to experimental psychology, chemistry, serology (the study of blood), genealogy, and contained a vast archive for books and the storage of all of his *Zählkarte*. He had assembled an impressive roster of experienced staff from across Germany. Of note was Alois Alzheimer, a former student from his days in Heidelberg. Designing his own pathology laboratory—a well-lit room with fifteen work spaces, tabletop microscopes, and a separate room for microphotography, on the third floor of the clinic—Alzheimer would become famous for his discovery of a "strange and serious" form of dementia. Kraepelin, in the eighth edition of his *Psychiatry* textbook, named it "Alzheimer's disease," a much-needed reminder that brain anatomy could offer something to the study of mental disease, even if manic-depressive insanity and dementia praecox were still unwilling to reveal their innermost secrets. Everywhere scientists cared to look—in blood sugar, measures of metabolism, gastric juices, nitrogen levels in exhaled breath, brain slices—there was no insight into the causes of depression.

It was away from the laboratory where Kraepelin and Alzheimer had their greatest influence on psychiatry. In addition to their written works, their lectures and workshops were legendary. "Whoever has listened to the lessons of Alzheimer and Kraepelin," the Italian physician Gaetano Perusini wrote in 1907, "wherever they came from, will not have found the journey too long." And wherever someone lived and worked, Kraepelin's *Psychiatry* textbook came to be seen as a bible of biological psychiatry. His diagnosis of manic-depressive insanity was a necessary update from notions of black bile and melancholia. But there were still heated debates over whether it was overly simplistic, and if there were other forms of depression that weren't depressive insanity. There were those who thought distinct disease states in psychiatry didn't exist and any attempts to classify patients were merely satisfying a basic human need for order. To Alfred Hoche, a German psychiatrist also working in Munich, the work of Kraepelin and his followers was flawed; "a kind of thought compulsion, a logical and aesthetic necessity, insists that we seek for well defined, self-contained disease entities," he wrote, "but here as elsewhere, unfortunately, our subjective need is no proof of the reality of that which we desire, no proof that these pure types do, in point of fact, actually occur." He added that the recent surge in excitement over classification, or "nosology," in psychiatry wasn't creating new understanding, but was like "clarifying a turbid fluid by pouring it busily from one vessel to another."

Kraepelin was his own worst critic. He cautioned that his classification system still didn't reflect a biological reality. "We are still so far removed from a real knowledge of the causes, phenomena, course, and termination of the individual forms [of mental disorder] that we cannot yet dream of a surely established edifice of knowledge at all," he stated at one of his famous *Lectures on Clinical Psychiatry*. "What we have formulated here is only a first sketch, which the advance of our science will often have occasion to change and enlarge in its details, and perhaps even in its principal lines." He was remarkably candid when it came to his work's shortcomings.

Kraepelin had tried his utmost to find a true map of mental disease and came out feeling like he hadn't done enough. As two historians of Kraepelin, Eric Engstrom and Kenneth Kendler, wrote, "Experience had shown that what at first appeared to be sharp clinical boundaries had become ever more blurred and that a 'thorough differentiation between normal and pathological conditions' was an impossibility." In lieu of certainty and fact, Kraepelin had created an introduction and a guide into the confusing, irrational, and often extraordinary minds of the mentally ill. Although psychoanalysis would come to dominate psychiatry by the middle of the twentieth century—especially in the United States—Kraepelin would have the last laugh.

In 1980, an emergence of "neo-Kraepelinian" philosophy resulted in the third edition of the *Diagnostic and Statistical Manual for Mental Disorders*, or *DSM-III*, a booklet created in the U.S. and used as a "bible" to mental illness around the world. As Edward Shorter, a historian of medicine from the University of Toronto, wrote, the "*DSM-III* represented a massive 'turning of the page' in nosology, and it had the effect of steering psychoanalysis toward the exit in psychiatry and the beginning of a reconciliation of psychiatry with the rest of medicine." Rather than trying to reveal a person's repressed memories or interpret dreams, psychiatrists argued that there were neat disease states—diagnoses—that could be tracked through time. In place of manic-depressive insanity, there was another all-encompassing diagnosis: major depressive disorder. But such a simplification belies the diversity of depression. Throughout the twentieth century, treatments would be developed that were specific to certain symptoms or severities of depression, a few beams of light to illuminate a once dark and endless forest.

In September 1915, Freud sent a letter to Karl Abraham, then working as a surgeon's assistant on the Eastern Front of World War I. Situated in Allenstein, a bucolic setting of lakes and trees, his characteristic stoicism was put to the test. "[I] have found confirmation of the solution of melancholia in a case I studied for two months," Freud wrote. Abraham was stunned. He couldn't believe that anyone, not even "the genius" Freud, had dipped into his field of expertise and found something that he had not. Freud's claim that he had solved the puzzle after two months of analysis seemed to belittle his years of careful observation, his thorough analysis of Giovanni Segantini, and two lengthy publications on this very topic.

As with many of his other theories, Freud had absorbed the ideas around him and added a subtle modification that he could call his own. Although hatred was a key part of depression, Freud wrote, it wasn't because of an absent or uncaring mother, as Abraham had argued. It was much broader. The causal factor was similar to mourning. But instead of the death of a loved one, the patient suffers from the loss of any type of "love-object," a family member, a relationship, or an unfulfilled expectation of the self. The mental energy that was once focused on this object, Freud wrote, was then branded by its departure. In depression, this negative energy was turned inward (not outward, as Abraham had proposed), straight back into the person's own mind—their "ego." This would become known as "retroflected hostility," a process that showed itself through the intense periods of self-reflection, critical judgment, and, ultimately, hatred. The depressed person, Freud wrote, begins to feel "worthless, incapable of any achievement and morally despicable. . . . [H]e reproaches himself, vilifies himself and expects to be punished. He abases himself before everyone and commiserates with his own relatives for being connected with someone so unworthy."

Reading through Freud's scribbled draft over and over, Abraham felt like his work had just been appropriated without credit. Love, loss, and hate—it was all uncomfortably familiar.

For the first decade of their friendship, Abraham was punctual in his correspondence with Freud. While the latter would often take days or weeks to respond, Abraham would reply almost as soon as he had opened the latest envelope. With this particular letter, however, he waited a month to reply. It wasn't because he was busy learning how to patch up injured soldiers on the Eastern Front. And it wasn't due to the delays of the postal service caused by the disruptions of war. It was because he was writing one of the longest letters he would ever send to Berggasse 19, Vienna.

When he finally posted his response to Freud's theory of depression, Abraham's letter was respectful and delicately balanced, much like the man himself. Although he hadn't experienced it firsthand, he knew that Freud was prone to volatile outbursts and swiftly made enemies of anyone who had contradicted his theories in the past. "I should like to remind you—not in order to stress any priority but merely to underline the points of agreement—that I also started from a comparison between the melancholic depression and mourning," he wrote. Fearing that even this was a little too bold, Abraham sheepishly added that, although he wished that his work were more clearly referenced, he was "able to accept all the essentials" of Freud's theory. "It seems to me that we ought to agree easily." Even those elements that he still couldn't accept, he wrote, "as you know, dear Professor, I am ready to re-learn."

Over a month later, Freud told Abraham that he had just finished writing the second draft of his grand treatise on depression. As requested, he credited Karl Abraham, "to whom we owe the most important of the few analytic studies on this subject." He wasn't saying that his theories were inspired by his colleague's research. Freud was merely informing his readers that Abraham had also done some work in this area.

Published in 1917, Freud's essay "Mourning and Melancholia" wasn't the first psychoanalytic interpretation of depression and neither was it the last. Nonetheless, it was a foundational text that crystallized a blurry set of ideas that Abraham had published and gave budding psychoanalysts the confidence to treat depressive disorders. Through the methods of free association and the analysis of dreams, thoughts, and memories, it was the psychoanalyst's job to reveal what love-object their patient had lost. Abraham, Freud, and their followers were paleontologists of the mind, digging through the unconscious strata of their depressed patients, hoping to uncover the old bones of a pathological monster.

By 1925, Abraham had fully adopted Freud's theory of object-loss and merged it with his former ideas of maternal abandonment. Depression, he wrote, is . . .

> connected to an event with which a person in their mental state at the time was not able to cope . . . a loss that caused a profound inner shock and was experienced as unbearable. It was always the loss of someone who was the center of the person's life. Loss did not necessarily mean death, but an intense relationship with another person was suddenly destroyed, for example by a disappointment that could never be set right. The feeling of total abandonment that resulted would then trigger depression. A disappointment of such magnitude could only happen in early childhood with regard to the mother and it would cause a desire for revenge that was hard to control. Hidden deep beneath that vengefulness was a nostalgia for the original mother, for the earliest sense of satisfaction at the mother's breast.

While Freud was focused on the Oedipus complex, a stage of psychosexual development that arose between the ages of three and five, Abraham pushed the origin of depression back into the earliest moments of human attachment. During the so-called oral phase, Abraham reasoned, the baby develops its first love-object: the warm, nourishing, and comforting milk and the breast from which they suckle. With this early love-object as a role model, humans are then primed to seek other attachments as they grow into adults. Healthy relationships that provide comfort, for example. But when these relationships break, the loss can feel like the world is ending, like a baby hungry for milk and screaming for its next feed. For those people who had passed through the "oral phase" of their infancy with frustration and disappointment, Abraham thought, this same feeling emerged once more in later life whenever a love-object was lost. Depression was the result.

"Abraham's [ideas] about a pre-Oedipal mother-son relationship," Anna Bentinck van Schoonheten writes, "gave a whole new direction to psychoanalysis." While Freud was still focused on the importance of father figures, penis envy, and death wishes, Abraham's disciples provided some of the most exciting avenues for psychological therapies to follow. From the 1920s onward, Melanie Klein, a former student of Abraham in Berlin, studied the behavior of newborns and young infants in order to understand the earliest stages of psychological development and how they might give rise to depression later in life. As one textbook on psychoanalysis and depression states, "[I]t was Melanie Klein who first elaborated the theory that this predisposition depends not on one or a series of traumatic incidents or disappointments but rather on the quality of the mother-child relationship in the first year of life. If this is of a type which does not promote in the child a feeling that he is secure and good and beloved, he is, according to Klein, never able to overcome his pronounced ambivalence toward his love-objects and forever prone to depressive breakdowns." In the 1950s, John Bowlby, a British psychoanalyst, would add to Abraham's and Klein's work and create his "attachment theory" of depression that made basic contact, clinging, and comfort essential resources for a person's mental health. It is important to remember that it wasn't Freud who first initiated this field of study. It was Karl Abraham.

Before the armistice of November 11, 1918, was signed and World War I ended, Abraham returned to Berlin from the sylvan hillsides and lake-filled valleys of Eastern Europe. After a few shock deaths in the family, he was shaken into a deep depression that turned his hair gray and his body thin. But he still remained in regular correspondence with Freud in Vienna, and continued to be his most loyal servant. Along with five other members, he wore a golden ring containing a small Grecian stone from Freud's personal collection of treasures and antiquities. After years of service, Abraham was twice elected president of the International Psychoanalytical Association and, from this vantage point, was a shrewd force in the promotion of Freudian psychology around the world. "Abraham will in future be recognized as one of those men of outstanding ability who helped fertilize and stir into movement the, till then, paretic science of psychiatry," the psychoanalyst Edward Glover wrote.

On June 7, 1925, Abraham wrote a letter to Freud while he was resting in bed. His throat was swollen and sore. His body temperature swung up and down with feverish ferocity. He believed that he had caught the flu from his recent sojourn to Holland and had brought it back with him to Berlin. In fact, he had a fish bone lodged in his throat that had become infected. In a time before antibiotics, such an unlikely and unfortunate injury could be lethal. Over the coming months, Abraham would exhibit the hallmark symptoms of septicemia: fevers, inexplicable recoveries, and even moments of euphoria. At those conferences that he managed to attend, he looked emaciated, a shadow of his former athletic self. When he was younger, Abraham was the first to conquer two mountains in Switzerland. He named the second peak after a climber who had failed in his attempt. Now, only in his late forties, he was taking a mountain railway in Bernese Oberland up to an altitude of 1,270 meters, this time in the pursuit of fresh air. Reclining in a deck chair at the Hotel Victoria, unaware that he was about to die, he read his favorite philosopher, Aristophanes, a writer of dark comedies in ancient Greece.

Upon hearing of Abraham's illness, Freud joked that he didn't know that he was capable of staying in bed. "I like to think of you only as a man continually and unfailingly at work," he wrote. "I feel your illness to be a kind of unfair competition and appeal to you to stop it as quickly as possible." Two years earlier, Freud had learned that the pain in his jaw was cancerous and spent large amounts of his own time recuperating from painful surgeries, radiotherapy, and frequent adjustments to his new prosthetic jaw.

Karl Abraham died on Christmas Day in 1925. Freud's last letter to Abraham's apartment in Berlin was to Hedwig, his colleague's wife of nearly twenty years and the mother of his two teenage children, Hilda and Gerd. Freud, writing in February 1926, was still unable to process his grief. "I have no substitute for him," he confessed to Hedwig, "and no consolatory words for you that would tell you anything new." The great translator of human misery was lost for words. Ernest Jones, an English psychoanalyst and fellow ring bearer, found the appropriate interpretation. "There is no way of meeting this blow," he wrote in a letter to Freud, "it cannot be dealt with, for nothing can ever cure it—not even time. The loss is quite irreplaceable."

Several months later, in October 1926, the field of biological psychiatry also lost a father figure. After a short bout of pneumonia, Emil Kraepelin was laid to rest in Heidelberg, the clement city where his family first found happiness in 1890. Buried in a graveyard that sat at the foot of a *Königstuhl* ("King's Chair"), the mountain that he had walked up and down so often in life, his name was engraved into a smooth lump of granite that looked like the upturned nose of a canoe. There was no date, no flashy accoutrements or religious maxims of any kind. Under his name, there was only a single sentence:

Dein Name mag vergehen
Bleibt nur dein Werk bestehen.

Your name may vanish, only your work will continue.

PART TWO

"The Biological Approach Seems to Be Working"

. . . the extent to which a treatment flourishes is directly dependent upon the specific features of the day's clinical landscape. In the long haul, viability is a matter of ecology, not virtue.

Jack Pressman

. . . for those patients who respond to a particular tranquilizer or antidepressant, old or new, that one drug is invaluable; it means for them the difference between sickness and health. . . . humans and not statistics suffer.

Pierre Deniker

Depression, which was once thought to be merely an unfortunate state of mind for which the depressed patient was largely responsible himself, is now recognized as a very common and painful disease—for which, fortunately, highly successful types of medical treatment have become available.

Nathan Kline

Fighting Fire with Fire

A decade before his death, Kraepelin's legacy had reached outside of Germany, crossed the English Channel, and settled in a new psychiatric stronghold in South London. After visiting Munich, Frederick Mott, director of the London County Council's Central Pathological Laboratory, was so impressed with what he saw that he returned to Britain intent on building something similar, a psychiatric hospital that was a hub of scientific research. He knew someone who would be interested in such a project: Henry Maudsley, a world-famous psychiatrist who wrote essays, novels, and rubbed shoulders with Charles Darwin and other members of the nineteenth-century intelligentsia. Maudsley thought it was a project long overdue. He offered £30,000 for the construction of a hospital that put science and research at the center of British psychiatry, a sizable sum at the time. By the end of the project, his donation had doubled. The language of psychiatry was changing from German to English.

Built in 1917 and looking more like a city hall than a mental institution, the Portland stone and red bricks of the Maudsley brought the treatment of mental illness into the community. There were few beds and most patients could come and go with as little as twenty-four hours' notice. This buzzing outpatients department was frequented by a diverse set of patients that would previously have been hidden away at home.

The first superintendent here, Edward Mapother, hoped to continue what Emil Kraepelin had started: to align psychiatry with medicine, a field of strict diagnoses and cures. The son of a former president of the Royal College of Surgeons in Ireland, Mapother followed in his father's footsteps and trained in medicine, specializing in neurology before working as a military doctor during World War I. As part of the Royal Army Medical Corps, he was positioned in northeastern France during the Battle of Loos, a two-week conflict in which Britain lost twice as many soldiers as the oncoming German army. Mapother saw firsthand what shock and stress could do to the mind, and how little he could do to help ease the pain. Some days, it felt like he shared the "thousand-yard stare" of these shell-shocked soldiers, peering into the abyss.

In the late 1920s, Mapother was at the center of a debate concerning the diagnosis of depression. More specifically, manic-depressive insanity, the all-encompassing diagnosis of Kraepelin. Was it a true reflection of the patients who were seen as outpatients and those who spent time in institutions? Some psychiatrists argued that there were actually two major groups of depression. The first was caused by environmental influences, either as childhood trauma, as many psychoanalysts posited, or as reactions to war or misfortune. This was known as "neurotic" or "reactive" depression. The second was more biological in origin and was known as "psychotic" or "endogenous" depression. (Confusingly, the terms *neurosis* and *psychosis* had previously been used to describe the complete opposite conditions. *Neurotic* was a term for physical nervous ailments such as neurasthenia. A psychosis was a disorder of the mind that often came with hallucinations. As psychiatry became anglicized, it was as if the language used became increasingly contradictory.) The terms were vague and often relied more on metaphor than scientific reasoning. One doctor, for example, explained that someone with neurotic depression "looked on a beautiful garden [and] obtained as much pleasure out of it as anyone else," while the psychotic "could see it was beautiful but . . . could not feel it."

In 1926, at the annual meeting of the British Medical Association held in Nottingham, the city of Robin Hood fame in the midlands of England, Mapother argued that there was only one depression: manic-depressive psychosis. It was a spectrum, a condition that varied "infinitely in degree," that each person could move up or down depending on the duration, severity, and potential recovery of their condition. Mild or severe, long or short, neurotic or psychotic, there "is no constant or specific cause, and no distinctive bodily change recognizable during life or after death," Mapother said. "The range of the term [depression] is a matter of convention." The only division that he could see was whether someone was institutionalized or not. People seen in the community were more likely to be called neurotic, while those in institutions were labeled as psychotic. "I can find no other basis for the distinction," Mapother said, "neither insight, nor cooperation in treatment, nor susceptibility to psychotherapy will serve."

Shy, introverted, and sometimes so nervous that he vomited before big presentations, Mapother filled the lecture hall with despair and frustration when it came to treatments. "A patient with serious emotional disturbance should be kept in bed as in the case of a feverish tuberculosis patient. . . . [O]pen air and sunlight are nearly as important in [depression] and massage is a useful substitute for exercise." A balanced and nutritious diet kept the worst of the weight loss at bay, while sedatives were prescribed to soothe troubled minds into a deep—but hardly refreshing—sleep. The treatment, in other words, was much the same as it was in William Tuke's days at the Retreat in York in the late eighteenth century. Over a hundred years had passed with little change. The one thing that *had* changed was the emergence of psychoanalysis, but Mapother was a vocal critic of its use. "[A]ctive psychotherapy, especially analysis of any kind, is merely harmful," he said.

It was a desperate situation, one that was made worse by the economic depression that encircled the globe for much of the 1930s, sapping funding from most studies into mental illness. "At the time," Heinz Lehmann, a German psychiatrist working in Montreal, Canada, wrote, "a clinician was almost helpless when faced with a depressed patient." A range of treatments—nitrous oxide inhalation, testosterone injections, X-ray irradiation, opium, the removal of teeth that were believed to be infected—had been tried and failed. Doctors and patients shared a feeling of hopelessness. Inside mental institutions, forced feedings and twenty-four-hour suicide watches were still common practice for the most severely ill patients. "Even in the more neurotic types of depression, and despite all the psychotherapy given, suicides were frequent, and patients often took months or years to get well no matter what one tried to do to help them," one psychiatrist working in London recalled. The only evidence-based finding came from *who* was most likely to suffer from depression. Based on the records from Emil Kraepelin's clinic, Helen Boyle, an English doctor who studied the thyroid and its associations with mental illness, said in 1930, "70 percent are women—a large preponderance of women."

In lieu of effective therapy, the staff at the Maudsley collected as much information on their patients as possible, hoping to uncover some link between their history and their current situation. Like psychoanalysts digging through the unconscious, biological psychiatrists reached back through the years in the hope of finding an explanation for a person's present. Ultimately, however, it was a means of keeping busy. After filling out thirty pages on a patient, one psychiatrist wrote that such an exercise "gave us a feeling that we were doing something for the patient by learning so much about him, even if we could not yet find any relief for his suffering."

The one success story in psychiatry came from an unlikely source: malaria. Injecting this infectious plasmodium into the blood of patients with the dreaded neurosyphilis, Julius Wagner-Jauregg, a mustachioed psychiatrist working in Vienna, found that this once-fatal disease could be cured if caught early enough in its course. Caused by an infection of syphilis that had corkscrewed into the delicate neurons of the brain (hence "neurosyphilis"), the disease would normally progress through a series of familiar stages, each a stepping-stone further into despair. First, a person's speech would start to degrade. Then muscles in the face and around the eye would begin to quiver and twitch uncontrollably. Finally, delusions of grandeur—the ownership of untold riches, unparalleled fluency in a foreign language, or hundreds of sexual partners—descended as patients rocked between depression and mania. Some people died believing they were immortal.

For over a century, a range of treatments had been tried. Low-calorie diets, bloodletting, leeching, cupping, purgatives, mercurial ointments, diuretics, proteins from tuberculosis, boiled milk: all failed. Once a diagnosis of neurosyphilis was confirmed, one doctor wrote, "[I]t is the duty of the physician to say at once that the case is without hope." But the fires of a malarial infection, Wagner-Jauregg found, could cleanse the body of this disease. Out of his first four hundred patients, 60 percent achieved remissions within two years and were able to return home and to work. In 1923, one doctor pushed this figure to over 72 percent. After eight to twelve rounds of fever, a three-day course of quinine—an antimalarial drug extracted from the bark of cinchona trees that grow on the slopes of the Andes Mountains in South America—rid the body of the *Plasmodium vivax* cells. Although milder than other types of malaria (such as *Plasmodium falciparum*), this so-called tertian malaria was still very capable of killing. Indeed, patients did die from this treatment.

In 1929, Wagner-Jauregg was awarded a Nobel Prize for his work, the first psychiatrist to receive the honor. Although antibiotics such as penicillin were soon introduced into the treatment of syphilis (preventing the disease from entering the nervous system in the first place), Wagner-Jauregg's legacy could still be found in the message his work sent to other biologically orientated psychiatrists: with daring effort and justified risk, even the most hopeless cases can be cured. With few other options than to sit and wait, treatments for depression—the embodiment of hopelessness—followed this example. "In this therapeutic and theoretical vacuum, almost any treatment was tried, providing it had the potential for treating large numbers of patients with a minimum of highly trained staff," Elliot Valenstein wrote in his book *Great and Desperate Cures*. "For many physicians, risking any therapeutic possibility was preferable to confessing helplessness."

Unfixing Thoughts

For most of her adult life, Becky was prone to extreme emotional outbursts. She screamed and wailed when she experienced the slightest disappointment in her day. She flailed her arms, knocking over anything that she could. Sometimes, she'd throw her own feces. Becky was, after all, a chimpanzee. But, one day in 1934, she suddenly stopped going—for lack of a better word—apeshit. She appeared to be a completely different chimp. She could endure endless disappointment, failing to open a box that contained treats hundreds of times in a row, and remain unfazed. In fact, she was more likely to be euphoric than angry. Fail or succeed, Becky seemed like a very content, happy chimpanzee.

Carlyle Jacobsen, a thirty-three-year-old psychologist from the Primate Research Center at Yale University in New Haven, Connecticut, presented Becky's story at the second International Neurological Congress at King's College London in August 1935. The significant change in her character, he explained, was related to a change in her brain. Along with his supervisor and director of the research center, John Fulton, Jacobsen explained how they had removed Becky's frontal lobes, the regions of the brain that sit just above the eye sockets. Her dramatic change in behavior, in other words, was a result of a physical modification in her brain. Compared to her former self, Jacobsen thought that Becky had joined a "happiness cult."

By this time in the 1930s, it was well-established that different parts of the brain had different functions. Like a children's jigsaw puzzle composed of several large pieces, scientists had started to separate what looks like a homogeneous mass of soft tissue into a complex organ with regions dedicated to memory, movement, and hearing. Even speech could be mapped to a specific blob of tissue—known as Broca's area—that was located on the underside of the temples. But did this "brain localization" theory hold true for traits as abstract and individual as emotion, temperament, personality? Indeed, a few remarkable case studies seemed to suggest that this was the case.

The most famous example, both now and in the 1930s, was the so-called American Crowbar Case. In September 1848, a twenty-five-year-old railroad worker named Phineas Gage was hammering his tamping iron (a meter-long rod that was used to pack explosives into a drilled hole) when a spark ignited and, in the blink of an eye, propelled the iron out the hole and through Gage's head. Landing thirty meters down the Rutland and Burlington Railroad, it was smeared with greasy bits of brain and blood. Although the tamping iron had entered his face just underneath the cheekbone and exited from the top of his forehead, a large chunk of his frontal lobes being destroyed in the process, Gage survived and seemed coherent soon after the accident. After his physical injuries healed, however, he wasn't the same person. His doctor noted that he was "fitful, irreverent, indulging at times in the grossest profanity (which was not previously his custom), manifesting but little deference for his fellows, impatient of restraint or advice when it conflicts with his desires." His friends simply thought that he was "no longer Gage." His employers found his change of character so marked that they refused to offer him another job. Unable to return to the railway lines, Gage became a curiosity of Barnum's circus.

Similarly intriguing insights into the function of the frontal lobes emerged after brain surgeries. In 1930, an American man known simply as "Patient A." had large sections of his frontal lobes removed as a result of a brain tumor. After the operation, he was reported to be emotionally rigid and often euphoric—two traits that seemed out of character when compared to his former self. Amazingly, the only thing that seemed to remain constant was his intelligence. Even without his frontal lobes, it was reported that he suffered no decline in IQ.

Egas Moniz, a sixty-year-old neurologist from Lisbon, Portugal, was sitting in the audience that day at King's College London, listening to Jacobsen's descriptions of Becky the chimpanzee. Tall and besuited with a shiny toupee covering his balding pate, Moniz asked Jacobsen whether such results might be used to develop new treatments for mental disorders. Could the frontal lobes be modified in some way to change the emotional extremes of psychiatric patients? Jacobsen replied that he hadn't given it much thought. He was an experimental psychologist, not a clinical psychiatrist. In truth, neither had Moniz until he heard this lecture. Still, he was excited to put such experiments into practice back home. He didn't have any chimpanzees to work with. But he knew where he might be given access to institutionalized mental patients.

Born on November 29, 1874, in the small coastal village of Avança, in northern Portugal, Moniz grew up hearing stories from his uncle about his country's colonial past, how this once-great empire had spanned the globe from Asia to South America and from Africa to Oceania. Moniz's own family history reached back five hundred years in Avança alone. But, as with the Portuguese Empire, these riches were slowly lost, and after the unfortunate and unrelated deaths of his parents, his brother, and his beloved uncle, Egas Moniz became the sole survivor and heir to a once powerful aristocracy.

After graduating from Coimbra University in 1899, Moniz accepted a position of lecturer at the medical school, trained in neurology in Paris, and became the Portuguese ambassador to Spain. During his political duties, he met the Nobel laureate Ramón y Cajal. Working in Barcelona, Cajal had published some of the most intricate and beautiful maps of our nervous system and, through a series of similarly intricate experiments, provided the best evidence for synapses, the tiny gaps that separate one neuron from another. He didn't know it at the time, but Moniz would follow in Cajal's footsteps; he, too, would be awarded a Nobel Prize, the first Portuguese scientist to receive the honor.

Moniz's first breakthrough in science came late in life. In his early fifties, working at the University of Lisbon, he published the first photo of cerebral angiography, an early brain-imaging technique that involved injecting an opaque fluid into the bloodstream and detecting its flow through the brain with X-rays. Areas of the brain where blood was blocked or diverted through other vessels indicated the presence of an anomaly such as a brain tumor. It was an exciting and state-of-the-art brain-imaging technique, just like the functional MRI scans of today. Single-minded, daring, and determined to add to his rich family history, Moniz published over one hundred articles and two books on angiography. He was hailed as one of the leading neurologists of the twentieth century, a person who other scientists could look to for guidance.

Moniz attended the 1935 congress in London to present his work into angiography. His photos were plastered over an entire wall. But, as he sat in the lecture hall and listened to Carlyle Jacobsen discuss Becky the chimpanzee, he thought that he had just discovered his next project. Instead of visualizing blockages in a person's brain due to cancerous tumors, he wondered whether he could introduce blockages of his own. "Seldom in the history of medicine has a laboratory observation been so quickly and dramatically translated into a therapeutic procedure," one psychiatrist later said.

On November 12, 1935, three months after the congress in London, a woman with severe depression and paranoia was placed under general anesthetic and prepped for her operation in a surgical room at Hospital Santa Marta, a long two-story building with a baroque tiled interior. With swollen hands from the gout that had painfully erupted when he was just twenty-four years old, Moniz lacked the dexterity for precise surgical procedures. But he guided his neurosurgeon, Pedro Almeida Lima, as best he could. The patient's hair had been shaved from the forehead to behind the ears. Pure alcohol was rubbed onto the bald scalp to avoid contamination. Then Lima bored a couple of holes through the unconscious woman's skull, each a few inches from the midline, like two upward-facing eye sockets. Swapping his drill for a sharp scalpel, he then cut through the liquid-filled cavity, known as the "dura," that cushions the delicate brain inside the skull. Draining the overflowing fluid and cauterizing the blood vessels, he was faced with the shiny pink folds of the cerebral cortex, the outer region of the brain that holds our most human characteristics: abstract thought, consciousness, the self. To access the bottom of the frontal lobes, an area that sat just above the back of her eye sockets, Lima used a hypodermic needle to reach through the cortex and squirted a few drops of pure ethanol to dehydrate, and destroy, any neurons in this region. Moniz called this site of destruction a "frontal barrier," one that he thought would release his patient from her fixed thoughts.

After the woman came to, Moniz declared the surgery a success. Her depression and restlessness had cleared and he said it was a "clinical cure." Over the next four months, Moniz and Lima performed similar operations on nine patients provided by José de Matos Sobral Cid, a professor of psychiatry at the Miguel Bombarda mental hospital in Lisbon. Nine of the patients (seven women and two men) were diagnosed with depression, six had schizophrenia, two had panic disorders, and one person suffered from what would later be called bipolar disorder. After the seventh patient, ethanol was swapped for a sharp metal instrument that Moniz called a "leucotome," after the Greek words *leukós*, for "white" (as in white matter), and *tome*, for a cutting instrument. About the size of a chopstick, the leucotome had a retractable wire loop near one end that, when rotated, made a circular cut in any soft tissue it touched. To sufficiently disconnect the frontal lobes from the lower parts of the brain, Moniz advised Lima to make four to six cuts, two or three on each side of the brain. This became known as the "core operation." The operation itself—a slicing of the white matter that sits just behind the frontal lobes—was called a prefrontal leucotomy.

Soon after the neurology congress in London, Moniz wrote that "true progress in this science [of psychiatry] will only be made with an organic orientation." A strict biological psychiatrist, he was no disciple of Freud's theories of the psyche, the ego, or the unconscious, that is. He was a strict biological psychiatrist. Mental disorders were rooted in the delicate concatenations of the brain, just as a broken leg is the product of a fractured bone. To Moniz, diseases such as depression and schizophrenia were the result of "fixed connections," pathways in the brain that were overused, like congested motorways. This biological traffic was replicated in a person's obsessions and certain thoughts or behavior. People suffering from severe depression, Moniz wrote, "live in a permanent state of anxiety caused by a fixed idea which predominates over all their lives." Adding that "these morbid ideas are deeply rooted in the synaptic complex which regulates the functioning of consciousness, stimulating it and keeping it in constant activity." To Moniz, this relationship between our anatomy and our actions revealed only one therapeutic option: "[I]t is necessary to alter these synaptic adjustments and change the paths chosen by the impulses in [an attempt] to modify the corresponding ideas and force thoughts along different paths." Could Moniz block or redirect the flow of pathological thoughts in a depressed person's brain? Could he free their fixed thoughts with a simple surgical operation? Just as Becky was made happy and stoic after surgery, Moniz argued that severing the frontal lobes from the rest of the brain could turn a depressive character into a member of Becky's "happiness cult."

Moniz later argued that he had been thinking about prefrontal leucotomies long before he had even heard of Becky the chimpanzee, but there doesn't seem to be any good evidence for this claim. Even a cursory glance through the scientific literature of the time would have revealed how the function of the frontal lobes was far from clear. The chimpanzee research from Carlyle Jacobsen and John Fulton at the Primate Research Center at Yale University, for instance, provided a valuable insight into the unpredictability of frontal lobe surgeries. Becky's pen pal, Lucy (also a chimpanzee), had the same operation, but was transformed from a calm character into a cantankerous, poo-flinging alter ego. It was the same operation with completely opposite outcomes. "In his review of the literature on the frontal lobes," the neuroscientist and historian Elliot Valenstein wrote, "Moniz simply extracted what was useful to his argument and ignored the rest."

On July 26, 1936, at the Parisian meeting of the Société Médico-Psychologique, one of Moniz's students explained how fourteen out of twenty patients had improved after their prefrontal leucotomy. Sobral Cid, the professor from Miguel Bombarda mental hospital, who was in attendance, argued that his patients were severely "diminished" and had exhibited a "degradation of personality" after the surgery. "As for the hypothesis of functional fixation by which [Moniz] explains the good results of his method," he added, "it rests on pure cerebral mythology."

In spite of such immediate firsthand criticism, leucotomies were soon performed in Romania, Italy, Brazil, and Cuba. But, even in these countries, the procedure was rarely used. Many psychiatrists were hesitant to take up such a radical and permanent approach to the treatment of mental disorders such as depression. Then, in the spring of 1936, an American psychiatrist from George Washington University opened a copy of Egas Moniz's first manuscript on leucotomy and thought that every page was bursting with a psychiatric revolution. His name was Walter Freeman, a goateed man who often walked with a cane, and he was as charismatic as he was controversial.

"The Brain Has Ceased to Be Sacred"

Walter Freeman was the kind of doctor who could fill a lecture hall on the weekend. Even though they could be socializing or sleeping, his students would even bring their dates to his classes on brain anatomy and psychiatry. He was both a lecturer and a performer. "In the gloomy and tight days of the Great Depression, Freeman's lectures substituted for entertainment," wrote his biographer Jack El-Hai. Standing at the front of the hall, Freeman would draw on the blackboard with both hands, not only to save time but as a dazzling accoutrement to his act. When he used patients from the university's psychiatric wards as living case studies, he treated each one as just another educational tool, largely insensitive to their feelings or confidentiality. To his audience, Freeman treated the bodies from the morgue and people from the hospital with the same level of respect, slicing through their brains with either a scalpel or his sharp tongue.

Often described as a genius, Freeman was an influential figure in psychiatry from a young age. In 1926, when he was just thirty-one years old, he was appointed as the head of his own department at George Washington University, an institution in the center of Washington, D.C. A year later, he became a secretary of the American Medical Association, an academic perch that allowed him to oversee and support certain developments in the field. Like Egas Moniz, he presented his own work into brain imaging at the neurological meeting in London in August 1935, the first time the pair met. A year later, after reading that Moniz had put his theory of "fixed thoughts" into practice in Lisbon, Freeman started collaborating with his own skilled neurosurgeon at George Washington University. Tall, clean-shaven, and taciturn, James Watts shared a passion for brain science and the potential of neurosurgery. A former student of John Fulton at the primate laboratories at Yale, he had met Becky the chimpanzee before her operation. In contrast to her usual cantankerous nature, she vaulted over a few rows of seats to greet him and wrapped her arms around his neck in a loving embrace. Fulton, meanwhile, wasn't so enamoured of Watts. His student's seriousness contrasted with his own jovial character. Yet he couldn't help but admire his skills as a surgeon. As he once commented, Watts had "the most beautiful pair of hands that I have ever seen at work at an operating table. It is really quite extraordinary."

Even the most skilled surgeon is only as good as his tools. Soon after they met, Watts and Freeman ordered a few leucotomes from the same French supplier that Egas Moniz used and started practicing his method of prefrontal leucotomy on pickled brains in the morgue. A promising collaboration was finding its feet in the chilly basement of the hospital.

In the autumn of 1936, after they were satisfied with their "coring" accuracy, they operated on their first patient. Alice Hammatt was a sixty-three-year-old housewife from Topeka, Kansas. Following her first pregnancy in her twenties and the death of the child, she had a long history of mental anguish. She struggled to sleep, felt constantly agitated and worried, and suffered from crushing bouts of depression. She was suicidal. Her main worry when it came to the operation was that her long curly hair had to be sacrificed in order to bore holes into her skull. After the operation, Freeman noted, she "no longer cared" about her appearance. In total, Hammatt had six cores cut into her brain, two more than Moniz's standard operations and an early sign that Freeman and Watts were set to develop their own method of prefrontal leucotomy. Indeed, they soon renamed the operation a "frontal lobotomy," as they didn't just cut through the white matter of the brain (as in a leucotomy), but also took cores from the frontal lobes themselves. (A "lobectomy" is a complete removal of the frontal lobes.)

Although hesitant to jump to conclusions, Freeman felt that Hammatt's operation was, on the whole, successful. "Whether any permanent residuals of frontal lobe deficit will be manifested is uncertain at the present time," he wrote, "but the agitation and depression that the patient evinced previous to her operation are relieved."

Although more commonly known for its use in chronic cases of schizophrenia, many of the first patients to receive lobotomies were suffering from what was then called involutional melancholia or agitated depression. They suffered from a devastating medley of delusions and morbid fears during their later years. People who had shown no sign of mental illness in their lives would, upon reaching their late forties or fifties, fall victim to a disease that was a common sight in mental institutions. As one author wrote in the 1930s, "[T]he involutional melancholic would be a thin, elderly man or woman, inert, with the head lifted up off the pillow. There were some sort of Parkinsonian-like qualities, mask-like face sunk deep into misery. . . . If you could get them to say anything, it would be something about how hopeless things were, how they were wicked, doomed to disease, death, and a terrible afterlife, if there was one."

Whether involutional melancholia was different from manic-depressive insanity was debatable. After all, similar delusions of guilt and wrongdoing could be found in younger patients. But there was something about this variety of depression that warranted its own classification. There was the late onset without any previous record of psychiatric illness. The endless hand-wringing. Even the personality of the two kinds of patients seemed widely different before their attack. While younger depressives were often described as outgoing, equable extroverts before and in between their depressive episodes, involutional melancholics were often introverted, sensitive, and lived day-to-day according to a strict routine. One author described their former selves as having "a narrow range of interests, poor facility for readjustment, asocial trends, inability to maintain friendships, intolerance and poor adult sexual adjustment, [and] also a pronounced and rigid ethical code." This obsessive compulsion for order was seen as an important predisposition to their later decline. After reaching middle age, their strict routines were thrown out of whack by the natural transitions of life. The kids grew up and left home. Friends and family fell sick and sometimes died. Divorce or stale marriages made them dwell heavily on the past, on the opportunities missed and the future that can never be. "The mind is occupied with the 'might-have-beens,'" one textbook from 1956 states.

Before he eventually incorporated involutional melancholia into the diagnosis of manic-depressive insanity, Emil Kraepelin made some characteristically astute observations of such patients under his care. "Remote and often insignificant facts are recalled, such as the stealing of fruit in childhood, disobedience to parents and neglect of friends," he wrote, "which now cause them the greatest anxiety. They are perfectly wretched about it." He noticed that such delusions had a strong religious element that emerged in repetitive mantras or agitated moments of personal reverie. "All wickedness is due to them," Kraepelin wrote. "[T]hey have desecrated the communion bread, or have spat upon the image of Christ. They are totally unworthy, should be buried alive. . . . [H]anging is too good." While thinking so little of themselves, involutional melancholics also raised their actions and former deeds to the highest level of religious wrongdoing. They were equally derogatory and self-important.

With their repetitive symptoms—hand-wringing, pacing, the preoccupation with a few recurring thoughts—involutional melancholics were the biological incarnation of Egas Moniz's theories. They were rigidly fixed in body and mind. And because they were often chronically ill, not cycling between depression and lucid moments, even the more drastic forms of treatment were deemed necessary to halt their suffering. Desperate for relief and in very real danger of death from suicide or starvation, the prefrontal lobotomy soon became the treatment of choice for involutional melancholics, a quick surgical operation that could release them from their morbid obsessions. "They all responded remarkably to prefrontal lobotomy," two psychiatrists working in Boston, Massachusetts, wrote in *The*

New England Journal of Medicine, the leading medical journal in the United States.

By November 18, 1936, just over two months after their first operation, Freeman and Watts had performed a total of six lobotomies. Again, their reports were a mix of positivity and caution over its use and potential misuse. "[T]he patients have become more placid, content, and more easily cared for by their relatives," they wrote. "We wish to emphasize also that indiscriminate use of the procedure could result in vast harm. Prefrontal lobotomy should at present be reserved for a small group of specially selected cases in which conservative methods of treatment have not yielded satisfactory results." Moreover, they warned, "every patient probably loses something by this operation, some spontaneity, some sparkle, some flavor of the personality."

From the outset, that is, neither Freeman nor Watts was under the impression that this was a cure for chronic depression or other mental disorders. A lobotomy couldn't bring a person back to health. It couldn't rewind time. But it could make people more peaceful, stoic, and content. Rather than suicidal, a patient was likely to be insensitive to their surroundings. A lack of emotional depth replaced constant worrying. A restless character was transformed into someone who sat in the same seat every day and stared out a window. "These patients are not only no longer distressed by their mental conflicts but also seem to have little capacity for any emotional experiences—pleasurable or otherwise," one author later wrote. "They are described by the nurses and the doctors, over and over, as dull, apathetic, listless, without drive or initiative, flat, lethargic, placid and unconcerned, childlike, docile, needing pushing, passive, lacking in spontaneity, without aim or purpose, preoccupied and dependent."

Family members—the people who often knew patients best before their operation—reported haunting transformations in their loved ones. "My husband may be better, but he is not the same," the wife of a schizophrenic patient said, "he is much inferior to what he was." A parent said that her daughter was "a different person. She is with me in body but her soul is in some way lost."

As with the chimpanzee experiments at Yale and the few examples of frontal lobe damage in humans, the operation's effect on personality was anyone's guess. "We could fairly accurately predict relief of certain symptoms like suicidal ideas," Watts later admitted during an interview in the 1970s. "But we could not nearly as accurately predict what kind of person [they were] going to be."

Late in 1937, Freeman and Watts replaced Moniz's coring method with what they called the "precision method." Instead of using a retractable wire loop to cut small spheres in the brain, they designed a cutting tool that looked like a butter knife crossed with a knitting needle. Flat and with a ruled edge to measure the depth of an incision, this tool enabled a clean cut rather than a rough core. (Their Washington-based manufacturer engraved both their names into their handles.) To guide his blade, Watts orientated himself with a series of landmarks on a person's skull. First, the squiggly line where the frontal bones of the skull fuse together (known as the coronal suture) was a natural marker for where to cut the two boring holes, each drilled along this line and just above the temples. Second, an arch of bone that sits above the nasal cavity provided an estimate of how far into the frontal lobes to cut.

This change in method was soon followed by a change in theory. In 1939, three years after the first lobotomy in the U.S., Freeman proposed that this operation didn't just disrupt "fixed thoughts" (as Moniz had proposed) but disconnected the frontal lobes from the emotional center of the brain, known as the thalamus. The more connections they cut, he thought, the less emotional oomph the thalamus provided to a person's thoughts. For the more severe or "incurable" cases of schizophrenia, Watts performed a "radical" lobotomy, disconnecting larger sections of the frontal lobes from the thalamus below. The common side effects—headaches, incontinence, confusion, and apathy—were also more likely and more severe. For the agitated, involutional melancholics, however, they performed a "minimal" or "standard" lobotomy, slicing a smaller section of the frontal lobes and still finding that the depression and anxiety often lifted. "Longstanding depressions disappear as if by magic," one doctor noted.

• • •

The immediate reaction to prefrontal lobotomy was as divisive as the operation itself. While some thought that it was the future of biological psychiatry, a much-needed addition to the paltry selection of treatments that it offered its patients, others thought that it was a reversion to the radical bloodlettings of yore. "This is not an operation, this is a mutilation," one psychiatrist said shortly after its introduction.

In April 1941, at a meeting of the American Psychiatric Association held in Cleveland, Ohio, a heated discussion on lobotomy took place around a single table. Walter Freeman, as expected, was at the center of attention, sitting on a six-person panel that discussed the merits and limits of neurosurgery, the umbrella term for any surgical operation on the brain. He started by defining what the frontal lobes were. They were linked to foresight, he said, "that is, with looking into the future and with the consciousness of the self." It was these traits, he thought, that lobotomy disrupted. A person lost interest with themselves, became disinterested in any prior delusions of guilt or persecution, two traits at the core of those depressions experienced by people who were prime candidates for lobotomies. Without foresight, patients lived in the present and failed to worry about the future. "The past and future seem telescoped for them into the present," the psychologist Mary Frances Robinson wrote. "[T]hey meet each situation as well as they can and then are ready for the next one."

Another panelist, Dr. Lyerly, then described twenty-four of his patients with involutional melancholia. Although they weren't always discharged, he said that over 70 percent had improved. "By 'improvement,' I mean that the patients came out of the severe mentally depressed states and became happy, cheerful and slightly elevated," he said. "The patient frequently wakes up with a smile on his face . . . and they were no longer such a problem in the hospital." As suicidal actions declined, forced feedings became a thing of the past. Everyone ate. And ate. And ate. While some argued that this increased appetite was due to the damage done to the brain, Freeman said that it was a by-product of their well-being. "[T]he gain in weight is merely an expression of the contentment with existence as it is," he told his fellow panelists.

M. A. Tarumianz, the superintendent of the Delaware State Hospital at Farnhurst, added to Lyerly's positive results. Shortly after Freeman and Watts created their "precision method," Tarumianz operated on eight patients with severe depressions. Between the ages of thirty-seven and sixty-six, seven had been suicidal and two had attempted to end their lives many times. After the operation, however, all of them had shown remarkable transformations. Two were even able to return to work.

The economic benefits of lobotomies seemed obvious. If institutionalized patients were able to return to their homes, their families, and even their former jobs, they were no longer an expense of the state. Trained psychiatrists, doctors, medication, and custodial care can quickly add up to costly medical bills. And since mental institutions were funded by taxpayers' money, everyone could benefit from any treatment that could reduce the inpatient population. With a back-of-the-envelope calculation, Tarumianz told his fellow panelists that lobotomies could save the average mental institution $351,000 every year.

But these calculations translated into an uncomfortable reality. Tarumianz's reduced costs of care didn't just encompass those patients who had been sent home, but also those who had died from their treatment. And with lobotomy this wasn't an insignificant number. With a skilled surgeon, between 2 to 4 percent of patients died during or shortly after the operation, usually from a major blood vessel being accidentally cut in their brain. But hidden away in unpublished cases, these numbers could increase to 10 or even 20 percent. Tarumianz used the conservative estimate of 10 percent mortality. This number—a fully grown male elephant in the room—was flicked aside as if it were an unimportant speck on case reports. "So far I have not seen a single case that has not responded favorably except in which death occurred," Tarumianz said.

Roy Grinker, a psychoanalyst from Chicago who had been trained by Sigmund Freud in Europe, added a voice of reason to this panel. Although he wasn't against the use of lobotomy, he refused to jump to any conclusions, given the lack of experience and empirical data. The economic costs were meaningless, he said. Any savings to the mental institution would simply be taken up by the patient's household or community. "In the long run it makes very little difference," he told his fellow panelists. His main concern, however, was the possibility that the patients would have recovered without any operation whatsoever. Although he didn't use the term, he was concerned about the placebo effect. "The amount of attention and interest in the long hospitalized [patient] is practically nil," Grinker said. "Now you have a new operation and you are enthusiastic about it. You bring your patient into the hospital ward. You operate on him. You supply him with nurses. You pay a lot of attention to him. You have brought to bear on this patient a tremendous amount of psychologic care which he did not have before and you get some results." How could anyone determine whether a successful operation was due to the lobotomy or all of the other elements of the treatment? "I have previously expressed my surprise that patients have been operated on after only a few months of mental illness," Grinker added. "It is obvious that even the natural process of recovery has not been allowed to take its effect. . . . But, once one cuts, there is no return."

Although similar critiques appeared in scientific journals in 1941, few of these words seemed to filter through to the public. In the early 1940s, journalists were largely positive when writing about prefrontal lobotomy. If newspapers and magazines were to be believed, a cure for mental disorders had just been discovered and Walter Freeman was akin to Jesus Christ, a bearded man dressed in white robes who could perform miracles. On May 24, 1941, a month after the conference in Cleveland, *The Saturday Evening Post* ran a three-page feature on Walter Freeman and James Watts. "The brain has ceased to be sacred," the science writer and editor Waldemar Kaempffert wrote. "Surgeons now think no more of operating upon it than they do cutting out an appendix." This wasn't a criticism of an overly gung ho approach to psychiatry. It was a statement of celebration, a comment on how far psychiatry had come in the treatment and understanding of mental disorders. The patients were the true benefactors of this daring approach. "A world that once seemed the abode of misery, cruelty and hate is now radiant with sunshine and kindness to them," Kaempffert wrote. Wearing face masks, surgical gowns, and pointing to an illuminated X-ray of a skull, Freeman and Watts were pictured as intrepid explorers ushering in a bright future of precision medicine.

In *Psychosurgery*, their 1942 textbook on lobotomies and similar operations, Freeman, Watts, and their colleague Thelma Hunt, an associate professor of psychology at George Washington University, dedicated their work to Egas Moniz, the man who "first conceived and executed a valid operation for mental disorder." In 1939, Moniz had been shot four times by a delusional patient in Lisbon. Although he survived, he was badly injured, permanently disabled, and went through months of arduous rehabilitation. He never fully recovered. In his letters to Freeman, he updated his younger disciple on his health. Already retired and approaching seventy years old, Moniz focused on his writings and only occasionally supervised the operation he had invented.

Psychosurgery was a pivotal turning point in the history of the prefrontal lobotomy. It solidified a growing realization in the psychiatric community that the procedure had moved away from its leucotomy roots to become the specialty of American psychiatry, of Walter Freeman and his surgical sidekick. Largely written in Freeman's flowing prose, the book provided a captivating journey from the skull trepanations of antiquity, through the theories of frontal lobe function, and into the prefrontal lobotomy and its origin at the International Neurological Congress in London, where Moniz heard the reports about Becky the chimpanzee. After performing eighty surgical operations between 1936 and 1941, Freeman concluded that the best results were found in depressions that came with severe agitation or fixed delusions. Overall, roughly two-thirds (63 percent) of such patients could be expected to improve after a lobotomy.

Miss E.G., a fifty-nine-year-old housekeeper who suffered bouts of depression every ten years, was a classic case study for the operation. She was also the first patient mentioned in *Psychosurgery*. Before the operation, she constantly fidgeted, rubbed her hands together as if massaging them with moisturizer. Her mind was elsewhere, her attention span short. And although she had no intention of killing herself, she wished that she would die soon. "The mood was one of enormous depression," Freeman wrote. On April 17, 1937, with her consent and that of her family and her priest, Miss E.G. was injected with morphine, an anesthetic, and given a lobotomy. "As the operation progressed she dropped off to sleep and snored gently," Freeman wrote. "Very little bleeding was encountered and the patient left the operating table in excellent condition."

Two weeks later Miss E.G. was discharged from the hospital. After the initial confusion and incontinence had cleared, her incisions healed and the scars were covered by a shock of new hair. "They tell me that I look ten years younger," she reported. "When I'm tired I still get the circles under my eyes, but they don't seem to be as deep as they used to be." Although she spent most of her time at her sister's home, continuing her work as a housekeeper, she enjoyed visiting the local cinema. Previously such an excursion was torture, the depression making any trip out meaningless and painful. "But last night I went and really enjoyed it," she reported to Freeman. "I forgot myself during the movie, and when I came out I felt refreshed rather than harassed like I used to."

Miss E.G. was content within her domestic lifestyle, but other patients were not so lucky. One person with depression died during the operation, a casualty caused by an accidental cut to the anterior cerebral artery that nourishes the frontal lobes with oxygenated blood. And, although Freeman never mentioned her by name, Rosemary Kennedy—the younger sister of the future president of the United States, John F. Kennedy—was permanently disabled after a botched lobotomy in November 1941. She wasn't named in *Psychosurgery*, but was simply labeled as one of the "poor" results. This was a tragic understatement.

• • •

As a wealthy businessman, investor, and politician, Joe Kennedy Sr. had built his life on his own success. But for more years than he liked to think about, he was at a loss with his eldest daughter. In many ways Rosemary was the perfect child—pretty, with long curly hair and a straight-toothed smile. But inside, Joe thought, she was a torrent of unsuitable emotions, a young woman who was set to tarnish the family name. She would not conform to the family's social norms; rather than being placidly well-mannered, she was rebellious and prone to staying away from home overnight. Joe thought that puberty had caused a change in her personality. But even as a baby Rosemary seemed different. Compared to her eight siblings, she didn't take to reading or writing so easily. While his other children excelled at school, Rosemary always remained a few years behind her age group. Her condition—what would today be termed learning difficulties, but was then known as "feeblemindedness"—was thought to have been a result of a lack of oxygen during the final stages of her birth.

Despite her disability, Rosemary rose to any occasion that she faced. When her father became the U.S. ambassador to Great Britain in 1938, she strolled through Buckingham Palace in the presence of royalty, turning heads as she walked past with her bright smile and trailing white dress. She had no problems in attracting the attention of the opposite sex. Like many young girls, she became fascinated by boys and started to enjoy their flirtations. As a pupil at a convent school in Washington, D.C., she would sneak out at night, freeing herself from the expectations of her family, and drink alcohol into the early morning. According to one biographer of the Kennedy family, the nuns at the school "feared that she was picking up men and might become pregnant or diseased."

It was when Rosemary turned twenty-one years old that her behavior became most alien to the Kennedys. She threw and smashed objects in the house. She would scream and storm up to her room after the slightest provocation. Once, she kicked her grandfather in the shin. On top of her irritability and anger, she was often depressed, a mental state that may have been rooted in the isolation she felt from her own family. Rosemary was one of nine siblings, and it was made obvious that she was not like them.

Her parents worried that her behavior was the result of some "neurological disturbance [that] had overtaken her, and it was becoming progressively worse." Without consulting his wife, Joe Kennedy Sr. spoke to a Boston-based doctor and asked whether the latest psychosurgery operations might "fix" his daughter. He was told that such action wasn't merited. Then he met with Walter Freeman and James Watts at their offices in Washington, D.C., a location not far from Rosemary's school. Watts told Joe that Rosemary suffered from "agitated depression" and that lobotomies had shown great promise in reducing the intense mood swings and outbursts of this affliction.

In November 1941, Rosemary was strapped to a table with only Novocain, a local anesthetic, for comfort. The pain was softened, but she was conscious throughout. The hard bone of her skull vibrated as the drill went in—twice, once on either side of her forehead. The experience was defined by "a grinding sound that is as distressing, or more so, than the drilling of a tooth." As James Watts pushed the leucotome in, Freeman knelt down next to Rosemary, asking her to complete simple cognitive tasks. Can you count backward from ten? Can you state the Lord's Prayer? What about "God Bless America"?

"We made an estimate on how far to cut based on how she responded," Freeman would later say.

After her scalp was stitched back together and the anesthetic had cleared, Rosemary's mental capacity was compared to that of a two-year-old. She couldn't walk properly and her right arm hung motionless at her side. She was incontinent. After a short stay at a private psychiatric hospital north of New York City, Rosemary lived the rest of her life in a small outbuilding—nicknamed "Kennedy Cottage" by her carers—at St. Coletta School for Exceptional Children in Jefferson, Wisconsin. "The nurse who attended the operation was so horrified by what happened to Rosemary that she left nursing altogether, haunted for the rest of her life by the outcome," writes Kate Clifford Larson in her 2015 biography of Rosemary.

Even with such blatant risks, lobotomies were soon performed in countries that were initially hesitant to accept this new treatment. The first lobotomies were performed in Britain in 1943. A year later, a few psychiatry departments in the Soviet Union started their own psychosurgery programs. As it spread, the operation was adapted by surgeons into a dizzying array of techniques. The angle of the incision, the region of the brain that was severed, and the type of tool used: all were modified to suit the patient or, more commonly, the latest theories of mental disorder. Some surgeons preferred to open up the skull rather than blindly cutting through two boreholes. Unswayed by this technique, Freeman wrote, "The surgeon sees what he cuts but does not know what he sees." While brain tissue was being frozen, cauterized, or sucked up like spaghetti, Freeman and Watts stuck to their trusted precision method. "We claimed ours was the best," Watts said, "but it's hard to know."

In the UK, Wylie McKissock, a leading surgeon from Atkinson Morley Hospital in Wimbledon, Southwest London, became a loud advocate for the Freeman-Watts lobotomy, performing over three thousand operations during his career. "This is not a time-consuming operation," he once wrote. "[It] can be done by a properly trained neurosurgeon in six minutes and seldom take more than ten minutes." Bore, prod, swivel, stitch—a personality changed forever.

In the mid-1940s, Walter Freeman replaced his "precision method" of lobotomy with a procedure that required little skill, few instruments, and zero surgical training. "[I]t proved to be the ideal operation for use in crowded mental hospitals with a shortage of everything except patients," Freeman wrote. He called it the transorbital lobotomy, but it would become more widely known as the "ice-pick method."

Unlike the prefrontal procedure that required two boreholes in the skull, the transorbital method entered the brain through the back of the eye sockets. Using a hammer and a metal ice pick, Freeman accessed the brain cavity with a forceful tap and then moved his blade back and forth, blind to what and where he cut. Its simplicity made the operation faster and, therefore, open to more patients. Its indiscriminate use spread throughout mental institutions and into the surrounding communities in the United States. Freeman would even go "on tour," driving over eleven thousand miles in a station wagon to perform his transorbital lobotomies. "Some of the 'black eyes' are beauties," he wrote in the same callous manner he had once shown in his university lectures. "I usually ask the family to provide the patient with sunglasses rather than explanations."

Freeman's mother once called him the "cat who walks by himself." He was independent, a loner, someone who thrived in his own company and didn't require any external input for motivation. But with transorbital lobotomy he was pushing his innate tendencies to their limits. Not just a maverick, he became a heretic. This procedure was not supported by even his closest peers. "What are these terrible things I hear about you doing lobotomies in your office with an ice pick?" John Fulton, the primate researcher from Yale, wrote to Freeman. "Why not use a shotgun? It would be quicker!" From its first use in 1946, James Watts refused to adopt or condone this method. "In both mental and physical disease, surgical intervention should be reserved for cases in which conservative therapy has been tried and failed or in cases where such treatment is known to be ineffective," he wrote. "It is my opinion that any surgical procedure involving cutting of brain tissue is a major surgical operation, no matter how quickly or [easily] one enters the intracranial cavity. Therefore, it follows, logically, that only those who have been schooled in neurosurgical technics, and can handle complications which may arise, should perform the operation." In response Freeman wrote, "I am equally insistent that it is a minor operation since it involves no cutting or sewing, and that it should be performed by psychiatrists."

Freeman had not only estranged himself from his closest colleague but the field of biological psychiatry as a whole. While he used his transorbital method on a whim, other leaders in the field had begun to acknowledge the need for a conscientious selection of patients before conducting any form of psychosurgery. "The more one employs and studies this method of operation, the more clearly one realizes that in individual, carefully selected cases it can yield much of value," Snorre Wohlfahrt, a psychiatrist working in Stockholm, Sweden, said in 1947. Selecting the wrong patient could lead to permanent damage or exacerbation of the initial problem.

How did such a radical treatment become so popular? Why did so many respected psychiatrists put their faith in such an irreversible, and ultimately unproven, surgical operation? It was as if they were probing every possibility and hoping that something—anything—would turn out to benefit their patients. And lobotomies certainly had the potential to change lives. It was an operation that could induce a state of harmony to a once hectic mind. "In several of the greatly improved cases the effect of the operation has been sensational," Wohlfahrt said. "Whereas they had previously been depressed, inhibited, egocentrically self-pitying individuals, or tormented by obsessive thoughts and confined in their activities, they have recuperated, become cheerful, active persons, freed from inhibitions and needless worry. . . . If time permitted, much might be said about these case histories and the revolutionary changes in character through which these patients have passed." The therapeutic goal for a lobotomy was centered on introducing a more manageable mental state and hoping that it was an improvement on what came before. (This approach was once likened to taking the "sting" out of mental disorders.) Although a patient's personality might be irrevocably altered, they were, hopefully, no longer a threat to themselves or their family. They could return home and might even return to work. "If the reward of the physician is the satisfaction of alleviating distress," one psychiatrist wrote, "we have felt that the operation brought this reward. Patients were more comfortable and families gratified." Rather than spending years on an overcrowded ward, tormented, sleepless, and kept in questionable conditions, the appeal of returning to their community would have been obvious to any patient (and their families). Plus, this operation was undoubtedly effective in reducing the number of institutionalized patients, a popular metric for success or failure in any medical treatment. Regardless of what state patients were in after the operation, psychiatrists and doctors believed that the more people out of mental hospitals, the better.

In 1949, Egas Moniz was awarded the Nobel Prize in Physiology or Medicine for "his discovery of the therapeutic value of leucotomy in certain psychoses." Controversial and hotly debated into the present, this honor no doubt helped to cement lobotomy as a real scientific discovery for the field of biological psychiatry. In reality, however, the wheels had already started to turn against this popular form of psychosurgery.

In 1953, Nikolai Oserezki, the director of the psychiatric institute at the I. P. Pavlov Leningrad Medical Institute, told a meeting of the World Federation for Mental Health that every lobotomy "violates the principles of humanity." Arguing that it was only good for turning patients into "vegetables" or "intellectual invalids," he voiced concerns that had been fueling active debate for years in the Soviet Union. Although only a few hundred operations had been performed since the end of 1944, there were many who advocated the use of lobotomies in chronic or otherwise incurable cases of mental disorder. Even so, based on the outspoken criticism and research of the psychiatrist Vasilii Giliarovskii, the Soviet minister of health signed a decree on December 9, 1950, that banned the procedure across the entire Communist state.

The initial positivity and hope that surrounded lobotomies was replaced with widespread caution or outright condemnation. An article published in *Scientific American* in 1950, for example, stated that "it is certainly an exaggeration to refer to all lobotomized patients as 'human vegetables,' as one author has done. Yet the results are serious enough to give us considerable concern." After the Soviet Union, many other countries banned the operation entirely. Its declining use in depression, however, was largely due to another treatment, a competitor that was invented two years after Walter Freeman had first heard of Moniz's prefrontal leucotomy. It was cheaper, required no surgical training whatsoever, and didn't merely flatten the emotional reactivity of agitated patients. But it wasn't a newly discovered drug that first revolutionized the field of biological psychiatry. It was the redirected flow of electricity from a light switch.

The Most Powerful Reaction

Ugo Cerletti, chair of the Clinic for Nervous and Mental Disorders at the University of Rome, was brought to the slaughterhouse on a rumor. In the spring of 1937—a time when Italy's fascist dictator Benito Mussolini was in crazed cahoots with Adolf Hitler—one of Cerletti's colleagues had told him that pigs were killed with quick bursts of electricity. "Electrical slaughtering," he called it. Welcomed to the slaughterhouse in Rome by the thick smell of manure and the din of snuffling grunts, Cerletti soon found that the rumor was false. The electric shock—125 volts delivered to the sides of the head with a large caliper-like contraption—instantly made the pigs freeze, fall onto the stone floor, and convulse as if they were having epileptic fits. Unsightly and brutal, this was the humane part of slaughter. It ensured that the pigs, a large white breed from England, were unconscious before their throats were cut. If left untouched, Cerletti noticed, they soon got back onto their hooves and trotted back to their pen pals. They seemed dazed, confused perhaps, but altogether unharmed.

"It occured to me that the hogs of the slaughterhouse could furnish the most valuable material in my experiments," Cerletti wrote. Swapping his laboratory coat for some grubby overalls, Cerletti, a broad-chested sixty-year-old with slick black hair and thick bushy eyebrows, picked up the large caliper-like contraption with both hands and placed it around a pig's head. A quick burst of electricity was all that was needed to induce the epileptic-like fit, he found. Only when he left the electricity running for over a minute—six hundred times longer than the usual tenth of a second—was the electric shock fatal. This was a huge margin of safety, Cerletti thought. "These observations gave me convincing evidence of the harmlessness of a few tenths of a second of application through the head of a 125-volt electric current, which was more than sufficient to ensure a complete convulsive seizure."

While Cerletti was shocking pigs at the slaughterhouse, his student, Lucio Bini, was busy at the university's clinic, a three-story, ocher-colored concrete building, studying how brief electric shocks affected the brains of dogs and other laboratory animals. The use of stray dogs in laboratory research, collected by the municipal dogcatcher, was normal in the 1930s, an era of biological science before the introduction of specific strains of mice and rats (today's go-to lab mammals). Peering down his tabletop microscope, Bini was looking for signs of nerve damage. Is that fibril out of place? Is that cell's nucleus disintegrated? Are there signs of hemorrhage? When he found time away from his other duties, Cerletti would pop in for a look. At the microscope, one of his students thought that he looked "like a falcon high in the sky, searching for prey on the ground below." Neither Cerletti nor Bini saw anything that might prevent this procedure from being used in humans. It was a huge moment for Cerletti. "[W]e had at our disposal, for the first time, the possibility of inducing in man, at will and without harm, convulsions," Cerletti wrote, "this most powerful reaction of the nervous system."

Epilepsy, and the convulsions that it can cause, winds a long and dark path through human history. Before it was given a medical name, this mysterious disease was explained in entirely supernatural terms. Demons, gods, or otherwise: the fits were the mark of the unearthly, a person who had been seized by an invisible spirit and left the confines of natural law. For their supposed curse, people with epilepsy have long been shunned, shamed, and isolated from their families, their communities, or both. At one time, some people even thought the convulsions were contagious and would step away from this so-called Sacred disease. Philosophers and physicians, however, inquisitive souls that they are, often stepped closer. For them, epileptic fits were a curiosity. Rather than a curse, many found them to be a potential cure for any concomitant diseases of the mind.

In his 1758 book, *A Treatise on Madness*, one of the first textbooks dedicated to mental illness, the English physician William Battie wrote, "[A]s ignorant as we are and perhaps shall always be of the reason, experience has shown that . . . one species of spasm however occasioned seldom fails to put an end to that other [malady] which before subsisted." In 1828, G. Burrows, a doctor working in London, found that two scruples of camphor oil—a gloopy liquid used for candle wax or embalming bodies—could induce a convulsive seizure in his patients with insanity. "A complete cure followed," he wrote. One of his contemporaries agreed, writing, "When the patient awakes from the epileptic attack his reason will return." In the 1930s, Ladislas von Meduna, a neurologist at the Brain Research Institute in Budapest, bolstered these broad observations down to a microscopic level. Slicing up hundreds of pickled brains from patients that had died from a variety of diseases, he found that those from epileptic and schizophrenic patients were opposites of each other. Those from epilepsy were packed with glial cells, the putty or "glue" of the brain's tangled web of neurons. In contrast, schizophrenic brains were deficient in these cells. "To explain the difference," von Meduna later wrote, "I developed the hypothesis that the [cause of] epilepsy has a stimulating effect upon the growth of the glia cells, while [that of] schizophrenia has an opposite, a paralyzing, effect of the glia system." In short, there was a "biological antagonism" between epilepsy and schizophrenia.

Although this theory would turn out to be wrong, it would still lead von Meduna to a grand therapeutic conclusion: if he could find a way to induce an epileptic fit, he might find a cure for schizophrenia, a disease that Emil Kraepelin had once deemed incurable.

As with malaria and neurosyphilis, could one disease treat another? Von Meduna tested various drugs on laboratory animals to see if any could reliably induce a convulsion. Strychnine, thebaine, nikethamide, caffeine, absinthe—all were unsatisfactory. In June 1934, he settled on camphor oil, the same convulsant that doctors had used centuries before his time. Injecting this dark and viscous substance into his patient's muscle, an epileptic-like fit would normally begin within a few minutes. Sometimes, however, it took half an hour—and the waiting was agony for the patient. Knowing that the fit was about to happen, but not knowing exactly when, left each person in a panicked state before they succumbed to the inevitable. The feeling would later be described as "impending doom." Sometimes the convulsion didn't even occur.

When the injection had the intended effect, the grand mal seizure fit progressed through its familiar stages. For the first few seconds, in what is known as the "tonic phase," the muscles of the body jerk into a state of contraction. The spine arches backward. The jaw clamps upward like a vise. With every sinew straining, vomit, urine, feces, and, for men, semen are sometimes ejected from the body. After this brutal beginning, the "clonic phase" takes over. For the next thirty to forty seconds, the limbs and trunk of the body start to gyrate as if the person is lying down in the back of a van being driven over a cobbled street. Frothing at the mouth is not unusual.

It wasn't a pleasant experience or a pretty sight. But von Meduna's patients seemed much healthier in the convulsion's wake. Out of the first twenty-four patients, he found that ten were rid of the hallucinations, delusions, and inexplicable paroxysms that defined their condition for so long. Publishing his results in 1934, von Meduna found that once-catatonic patients started to get out of bed, dress themselves, eat food without forced feedings, and—for the first time in years—start talking. Soon after this, he found a replacement for camphor oil, a drug known as Cardiazol (marketed as Metrazol in the U.S.). Faster in its action, it reduced the level of anxiety before the convulsion. This "convulsive therapy" would soon encircle the globe, spreading from Budapest to other countries in Europe and into the Americas. Feeling like he had just found the source of the Nile, von Meduna wrote, "[A] new art and possibility of cure opens up for us."

After a few of his students returned from a training course in convulsive therapy in Vienna in 1936, Ugo Cerletti had one question: Why didn't they use electricity instead? Before he visited the slaughterhouse, he had been using short sharp bursts of electricity to induce epileptic seizures in his laboratory animals for years, studying the effects on the hippocampus, a region of the brain involved in memory. It was a project driven more by curiosity than any urge to develop a new medical treatment. "I never thought that these experiments would have any practical application," he wrote. But the link between electricity and convulsive therapy seemed too obvious to ignore. After centuries of using powerful drugs like camphor oil, he wondered whether his research could offer something more humane.

In March 1938, police officers brought a new patient into Cerletti's clinic. In his late thirties or early forties, the man had been found wandering through the central station in Rome in a confused state, muttering to himself in a language that no one else understood. As one of Cerletti's students later wrote, he "was unemotional, living passively, like a tree that does not give fruit. . . . We concluded that we were dealing with a mentality that was completely unraveled, and gave little hope, even for a partial recovery." This patient was exactly what Cerletti had been waiting for. He asked Bini to set up their electroshock machine—a shoebox-sized, cream-colored cube of metal with a few dials, knobs, and switches on its top surface—in a quiet back room on the second floor of the clinic. Used by a laboratory technician who took naps between shifts, it was empty but for a small metal-framed bed in the corner. It was unlikely that they would be disturbed or, as Cerletti feared, discovered by his peers. Just in case, he asked one of his younger students to poke his head out the door now and again, just to check if anyone might be coming.

The air in the room was thick with anticipation. In a few seconds, Cerletti's reputation could be destroyed. A former student of Emil Kraepelin and Alois Alzheimer in Munich, Germany, Cerletti was a gifted teacher in his own right and, upon returning to Italy, was nicknamed "the Maestro" for his didactic approach to psychiatry. He exuded warmth and empathy, a humanism that later became known as the "Cerletti atmosphere." This academic standing took decades to amass. But it could all be lost in an instant. Cerletti knew what happened in this disused room was on his head. What if the patient died? What if he and Lucio Bini had missed something in their brain autopsies? He had been concerned about the answers to such questions ever since he first heard of von Meduna's convulsive therapy. Inspired by the "youthful enthusiasm" of Bini, he also couldn't help thinking that this could be the spark that fuels a revolution in biological psychiatry.

Bini, a portly thirty-year-old with a thick mustache, stood next to the machine he had helped build. Plugged into a nearby light switch, a couple of thick black wires snaked into a device that looked like a huge pair of headphones. Inside each of two thick pads of brown cloth was an electrode, a flat piece of metal where the electric shock would jump from the wire into the patient's skull. Dipped in a saline solution, these pads were placed over the patient's temples and, when all was set, Cerletti gave the signal for the treatment to begin. As the switch was flicked on, an automatic stopwatch inside the machine allowed the electricity to flow for a fraction of a second. Almost immediately, the patient's body tensed up, he let out a groan, and his back arched upward from the bed. His mouth opened wide and then clamped down on a piece of pipe wrapped in rubber that protected his tongue from his teeth. After a few seconds, the tonic phase of the convulsion ended and the patient progressed into the clonic phase. His legs and arms thrashed wildly.

Then he turned blue. He wasn't breathing and his blood and brain were being deprived of oxygen. The room was silent except for Bini's voice. He had started counting. Soon, his colleagues were counting along with him. It became a ceremonial chant of sorts, one that they all hoped would culminate with a sign of life. At the forty-eighth second—a period that seemed like an eternity—the patient sighed heavily and everyone in the room took a long, deep breath. It wasn't particularly hot in the room, but one student, Ferdinando Accornero, noticed that his forehead was beaded with sweat. "Now the patient was breathing regularly, was sleeping, and was calm," Accornero wrote. "We glanced at each other; in our eyes, there was a new shining light." Cerletti confirmed that the session was a success.

Although it is often cited as such, this wasn't the first time this patient received electroshock therapy. The details of the first session—what happened, who was there, whether it worked—have been lost over subsequent decades. Memories are fallible. Biographies are often biased. And stories are reshaped every time they are told. For a touch of clarity, it is necessary to go back to the beginning: to Lucio Bini's original notebooks. Reading through these documents, Roberta Passione, a historian from the University of Bologna, Italy, discovered a slightly different story. In black scribbles across a small notepad, Bini recorded that after the first eighty volts for one-tenth of second ended, the patient was singing at the top of his voice. The subsequent shocks were then increased to half a second and then three-quarters of a second. Voltage remained constant, and so did the result—there was no convulsion. "The patient is freed, gets up immediately, walks quietly back to the ward, talking in his usual loose way," Bini wrote. "In the afternoon and the following day, he did not present any phenomenon of particular mention." He hadn't been harmed. But he also wasn't showing any signs of improvement.

Over the next few weeks, however, this first patient would be given eleven sessions of electroshock therapy—eleven successful convulsions—and would show such an improvement that he'd be discharged from the university's clinic on June 17, 1938, returning home to his wife in Milan and continuing his work as an engineer. Three months later, however, his symptoms slowly reappeared. In bed, he would whisper to himself and begin to distance himself from his wife. It was the first sign that this was no miracle cure. Electroshock therapy could boost a patient back to health, but it couldn't keep them there forever.

As Cerletti and Bini gained more experience with this new form of convulsive therapy, they began to share their results and methods. On May 28, 1938, just over two months after their first patient had been brought into their clinic, they presented their observations at a meeting of the Royal Medical Academy of Rome. Cerletti began by describing the previous form of convulsive therapy: the Cardiazol method of von Meduna. It was costly, involved a painful injection, and the fear of impending doom before the convulsion understandably made patients hesitant to be treated a second time. Even after the convulsion had passed, the remaining Cardiazol in the patient's body caused an agitated state that could last for a number of hours. Electroshock didn't come with these drawbacks. "Loss of consciousness instantaneously follows delivery of the current," Cerletti told his audience. "Afterwards, there is a phase of muscular relaxation with stertorous breathing. The state of unconsciousness lightens gradually and the patient becomes more alert, relaxes the jaw, moves the eyes and begins to answer to verbal stimuli. After five minutes he can talk but remains clouded. After eight to ten minutes the patient is completely himself. If not disturbed, he falls asleep and after a few hours wakes up completely restored.

"The advantage of this method," he added, "is the immediate and absolute loss of consciousness of the subject, lasting for all the length of the treatment. The patients, if asked about their experience, state they do not remember anything but having slept."

Back in the clinic at the University of Rome La Sapienza, scenes of secrecy transformed into loud celebration. Three to four times a week, one of Cerletti's assistants, Spartaco Mazzanti, would blow an out-of-tune trumpet through the wards to signal that the next session of this revolutionary treatment was about to start. The once disused back room on the second floor became packed with curious onlookers, students, and a suite of psychiatrists recording the heart rate, blood pressure, brain activity, and other quantifiable measures of a patient's health.

Despite communications between fascist Italy and the rest of the world being stymied, news of Cerletti's work nevertheless reached across the Atlantic. "Although physicians were reluctant to earmark the new treatment as particularly beneficial for any specific type of insanity," an article in the *New York Times* stated in June 1940, "it was said that it had been used successfully in [the] treatment of schilzophrenia [*sic*]." This was already an outdated conclusion in Cerletti's clinic in Rome. In his first manuscript on the topic, titled "L'Elletroshock" and published in the Italian journal *Rivista Sperimentale di Freniatria* in 1940, he wrote, "Even more brilliant results than in schizophrenia were obtained in manic-depressive insanity, particularly in depressive episodes." From the 180 or so patients who had received electroshock therapy in Rome since April 1938, he found that depressed patients required fewer convulsions to bring about a recovery. In some cases, only four sessions, spread over a week or so, were all that was needed. In comparison, schizophrenic patients might need twenty sessions or more. "We seem to be able to state that the morbid [depressive] episodes are actually halted or at least considerably shortened in their presumable duration," Cerletti concluded. "A series of complete remissions was obtained."

Ugo Cerletti's detailed manuscript on electroshock therapy—a product of two years of research, detailed methodology, and over 180 patients—was seldom read outside of Italy. Even today, it hasn't been translated from its original Italian. Appropriately for the "Maestro," it was his cherished students and colleagues who helped spread his invention around the globe. And none of his former students had a larger influence than Lothar Kalinowsky, a German neurologist who witnessed the second patient to ever receive electroshock therapy, in 1938, and vowed to never see it again. That evening, when Kalinowsky arrived home to his wife, Hilda, he looked as "white as a sheet." Perhaps it was the involuntary, high-pitched yelp as the air was forced from the patient's lungs by the brief electric shock. Or it could have been its speed, an instantaneous fit triggered by the flipping of a switch. And while the chemically induced convulsions weren't too dissimilar from an injection of other drugs such as anesthetics, this electrical procedure seemed altogether unnatural, forced, manufactured by a machine. Whatever his reasons, Kalinowsky soon saw the potential of this new treatment and broke his vow, performing one thousand sessions during his time in Rome. "It is amazing to see most depressions of various depths and duration clear up with the same number of three to five treatments," he later wrote. The great value of electroshock therapy, he continued, "cannot be questioned when we realize how much human suffering is prevented by shortening the [depressive] episode."

After working directly under Ugo Cerletti for five years, he was forced to move in 1938, as Mussolini signed the "Pact of Steel" with Nazi Germany, joining the two nations together like two metals melding into a formidable alloy. Upon hearing that all Jewish émigrés from Germany were to be apprehended and imprisoned, thirty-seven-year-old Kalinowsky and Hilda rushed onto the next train to Zurich, Switzerland, leaving behind their home, their possessions, and even their two children. The girls would be picked up from school by family friends and taken to a safe (non-Jewish) household. Kalinowsky had planned for this. In 1933, he had fled from his home in Berlin as fascism took root following the Reichstag Fire. With anti-Semitism spreading across Europe like a poisonous vine, emigration was always likely.

Inside his small briefcase, Kalinowsky packed what would be his future. Tucked away among his clothes and other essentials was a blueprint of Bini's electroshock machine. As Kalinowsky traveled from one city to another, he would become a psychiatrist, a salesman, and an international champion of electroshock therapy. In Paris and Amsterdam he showed Bini's designs to directors of mental hospitals and helped guide their construction and subsequent use. Not everyone was sold on the idea, however. Aubrey Lewis, an Australian psychiatrist and the director of the Maudsley Hospital in London, met Kalinowsky with stern indifference. But slowly his peregrinations spread the work of Cerletti and Bini into Europe and beyond. In 1940, Kalinowsky arrived with his family in New York. Finding work at several mental hospitals—including the New York State Psychiatric Institute, a largely psychoanalytical institution on the Upper West Side of Manhattan—Kalinowsky attracted psychiatrists from all over the United States. For his central role in its dissemination, one historian called Kalinowsky the Johnny Appleseed of electroshock therapy.

There was immediate resistance to electroshock therapy in the United States. Since the late nineteenth century, the only other method of applying electricity to a person's skull in this country wasn't an attempt to cure a patient but to kill a prisoner. The electric chair was an infamous form of criminal punishment. In reality, electroshock therapy and the electric chair were as different from each other as licking a nine-volt battery is from being tasered in the face. While the electric chair can deliver up to 2,000 volts to the head and legs (flowing through the heart and causing a cardiac arrest), Bini's electroshock machine—and other machines like it—only delivered between 70 and 120 volts to the temples. (A small fraction of this actually reached the brain.) The former is switched on for upward of a minute, while the latter only lasted for a few milliseconds. Despite this early misinformation, electroshock therapy was introduced into the U.S. in early 1940. As Renato Almansi and David Impastato, two psychiatrists working in New York at the time, recalled, "[The] doubts and obstacles were soon dissipated and replaced with typical American enthusiasm."

Not all depressions responded equally well to electroshock therapy. If someone seemed to be living in a confusing state of slow motion, suffered from anorexia, woke up early in the morning without much sleep, and had symptoms that gradually improved throughout the day, then electroshock therapy was almost guaranteed to help. If delusions or psychosis were present—as had been one of the defining features of involutional melancholia—then figures near to 100 percent effectiveness were reported. "The fundamental fact remains that a new uniformly successful treatment, introduced into psychiatry, effectively terminates depressive psychoses," one psychiatrist said in November 1944, "whereas all previous methods have failed to produce such spectacular results." Another psychiatrist wrote that it was nothing short of a miracle: "Patients with severe depressive disorders and stuporous states that had disconnected them from the real world for months and even years came back to life." Like a forest fire vital in liberating the seeds from inside tough pine cones, a quick burst of electricity—and, more important, the convulsion that it caused—seemed to reanimate a person from the thick psychological shell in which they were encased.

Although it was tried, electroshock therapy was rarely as effective in milder depressions that were found in the community. This was puzzling. Why would a medical treatment be more effective in the worst cases, while hardly influencing the milder versions? Imagine a drug that kills cancer only when it has metastasized around the body and not before. It didn't make much sense. The explanation, many psychiatrists thought, was that electroshock therapy was specific to severe depression because it was a distinct disease state. This was "endogenous depression" (coming from within), the more biological form of the disease and a distinction that would continue to be used for much of the twentieth century. The idea that manic-depressive insanity was a spectrum—the argument that Edward Mapother had staunchly held at the Maudsley in the 1920s—was starting to break apart.

The other form of depression, the one that didn't respond to electroshock therapy, was termed *reactive depression*. Rather than being biologically determined, this form of illness was related to stressful life events, personality types, and a more responsive (or reactive) mental state to changing conditions. If the stresses of life were removed or ameliorated, people with reactive depression had a good prognosis. They were likely to recover. Those with endogenous depression, meanwhile, would remain tormented no matter how good their lives seemed to everyone else. These two categories weren't perfect reflections of reality, and a great debate between biological psychiatrists ensued for much of the twentieth century regarding their accuracy and clinical utility. There were huge gray areas between the two. Endogenous depressions could be triggered by environmental stressors such as divorce, unemployment, and the death of loved ones. Reactive depressions might not change even with the greatest reversals in fortune. But this division was nonetheless useful in predicting who might respond to the different treatments available. Without biological markers from brain slices or blood samples, knowing that electroshock therapy worked in more severe, often delusional, depressions was a boon to psychiatry. It meant that a life-threatening condition now had a good prognosis. "The greater the loss of contact with reality," Kalinowsky wrote, "the better the treatment."

Psychoanalysts were never in doubt that there were two major types of depression. In 1917, even Sigmund Freud wrote that so-called psychotic depressions—those with delusions and a better response to electroshock therapy—had some basis in biology. It wasn't purely a disorder of unconscious conflicts. The idea of at least two main types of depression was widely shared throughout psychiatry.

The main difference was in the treatment they received. Whether someone was given electroshock therapy or sat through endless sessions of psychoanalysis depended less on their depression—its symptoms, severity, or supposed biology—and more on which office they walked into. The psychoanalyst believed that patients required insight into their unconscious conflicts and repressed desires, a deep dive that no amount of electroshock therapy could provide. As a report from the New York State Psychiatric Institute noted, "There was a tendency for the residents to prefer to delay all forms of organic [biological] treatment until extensive trials of psychotherapy were exhausted." But, once exhausted, there was no arguing with the efficacy of electroshock therapy. As one biological psychiatrist said, "The analysts denigrate me, but when their mothers get depressed, they send them to me."

Legacy

In 1938, the year that electroshock was still being performed in secrecy, Sigmund Freud's life was threatened by the same fascist ideology that had forced Kalinowsky to leave Berlin. Although he was an atheist, Freud was raised in a Jewish household and married to a Jewish wife. Because of the violent anti-Semitism in the Nazi Party, psychoanalysis was labeled as a "Jewish science" and, therefore, worthy of destruction. In Berlin, the Gestapo removed Freud's books from the public library and burned them in the street. In his hallmark wit, Freud wrote, "What progress we are making. In the Middle Ages they would have burnt me, nowadays they are content with burning my books." After his home was raided in Vienna, Freud decided to leave the city in which he had lived and worked for seventy-nine years. When he was just a toddler, his mother and father had fled from anti-Semitism in Germany and settled in Vienna. Now, with his family seemingly under threat, it was Freud's turn to flee.

After a brief stopover in Paris, he took the night ferry to England and found a welcoming party of friends and family at Victoria Station in London. Although he had hoped to see out his last days in Vienna, he had always admired England from afar. In August 1939, a month before Germany invaded Poland and World War II began, he sat in his study, surrounded by his antiquities and furniture that had been delivered from Vienna. Experiencing firsthand the violence and hatred of Nazism, he hoped that he wouldn't live to see its domination in Europe. Gazing out at the flowers that lined his accomodation's garden, Freud knew that he wouldn't see this latest war unfold.

His enemy—the squamous cell carcinoma, which had first been diagnosed in 1923—had defied years of painful surgical operations, electric cauterization, and radiation therapy. After his last surgery in early 1939, he wrote to his friend Marie Bonaparte, "My world is again what it was before—a little island of pain floating on a sea of indifference." On September 22, 1939, Freud's island of pain was flooded with three centigrams of morphine. He fell asleep and didn't wake up. He took his final breath early the next morning. Unlike four of his five sisters who were murdered by the Nazis, Freud chose euthanasia.

Psychoanalysis would grow to its zenith after his death. By the middle of the century, it was said that a psychoanalyst sat at the top of every psychiatry department in the United States. Not only was this field of the unconscious and repressed memories used in treatment, but also in classification, the field most commonly associated with Emil Kraepelin. In 1952, the first *Diagnostic and Statistical Manual for Mental Disorders* (*DSM-I*) was published and was based on psychoanalytic concepts. Rather than neat clinical groups like Kraepelin had advocated in his textbooks, each patient was often classified according to a supposed underlying cause from the depths of the mind. Depressive reaction, for example, was a "psychoneurosis" caused by repressed anxiety. In such patients, the *DSM-I* reads, there was "no gross distortion or falsification of external reality." Depressive psychosis, meanwhile, was, as Freud had suggested, a more biological disorder that often came with delusions and hallucinations, and was rarely tied to a stressful life event. These groupings would remain popular, especially in the United States, for the next three decades.

It was a fatal twist of irony. In 1940, as electroshock therapy was providing a few remarkable scenes of recovery across the world, biological psychiatry was also leading thousands of mentally ill people to their doom. Emil Kraepelin's popular *Psychiatry* textbook was also popular within the ranks of the Nazi Party. In its most expansive eighth edition, published in three volumes between 1909 and 1912, Kraepelin wrote, "The relatives have often suffered from the same form of the disease." Depression begets depression. This idea of "defective heredity" aligned with the themes of eugenics, the field of science that had swept through Britain and the United States and led to the mass sterilization of thousands of people with disabilities or severe mental illnesses. Men underwent vasectomies. Women had their tubes tied or their ovaries blasted with ionizing radiation. Only in Germany, however, was eugenics taken to its most extreme ends. Before the Holocaust claimed the lives of millions of Jews, thousands of mentally ill inpatients were murdered. Those with "manic-depressive insanity" were killed because of a diagnosis, one that Emil Kraepelin had pushed into the zeitgeist.

In 1916, a colleague of Kraepelin's in Munich, Ernst Rüdin, had provided the first piece of solid evidence that mental disorders were heritable. If Kraepelin was a tireless hoarder of patient reports, Rüdin was an avid collector of family histories. Both were obsessed with big data and even bigger ideas. Rüdin had collected information on the families of 701 patients at the Munich clinic who were diagnosed with dementia praecox. Using large sample sizes and the latest statistical methods, he provided the first conclusive evidence that severe mental disorders were often passed on between parent and offspring. It ran in families. He even created an equation that could estimate the likelihood of the disorder emerging in the future: the "empirical heredity prognosis."

After dementia praecox, Rüdin turned to manic-depressive insanity. Once again, he collected detailed information from the families of 650 patients who had been diagnosed with the disease since 1904 at the Munich clinic. He used the same statistical methods as before, refining them where he could. But he didn't get the same result. There wasn't a strong heritable component. No matter how he played around with his data, it was a weak connection at best. "[Rüdin's] demands for negative eugenic measures against patients with [mood] disorders and their families could not be justified on the grounds of the heredity figures he had calculated," Hubertus Himmerich and his colleagues from the University of Leipzig wrote in 2017.

But Rüdin never published this study. As a cofounder of the

Society for Racial Hygiene and an editor of the *Archive of Biological Hygiene for Race and Society*, his findings didn't fit with that of his—or his country's—worldview. Manic-depressive insanity remained in the realm of heritable diseases along with dementia praecox, learning disability, epilepsy, Huntington's chorea, and hereditary blindness. All were eventual targets of extermination. Emil Kraepelin may have created the classification system on which psychiatry was built, but, by claiming that "defective heredity" was the key ingredient of mental disorders, he also provided the basis for some of the greatest tragedies in human history.

Like the British mathematician Francis Galton, who coined the term *eugenics* in 1883, Adolf Hitler was a radical Social Darwinist. "The volkisch state must see to it that only the healthy beget children," he wrote. "Here the state must act as the guardian of a millennial future. . . . It must put the most modern medical means in the service of this knowledge. It must declare unfit for propagation all who are in any way visibly sick or who have inherited a disease and can therefore pass it on." "National Socialism," another Nazi leader added, "is nothing but applied biology."

Although eugenics had its critics—the psychiatrists Oswald Bumke in Germany and Henry Maudsley in Britain, for instance, cautioned how complex traits such as mental illness couldn't be so simply selected against—over 410,000 people with mental or physical disabilities were surgically sterilized in the 1930s.

As the German military pushed eastward into Poland in 1939 and then up to Russia the year after, the residents of entire mental institutions were herded up and gassed, shot, drugged, or blown up with explosives before being buried in mass graves by Polish prisoners (who were then shot). Between 1939 and 1941, a network of six killing centers was set up across the expanding German Reich as part of the Aktion T4 program for the extermination of the mentally ill and the disabled. In less than two years, an estimated seventy thousand people were murdered. At Hartheim Castle, "one of Upper Austria's most beautiful and most significant Renaissance castles," over eighteen thousand people were killed, most by carbon monoxide poisoning inside a tiled room with showerheads in the ceiling. Their remains were burned into ash. This ash was tipped into the river or simply scattered on the surrounding castle grounds. In 2001, these human remains were discovered at Hartheim Castle and buried in an ecumenical ceremony. The ash inside each urn was a nameless mixture of people that all shared the same fate.

Aktion T4 was soon put to a halt. Relatives learned about the fate of their parents, siblings, or children. The sweet smell of burning bodies filled the air around killing centers and locals noted how the large gray buses were always dropping people off at these sites but never picking anyone up. (Staff at Hartheim Castle used these buses to go to the cinema in nearby Linz.) Bookkeeping errors—such as the death certificate of a man who'd had his appendix removed long ago stating that he died from appendicitis—made the widespread rumors of killing centers come into a shocking reality. But the killing didn't stop there. Although Aktion T4 was officially terminated in August 1941, the Nazi Party found other ways to kill those people infamously labeled "life unworthy of life." As World War II continued, supplies of food and medicine were shuttled to the front lines and mental asylums were left empty. An unknown number of thousands died from starvation and disease. This was known as "decentralized euthanasia," or "wild euthanasia."

In 1944, a year before World War II ended and the true devastation of concentration camps was revealed, Emil Gelny, the director of two psychiatric hospitals in Austria, killed three hundred mentally ill patients with an electroshock machine. But this wasn't electroshock therapy. He had bought an Elkra II, a black machine that looked like a portable speaker on a small Christmas tree stand, and adapted it so the voltage could be increased, the duration of the shock extended from milliseconds to minutes, and added four more electrodes that could be attached to a person's wrists and ankles. Patients were individually treated with electroshock therapy by the routine method (two electrodes on the sides of the head), and then, while they were unconscious, the four electrodes were attached and the machine was switched on. They were dead in less than ten minutes. Gelny performed his form of euthanasia for visiting psychiatrists, detailing how it was preferred to injecting sedatives, as it was cheap, quick, and patients died thinking that they were receiving electroshock therapy, a common treatment in psychiatric hospitals at the time. This was later referred to as "the darkest chapter in biological psychiatry."

Although Emil Gelny's euthanasia was an extreme case, even the standard practice of electroshock therapy could easily be misused. In its early decades of use, electroshock therapy was adopted as a method of control, a means of punishing unruly or loud patients. Although it was relatively painless and a person lost consciousness before the convulsion, patients would become drowsy, confused, and quieter in the wake of the treatment. In what became known as "regressive shock therapy," treatments were given to some chronic schizophrenic patients several times a day for a week or more, a terrifying frequency that made a patient "so regressed that he is bedridden, does not know his name, is doubly incontinent, and is unable to swallow." More commonly, electroshock therapy was given to people with mental disorders who would never benefit from its use. But, for this story, a journey in and out of depression, its potential benefits were clear to anyone who used it properly. In the right hands and on the right patients, it was amazing, revolutionary, a treatment that saved lives. "The recovery of the severe, long-standing, agitated depressions makes a profound impression," one psychiatrist said in 1941. "The morale of the personnel [at the hospital] has been improved by the feeling of active service and treatment for such a large percentage of patients." With electroshock therapy, between 80 and 90 percent of patients were lifted from their depression, a figure that remained largely constant as Lothar Kalinowsky traveled from one country to another. Without electroshock therapy, one psychiatrist working at the Maudsley Hospital in London recalled in 1978, "I would not have lasted out in psychiatry, as I would not have been able to tolerate the sadness and hopelessness of mental illnesses before the introduction of convulsive therapy."

Kalinowsky had heard or read reports of ten thousand sessions of electroshock therapy, performed in doctors' offices, in patients' homes, and even in hotel suites. Despite such widespread use, he found that there had been no serious complications. "To date in the hands of various investigators," he wrote, "the electric shock method seems relatively harmless." As one of his contemporaries put it: "[I]n an era in which hypothermia, hyperthermia, lobotomy, and lobectomy are being carried out for the treatment of mental diseases, electric shock is, relatively speaking, a mild approach."

Cerletti's Monster

While lobotomies were still being used for more chronic states of depression, particularly those of elderly patients, the use of electroshock therapy largely removed the need for this drastic surgical operation. The exceptions to this general rule were soldiers returning from World War II. After joining the war effort in December 1941, thousands of American men were flown back to their families and their homes, but they often weren't the same as when they'd left. They were panicked and fearful. Some talked to themselves or believed they were being pursued. Nightmares reached into reality. Others were depressed and unable to resume the active lives they had once lived. Electroshock therapy was used and frequently found wanting. Deep anxieties and neuroses, fears and frights, were never part of its forte. (As Kalinowsky once wrote, "It cannot be emphasized enough that contrary to psychotics, some neurotics may be harmed by [electroshock therapy]. Anxiety, as the most frequent symptom in neurotics, is often aggravated.") These veterans needed another option.

"By August 26th, 1943, many relatives and guardians of patients who had failed to improve with shock therapy, began to demand more rapid and drastic treatment," wrote Fred Mettler, a neurologist from Columbia University in New York City. Just as the spouse of a cancer patient might request the details of the latest drug trial, families of war veterans were willing to try anything that might bring their fathers, husbands, and sons back. In response, the Veterans Health Administration issued information about lobotomies, stating they were a "last resort" for those who had failed to respond to electroshock therapy.

It was these patients that Mettler would use to try and understand whether the vague procedure of lobotomy could be made more precise, to disentangle which parts of the brain could be excised for greater effect. "Could we find, for example, that only one cortical area need be removed to achieve a particular result?" he asked. "[Could] an essentially uncertain operation be reduced to a proper surgical technique and its unnecessary damage eliminated?" It was such investigations that would start to separate the technique of a prefrontal lobotomy in a list of other surgical techniques—cingulotomies, topectomies, amygdalotomies, and tractotomies—that would continue to be used for the rest of the twentieth century. A survey from forty-four neurosurgical centers in the UK in 1978, for example, found that 431 such operations were performed over a two-year period. Depression was the most common diagnosis recorded, accounting for 63 percent of patients. With greater medical safeguards such as the UK's 1983 Mental Health Act, which stipulated the use of neurosurgery only after the consent of the patient, their psychiatrist, and two other representatives from the Mental Health Act Commission, such treatments would continue to decline in popularity, being used only in cases of depression, obsessive-compulsive disorder, or anxiety that were chronic and intractable.

From this history of greater precision, a technique that used electrical stimulation rather than surgery emerged. Today, deep brain stimulation (DBS) is a promising treatment for people with severe depressions who have failed to respond to all other options. As Jack El-Hai, a biographer of Walter Freeman, wrote in 2005, "Freeman's career may seem bizarre to many people today, but it is playing out again—with better technology, a better understanding of the functioning of the brain, and better ethical guidelines—in the burgeoning world of the new psychosurgery." We will travel to this new world later in this story.

• • •

Although it wasn't as drastic or permanent as a prefrontal lobotomy, the side effects of electroshock therapy in the 1940s were still severe and numerous. Fractures in the bones of the limbs, spine, and pelvis were found in around 23 percent of patients (for chemically induced convulsions this figure could be as high as 43 percent). This was the most pressing concern during the treatment's early years. "If shock treatment is to survive," one author wrote in 1940, "the incidence of fracture complications must be reduced to a minimum." Most commonly, the middle region of the spine suffered from compression fractures as a result of the back muscles contracting powerfully in the first few moments of the convulsion. Although these were often hairline fractures, the type only seen on X-rays and which healed quickly, they were still an undesirable side effect, both for the patient and the image of electroshock therapy. Nonetheless, some patients were willing to sacrifice their skeleton for their mental health. As one study reported, "The patient felt greatly improved and was content to have his shoulder broken rather than suffer the depression."

By 1941, Lothar Kalinowsky had developed a quick fix that substantially reduced the risk of fractures: sandbags. By placing three heavy sand-filled sacks underneath the spine of his patients, he enforced a posture—known as hyperextension—that stretched the internal edges of their vertebrae away from each other. As it was these edges that were most likely to be compressed, this position could reduce the risk of fractures significantly. Using X-rays to scan the vertebrae of sixty patients before and after their treatment, Kalinowsky found no evidence of damage.

Sandbags were far from a fix-all, however. Many people were still unable to benefit from electroshock therapy, as the risks were still too high. The elderly, for example, often had bones too brittle to undergo a full-blown grand mal seizure. If they did, their hips and limbs might snap like dry twigs. In such cases, even the best-placed sandbags would be insufficient.

Then there was the memory loss. Although often temporary, this was hugely disorientating for a large percentage of patients. Some forgot what they did in the days preceding their electroshock therapy session. Others lost years of their life. Even memorable events such as their own wedding, family holidays, or military service could be forgotten. In the 1940s, there was a theory that this "retrograde amnesia" wasn't a side effect, but was actually the curative element of the treatment, as if patients simply forgot that they were ill. (This is now known not to be the case; there is no relationship between memory loss and the treatment's effectiveness.)

A minority of patients lost more than memories. Studies in the 1940s found that around 0.06 percent of patients died during or following their course of electroshock therapy. One psychiatrist wrote that this was "negligible," especially when compared to the figures of 2 to 20 percent for lobotomies. The most common cause of death was heart failure, but there was one account of a ruptured intestine in a twenty-eight-year-old "aggressive schizophrenic." Although many deaths couldn't be directly attributed to the treatment (they may have been coincidental), this last example seemed causative. "There would not appear to be any reason for the condition other than the electrical treatment administered to the deceased," the pathologist wrote. These potential complications needed to be taken into account for each patient and their underlying conditions. "A psychiatrist, because he so often has to deal with patients incapable of making their own decisions, has an even greater responsibility to bear in this respect and must accept it."

As antibiotics were being mass-produced for the first time, making once-lethal infections vanish within a few weeks, electroshock therapy became a staple of psychiatry. By 1944, the first controlled study into its efficacy had been published. Kenneth Tillotson and Wolfgang Sulzbach, two psychiatrists working at McLean Hospital in Belmont, Massachusetts, compared the rate of recovery in seventy depressed patients who were given electroshock therapy to sixty-eight comparable patients who received the standard care (essentially no care) within the hospital wards. With this simple difference—one group with and one group without electroshock treatment—they hoped to investigate a well-known complication in any treatment for depression: patients often recovered on their own. "The efficiency of convulsive treatments is frequently minimized or entirely disputed on account of the generally favorable prognosis in depressions," Tillotson and Sulzbach wrote. "Our evidence, however, argues against this."

Over four years of study, not only did 30 percent more patients recover in the electroshock group compared to controls (80 percent vs. 50 percent), but they were also half as likely to relapse back into a depressive state a year in the future. (A similar study also published in 1944 presented figures of 90 percent for electroshock and 40 percent for controls.) Perhaps most spectacular were four patients who had suffered from unremitting depression for upward of fifteen years who then recovered in a matter of weeks or months. Two years later, they were still depression free. And although short-term memory loss was still a common complaint, over long periods of time electroshock therapy actually had cognitive benefits, especially for elderly patients. Rid of the sluggishness of depression, they "displayed a far more efficient intellectual as well as emotional adaptivity to their environment than ever before in their life," Tillotson and Sulzbach wrote.

While Kalinowsky saw electroshock therapy as a safe, effective, and profitable profession, his former supervisor, Ugo Cerletti, remained in Rome and grew ever more critical of his creation's wanton usage. "[I]n private practice," he told an audience at the first Congress of the Italian Society of Psychiatry since the end of World War II, "progress cannot be made . . . since not all the apparatus, the tools and the analyses which are commonly available in a modern clinic can be used." More than just a lack of equipment, he added that the financial incentives of treating private outpatients was tempting many budding researchers away from curiosity-driven medical research.

As with Mary Shelley's Dr. Frankenstein, Cerletti thought that he had created a monster that wasn't only damaging his profession but also the patients he felt a duty to help. "I came to the conclusion that we must get away from the use of electroshock," he later wrote. "When subjecting unconscious patients to such an extremely violent reaction as these convulsions, I had a sense of illicitness and felt as though I had somehow betrayed these patients." Worried that he had also inadvertently aided in the demise of laboratory research, Cerletti, now approaching seventy years old and with his eyebrows bushier than ever, worked even harder to undo the damage he had done.

He didn't lose faith in the efficacy of electroshock therapy. He was as convinced as anyone of its therapeutic potential, especially for certain severe forms of depression. He just didn't think that a convulsive treatment always required a convulsion. Surely all the thrashing of limbs, crunching of vertebrae, and clamping of jaws wasn't necessary if the brain was the focus of an epileptic fit? Why should the body go through so much trouble as a side reaction? For Cerletti, the path forward was as clear as the stained slices of brain tissue under his microscope. "The future lies in the domain of biochemists," he wrote.

As a student at the University of Turin in the late nineteenth century, Cerletti had failed chemistry. Even when he retook the test, he still only scraped through. Even so, it would be this world of molecules, test tubes, pipettes, and graded solutions to which he turned in his later years. Cerletti became obsessed with a family of chemicals that he called *acroaganines*. Taken from the Greek words *akros* (extreme) and *agon* (struggle or defense), they were the chemical basis of electroshock therapy, he thought. Just as an immune response leads to the release of antibodies, he proposed that a convulsion released its own "highly vitalizing substances with defensive properties" against depression and other mental illnesses.

As with his early tests into the safety of electroshock therapy, Cerletti looked to pigs for answers. He returned to the slaughterhouse in Rome. After administering electroshocks several times before they were killed, Cerletti would extract the solution that bathed the pigs' brains. It was these foul few milliliters, gloopy and sickly yellow, that, Cerletti thought, contained his coveted *acroaganines*. Once distilled, he hoped to inject this solution into his patients, removing the need for electricity or a convulsion. Could these brain chemicals lead to a less catastrophic form of therapy? Would this discovery attract the next generation of psychiatrists back into the laboratory, a place where medical revolutions emerge from such daring theories?

Young or old, Cerletti didn't succeed in chemistry. *Acroaganines* were in the same chemical family as fairy dust and black bile—they didn't exist. His final experiments were all dead ends.

Psychoanalysts had their own theories about electroshock therapy. Fear of imminent death, punishment from a parent-like figure, enforced regression into an infantile state: all were proposed as potential mechanisms for this treatment's efficacy. Ultimately, Edith Weigert, a leading psychiatrist in the U.S., wrote, "the effect of the different [shock] treatments is a breaking through of the patient's autism, making him more affectionate, more interested in the outside world, more manageable, more sociable." Although it was an invention of biological psychiatry, psychoanalysts still turned to electroshock therapy as a tool that could "open up" their patients and make them more amenable to psychotherapy. "[W]hatever doubt may exist about the dynamics of improvement," one psychotherapist wrote in 1943, "there can be no doubt of the empirical fact that electro-shock therapy is an important contribution to the art and science of mental healing."

From its underlying biology to its effects on the mind, no one knew how electroshock therapy worked. "It is true that no adequate theory for [electroshock therapy] is available," Kalinowsky wrote in 1949, "but this is no valid reason against its application. The patient who needs help, cannot wait until empirical knowledge has found a satisfactory theoretical basis." To him, the important thing wasn't how it worked. Just that it did.

Although it came with disabling side effects, Kalinowsky argued that any psychiatrist who refused to employ this treatment for the most severe mental disorders was guilty of medical malpractice. Refusing to adopt a treatment that can relieve months of suffering in patients, he argued, was like withholding food from someone who is starving. "The surgeon does not refuse a necessary operation because of its impending risks," he wrote in *Somatic Treatments in Psychiatry*, a textbook he coauthored with a colleague at the New York State Psychiatric Institute. "Mental disorders are as destructive as a malignant growth and far more terrible in the suffering they may cause. Risks are therefore justified."

Ugo Cerletti was right about the potential for biochemistry in electroshock therapy. It just didn't come from *acroaganines*. As early as the late 1930s, muscle relaxants were removing the need for the violent convulsions that electroshock therapy induced. With an injection of curare (pronounced "coo-rah-ree"), the muscles of the body would remain relaxed while the brain still went through the classic electrical fluctuations that defined the clonic and tonic phases of a grand mal seizure. But it was a dangerous practice that blurred the lines between life and death. Made from the boiled stems of climbing plants, curare had been used to kill prey in the Amazon rainforest since time immemorial. Painted onto spear tips or arrowheads, this viscous paste quickly entered the bloodstream and stopped the muscles that surround the lungs from expanding and contracting. In a few minutes, animals as large as deer would fall to their knees and suffocate. At one-tenth of a lethal dose, however, curare was used to reduce the violent contractions of convulsive therapy. Paralyzing the muscles for fifteen to twenty minutes, the convulsion was restricted to the brain and hidden from the body. With artificial respiration reducing the risk of suffocation, the problem of fractures became a distant memory.

Lothar Kalinowsky refused to use curare. "I had my first and only fatality . . . in a patient where I used curare," he confessed years later. Although fatalities were rare, there was no doubt that, in the wrong hands, curare could paralyze a patient's breathing or even their heart to such an extent that artificial respiration was not always available or effective. To Kalinowsky and others, curare was "more dangerous than the complications it is supposed to prevent." Symptomless fractures were better than accidental suffocation.

Curare was nevertheless an insight into a safer future of electroshock therapy. It shined a light on the essential organ of a convulsion: the brain. The body needn't move at all; as long as the electroencephalogram (EEG)—a readout of brain activity from electrodes placed on the head—showed the steep scribblings that define a grand mal seizure, then the session could be deemed a success. Such "modified electroshock therapy also injected a sense of hope for those people who were once thought to be unsuitable for this treatment—people with medical problems such as osteoporosis, slipped disks, and previous fractures. Even the brittle bones and dwindling reserves of cartilage of the elderly were no longer contraindications for this treatment. This was a boon not only for psychiatry but for medicine in general. Late-life, or "geriatric," depression not only coincides with other chronic diseases such as diabetes, cancer, hypertension, and cardiovascular problems, but can exacerbate them. It makes cancer more likely and more lethal. It reduces a person's adherence to pills that help keep diabetes at bay, thereby reducing their efficacy. And people with depression are far more likely to die from heart complications compared to someone who isn't depressed. In late life, that is, untreated depression is a dangerous and potentially lethal co-traveler. As early as the 1940s, however, electroshock therapy had the potential to shunt the aging process toward a healthier path.

In 1952, a time when psychiatrists were still learning how to harness the potential of electrically induced convulsions, a safer relaxant entered the field of medicine. Appropriately for this story, a chemist working just a stone's throw from Ugo Cerletti's laboratories in Rome began his work on a fast-acting, rapidly dissipating muscle relaxant that was initially termed "curare of short action." Succinylcholine (its formal name), or "sux," was a vital development in the treatment of depression. Even Kalinowsky accepted this new drug and pushed his sandbags out of his methods and his private practice. Along with general anesthesia and artificial respiration (a requirement for any use of a general anesthetic), sux helped transform a dated shock therapy into a modern medical procedure. To complete the break from the past, electroshock therapy went through several rebrands. Over the coming decades, it would be referred to as "electroplexy," "cerebroversion," "central stimulation (CS)," "brain stimulation therapy (BST)," and, wait for it, "central stimulation with patterned response (CSPR)." But the name that stuck was "electroconvulsive therapy" and its famous initialism, ECT, a revolutionary life-saving treatment that had the potential to both help and harm mentally ill patients.

From the 1960s onward, however, it would nearly drop out of psychiatric practice altogether. "It was as if penicillin had somehow vanished from the medical armamentarium," wrote Edward Shorter and David Healy, two medical historians, in their 2007 book, *Shock Therapy*, "and a generation's memory of its very existence had been somehow erased."

The Psychic Energizers

It had smooth white fur and a little pink nose that wiggled a set of whiskers from side to side, and it lived in a wire cage. It looked like every other mouse Nathan Kline had ever seen. But as he watched this particular mouse, Kline, the director of psychiatric research at Rockland State Hospital in New York, noticed that it was behaving differently from most of its laboratory kin. It was more active, more aware of its surroundings than a run-of-the-mill mouse. It seemed to buzz with energy, powered by some invisible force within the confines of its enclosure. If it had a wheel it might have been able to run nonstop for hours without food or water. This mouse—nameless, just one of hundreds—was just what Kline had spent years searching for.

Kline, a smartly dressed man with thick horn-rimmed glasses who was once described as having a "twinkle in his eye that spoke of something extraordinarily curious," was visiting the laboratories of Charles Scott, a leading pharmacologist in the U.S. Before his arrival, the mouse had been injected with reserpine, a drug extracted from the root of a plant that had been used in Indian medicine for centuries. Known as sarpaganda, or snakeroot, it was used for a variety of ills. Snakebites and scorpion stings. Asthma. High blood pressure. Insomnia and insanity. Mahatma Gandhi used snakeroot to "attain states of introspection and meditation." Kline was very familiar with reserpine. In 1954, two years before his visit to Scott's laboratory, he had helped introduce the drug into the U.S. as a treatment for calming the manic, often destructive highs of some of his patients. On one of the psychiatric wards where Kline first trialed the drug at

Rockland State Hospital, it became much quieter and the furniture, windows, and people weren't in need of repair as often. Even the hospital's glazier noticed that he wasn't spending as much time working on that particular ward.

Reserpine was one of the first "tranquilizers," a class of drugs that didn't come with the lethargic side effects or addictive qualities of strong sedatives such as barbiturates and opium, both of which had been used for decades in mental hospitals.

In 1952, a similar chemical trialed at Sainte-Anne Hospital in Paris, France, known as chlorpromazine (marketed as Thorazine in the U.S.) was found to be even more effective in the treatment of schizophrenia than reserpine. Thomas Ban, a clinical psychiatrist and historian, was working in Austria when chlorpromazine was first released. He thought that it was nothing short of a miracle drug. "I could sleep through the night, and patients whom I never really thought would come back, came back," he recalled, his face beaming with delight at the memory. "It was unbelievable. The whole psychiatric establishment didn't believe it."

The effect of these two drugs—known as neuroleptics or, later, antipsychotics—was far greater than fewer smashed windows on hospital grounds and a better night's sleep. For the first time in human history, severe, life-threatening mental illnesses could be targeted with drugs, allowing outcasts of society to become outpatients with a prescription and regular medical supervision. For the first time in their adult lives, people were able to live in their community, in a home or halfway house, outside the walls of a mental hospital. "This was, I think, one of the most exciting periods that I've ever had in my life," Joseph Barsa, a stocky, balding, and conservative colleague of Nathan Kline at Rockland State Hospital, recalled. "People who had been in the hospital for twenty or thirty years began to improve, and we were able to discharge them. And the thing is, the joy, the excitement, and the enthusiasm spread throughout the staff. It was a different environment."

Reserpine wasn't perfect and sometimes didn't work at all, but it nonetheless provided a shift in how Kline thought about the treatment of mental illness. He started to wonder whether there was a chemical opposite to such tranquilizers. "If one compound could swing the emotional pendulum down," he wrote, "then there should be another compound that could swing it up." He had no idea what this hypothetical chemical would look like, where it would come from, or how it would work. He just knew it had to exist.

In the spring of 1956, three years after his first trial with reserpine, he thought he had found it in Scott's laboratory in New Jersey. Even though the mouse had been injected with reserpine, a tranquilizer, it was full of energy. The mouse should have been dozing and fatigued. It should have been stationary, not endlessly seeking. This paradox had a chemical explanation. The day before Kline's visit, the mouse had been given a different class of drug. It was called iproniazid, a chemical that seemed to reverse the depressing effects of reserpine. Kline thought that this could be the upward swing of his emotional pendulum.

From their source, reserpine and iproniazid were very different drugs. While the former was extracted from the root of a plant that blooms in white flowers across India, the Himalayas, and throughout most of Indonesia, the latter was a product of Nazi rocket fuel. One was a chemical wrested from nature, the other forged in modern warfare.

Hydrazine is a clear, highly combustible liquid that was used by the Nazis toward the end of World War II, propelling their V-2 missiles into Britain and France. After the war ended, pharmaceutical companies in the U.S. bought these redundant reserves at low cost, poring through their properties for anything that could be put to use in the postwar world. Needless to say, it was dangerous work. A single spark could fuel an explosion. But great risk often reaps great rewards. In 1951, one chemical showed promising antitubercular activity in the petri dish. It was called isoniazid, soon marketed as Rimifon. Then, by adding a clump of carbon and hydrogen atoms to this drug, Herbert Fox, a chemist working at the pharmaceutical giant Hoffmann–La Roche, created iproniazid. It was his finest creation, superior to its mother molecule, and succeeded where many of the strongest antibiotics—such as streptomycin—had failed. It killed *Mycobacterium tuberculosis*.

From Fox's laboratory in New Jersey, the drug was sent around the U.S. and given to various animals infected with tuberculosis. Guinea pigs, rabbits, rhesus monkeys: all showed promising signs or even a complete recovery from their infection. When the same result was found in humans, the mass media exploded with hyperbole. "WONDER DRUG FIGHTS TB," the *New York Post* wrote in two-inch type on its front page. During a fifteen-minute radio broadcast that was aired across twelve American networks in February 1952, listeners were told:

> Tuberculosis—the disease which destroyed more than five million lives last year, whose indiscriminate path through man's recorded history has filled more graves than war, famine, or pestilence—has been stopped. Scientists and clinical physicians have reported "amazing results" in the treatment in New York hospitals of some 200 tuberculosis patients . . . for whom all hope had been abandoned.

There were also reports of patients feeling a little too good, euphoric even. At Seaview Hospital on Staten Island—a four-million-dollar complex that the *New York Times* called "the largest and finest hospital ever built for the care and treatment of those who suffer from tuberculosis"—patients started to feel better the day after taking the new pills, even though their lungs were still pocked with painful lesions filled with bacteria. "A few months ago, the only sound here was the sound of victims of tuberculosis, coughing up their lives," one news article stated. After iproniazid, however, formerly terminally ill patients were "dancing in the halls tho' there were holes in their lungs."

"No bed cases remain," the doctors at Seaview confirmed, "the wards have a completely new appearance." Each room was a scene of celebration.

Intrigued, pharmacologists soon started to pick apart what these drugs actually did inside the body. Even at a molecular level, iproniazid and reserpine were opposites of each other. While reserpine reduced the levels of serotonin in the brains of rabbits and rats, iproniazid increased them. But what did this mean? In the mid-1950s, little was known about serotonin's role in the brain. After all, it was only in 1952 that Betty Twarog, a twenty-four-year-old PhD student working at Harvard University, first found evidence of this molecule in brain tissue. (Independently, researchers in Edinburgh discovered serotonin in brain tissue and presented their work to the British Pharmacological Society in July 1952, a month after Twarog submitted her paper.) Prior to this, serotonin was studied in the blood serum (hence "serotonin") and the gut (where 90 percent of serotonin is produced). But even without any definite explanations, the modulation of serotonin became a hot topic of research. Was serotonin of prime importance in mood regulation? Did the tranquilizing effects of reserpine and the euphoria of iproniazid center around this single molecule? If so, when Nathan Kline thought that there might be a chemical opposite to his tranquilizers, he didn't expect it to be so neatly packaged.

If it could reverse the depressing effects of reserpine in mice and make critically ill patients dance, could iproniazid also reverse mental depressions? Shortly after he saw the hyperactive mouse in New Jersey, Kline tested iproniazid on some of his resident patients at Rockland State Hospital. Along with his colleagues Harry Loomer and John Saunders, Kline chose seventeen schizophrenic patients who were also suffering from severe bouts of depression. One such patient, a forty-two-year-old woman known as "E.S.," who had been hospitalized for twenty years and hadn't responded to convulsive therapy or reserpine, was "quiet, withdrawn, and unresponsive to questions." Her attendant said that "she just wasn't there." She remained this way, silent and in the dark, for the first five weeks of her iproniazid treatment. During the sixth week, however, some invisible barrier seemed to have been lifted between herself and those around her. "She is loud and talkative," Loomer reported, "some of her verbalizations are quite sensible." In total, 70 percent of inpatients showed some sign of improvement after taking iproniazid.

This was a promising start. But the most remarkable findings were to be found outside the hospital's walls, across the Hudson River, and down into the lively heart of Manhattan, a place where depression seemed to thrive. Every afternoon at around four thirty, Kline drove his black Thunderbird soft-top into the city to meet his private patients, sometimes staying until after 11:00 p.m. His offices were on the ground floor of an East Side greystone that contained so much artwork, it was likened to a "metastasis from the Museum of Modern Art." With five nurses and three psychiatrists, four patients could be seen every hour, a turnover that made Nathan Kline's clinic the McDonald's of clinical psychiatry. Over two-thirds of his patients were suffering from depression. Although, as Kline later wrote, "[s]uffer is often too mild a term for the wracking agony and raw pain [of depression] which cries out for relief." In early 1957, the cries of many of his patients were met for the first time—with iproniazid.

"You see, Doctor," one patient told Kline, "the pills have worked. . . . I am not depressed or frightened anymore. I go everywhere and talk to everyone. I am full of energy. . . . I knew that I was in for another year of depression. I shall be eternally grateful." The depression of another patient, a thirty-year-old housewife who had been given seven years of psychoanalysis, started to lift for the first time after taking iproniazid three times a day for three weeks. Out of thirty-one patients, Kline found that thirty "responded with complete remission of symptoms." If such reports could be replicated elsewhere, nothing in the history of psychiatry could compare to iproniazid.

"There is a revolution in psychiatry," wrote Earl Ubell, the science editor of the *New York Herald Tribune*. "The biological approach seems to be working."

Kline called iproniazid a "psychic energizer," the chemical opposite of tranquilizers such as reserpine. Its real name, however, is monoamine oxidase inhibitor, or, in short, MAO inhibitor. It's not as catchy, but is certainly more accurate. A monoamine is a single molecule that is derived from an amino acid, the building blocks of proteins that make up the bulk of every living thing from microscopic bacteria to massive blue whales. Serotonin, for instance, is a monoamine created from tryptophan, an amino acid found in most protein-rich foods such as tofu, chocolate, and mushrooms. Dopamine and norepinephrine, two other well-studied monoamines, stem from tyrosine. Each monoamine is a crucial signaling molecule in the body's tissues, sending messages from one neuron to another, or from a neuron to a muscle or an organ. Your every thought, memory, and move is founded on a well-choreographed dance of monoamines.

Let's re-create their dance floor. Imagine clenching your fists in front of you and moving them toward each other, right hand to left hand and knuckle to knuckle. Leave a small gap in between. Don't let them touch. This is a gigantic and hugely simplified model of two neurons meeting in your brain. The gap between your knuckles is important because that represents the synapse, a place where monoamines flow from one neuron (say, your right hand) to the other (your left hand). At this junction, the electrical signal that flowed through this neuron is translated into a complex chemical signal. Whether it's serotonin, dopamine, or any of the other monoamines found in the body, vital information is relayed through the brain's neural networks in an instant. For this reason, monoamines are called "neurotransmitters," chemicals that transmit information between neurons. Our brains aren't just electrical organs (as was thought up until the middle of the twentieth century), they are powered by chemistry, too.

All chemical reactions need an off switch. Whether it is temperature regulation, or the amount of oxygen in the blood, everything needs to be kept in check. This is called homeostasis, roughly meaning: the same thing (*homo*) kept constant (*stasis*). For monoamines in the body, this control comes from an enzyme, a relatively huge lump of protein called monoamine oxidase that can grab hold of serotonin and norepinephrine and temporarily disable them by altering their chemical structure. MAO puts an end to whatever message the monoamine was sending. Without this off switch, the brain would be awash in a confusing abundance of signals, all vying for space in the synapse and causing an overload for our everyday consciousness.

To different degrees, iproniazid stops this off switch from working (hence, MAO inhibitor). By preventing the MAO enzyme from doing its job it keeps monoamines such as serotonin and norepinephrine as they are—signaling molecules in the brain's synapses. (The first monoamine oxidase, discovered by Mary Hare, a chemist at the University of Cambridge, in 1928, interacted with tyramine, another monoamine.) It's a cascade effect: by blocking the enzyme that normally breaks down serotonin and norepinephrine, these monoamines are free to increase in number. Iproniazid in the brain is like the introduction of wolves into Yellowstone National Park. These large predators indirectly increase the growth of grass and trees by reducing the feeding activity of deer. Iproniazid, by similarly blocking the feeding activity of MAO, increases the amount of monoamines in the brain, the fundamental basis of our own internal ecosystem. At the right dose, a new homeostasis—one with slightly higher levels of serotonin and norepinephrine—can take root.

With iproniazid and reserpine, an energizer and a tranquilizer, Kline dreamed of a day when emotions could be tinkered with like body temperature or blood sugar levels, guiding people away from pathological extremes toward a healthy norm.

Drug therapy didn't sit well with psychoanalysis, the dominant psychiatric worldview at the time. The theories put forward by Karl Abraham, Sigmund Freud, and Melanie Klein were still seen as the one true insight into the cause and treatment of depression. Nathan Kline, originally trained by a leading psychoanalyst in the U.S., was well versed in the latest revisions of this theory that focused on loss and feelings of deprivation that were first sown in childhood. "One supposed mode of development [for depression] is that a child somehow fails to receive from the parents the love and support it needs," Kline wrote. "The child resents this bitterly but cannot express it openly because of guilt and so turns the anger inward," adding, "In effect, the child enters into a kind of subconscious alliance with the parents, rejecting himself as he believes they reject him and forming toward himself feelings of inadequacy and unworthiness. Thus created is a pattern of responses that becomes deeply embedded in the individual's personality. He will be plunged into depression, so the theory goes, whenever some stress situation brings out his buried feelings of rejection and failure."

Depression was the emotional aftershock of object-loss and the conflicts that lay hidden in the deep strata of the unconscious. Only by extracting them—through hours of analysis "on the couch"—could a patient start a process of recovery. It could take years and thousands of dollars until they were freed from their affliction. There was no place for iproniazid in this technique. If any benefits were reported from drug therapy, psychoanalysts argued, it was because they simply masked the deeper causes of depression, just as drinking alcohol can be effective at covering up stress. Such a deep-seated conflict as depression wasn't amenable to simple changes in chemistry. To say otherwise was close to heresy, a viewpoint that could end a once-promising career in psychiatry. "So intense were the feelings that some good friends and respected colleagues took me aside to warn me that I was making a fool of myself," Kline wrote. "I was told that if I persisted in pursuing such an eccentric course, it could jeopardize my career.

"Such warnings, of course, were both well meaning and sincerely motivated," he continued. "They came from men who believed drug treatment of mental disturbance was, by definition, a dangerous quackery." As one of his contemporaries later remarked, "No one in their right mind in psychiatry was working with drugs. You used shock or various psychotherapies."

Kline grew up along the boardwalk and busy streets of Atlantic City, New Jersey, a place that seemed to be—and often was—governed by its own rules. Under the jurisdiction of Enoch "Nucky" Johnson, a man who, according to the *Atlantic City Press*, "had flair, flamboyance, was politically amoral and ruthless," the city had turned into a hub of drinking, sex work, and gambling during a time of prohibition and economic austerity. "We have whisky, wine, women, song and slot machines," Johnson once said. "I won't deny it and I won't apologize for it. If the majority of the people didn't want them they wouldn't be profitable and they would not exist. The fact that they do exist proves to me that the people want them." It was as if Atlantic City had broken away from the mainland and floated adrift from the rigid work ethic that founded the American dream.

For Kline, a resident here, the endless mesh of entertainment and hedonism wasn't a holiday. For all he knew it was life itself. Although the cast in Atlantic City was in constant motion—people moving in and out like the ocean's tide—the characters remained the same. Drunken, disorderly, criminally extravagant. "I thought everyone lived that way all the time," Kline later wrote, "and, naturally, I did the same. Money was meant to be spent, objects were in stores in order to be purchased, and life was there in order to be lived." As his teenage interests in literature and poetry shifted toward medicine and psychiatry, he knew that he would never outgrow his childhood in Atlantic City.

Kline seemed to make up his own rules and live by them. When the medical director of a pharmaceutical company rejected his bid to test iproniazid, for example, Kline circumvented his decision by inviting the company's president to lunch at Theodore's Restaurant in New York City and, "in the flush of excellent food and wine," was able to sweet-talk his way to what he wanted. He was known for his womanizing and would separate from his wife after nine years of marriage. He intentionally antagonized his peers at international conferences and would give experimental drugs to mentally ill patients without consent or approval from medical ethics committees. In short, he lived life as if he were playing a game and he was the protagonist.

Nathan Kline was no saint. When it came to the treatment of depression, however, he would do more than anyone to push its treatment into the mainstream, not only providing relief for millions of people, but changing the scientific perception of what depression is.

During a time of psychoanalytical obsession, he stuck his neck out for biological psychiatry and seemed to relish the attention that it brought him. The front cover of his best-selling book, *From Sad to Glad*, made his chutzpah plain to see: "Depression: You Can Conquer It Without Analysis!"

The Shoes That Prozac Would Fill

Ten years before Nathan Kline gave his patients iproniazid, another class of drugs was widely used to treat depression. Originally created as a potential decongestant to replace ephedrine (a chemical extracted from the ephedra plant that had been used in Chinese medicine for millennia), Benzedrine sulfate was shown to make withdrawn, depressed patients at the Maudsley Hospital in London more outgoing, even chatty. "Almost everybody showed an increased tendency to talk, but the effect was most striking in depressive patients," the psychiatrists wrote. "[T]hey overcame their [depression] and several of them talked spontaneously to other people for the first time since their admission." As they spoke, they felt better about themselves and the world around them. One patient declared that she was an "exulted being." Another compared Benzedrine to a double shot of whiskey, a feeling of "energy and self-confidence." The main difference, he added, was that whiskey was only enjoyable when he was feeling healthy. When he was depressed it made his situation even more miserable. But Benzedrine didn't.

In 1937, a study from researchers at the Mayo Clinic in Minnesota found that, out of thirty patients who were given Benzedrine sulfate to take with breakfast and before lunch, twenty-one showed almost immediate response. "As soon as I started taking the benzedrine the depression and feeling of fear left me at once," a forty-five-year-old man who had suffered from bouts of depression since 1918 said. "[I]n other words it made me feel like I have felt before under normal conditions when I have been at my best of vigor."

Benzedrine sulfate is an amphetamine, a drug in the same family as ecstasy, speed, and crystal meth. In the late 1930s, however, it was most likened to caffeine, as if patients were being lifted from a depression with a particularly strong coffee. "Benzedrine is a stimulant," Eric Guttmann, a researcher at the Maudsley, told the Royal Society of Medicine in October 1938, "and although it does not appear to alter a depression fundamentally, it can lead to a symptomatic improvement which may tide the patient over a critical period, or it may provide that final impetus which is so often necessary in a depression that drags on." Drugstores in the U.S. and pharmacies in Britain stocked the growing variety of amphetamines that followed in Benzedrine's footsteps, offering relief from narcolepsy, fatigue, and general tiredness. University students could purchase such "Pepper-upper (Pep) pills" and stay awake for days, allowing prolonged periods of study that were previously impossible. Soldiers on both sides of World War II gulped down amphetamines to keep their minds alert, their eyes focused, and their bodies ready to fight, even after sleepless nights or long marches. While the British armies allocated Benzedrine to their infantry and aviators, the Nazis preferred Pervitin, a brand name for methamphetamine (a form of crystal meth).

Although there were warning signs from the beginning—there were reports of students fainting and dying after taking Benzedrine to keep them awake during exam season—amphetamines were marketed as a treatment for "mild depression" in 1939. The pharmaceutical company that produced Benzedrine—Smith, Kline & French—forced its product into the mainstream. One poster from 1945, for instance, shows a besuited man standing proud, his chest pushed outward and his hands on his hips. He is smiling and gazing into the distance. Behind him, his enlarged countenance is in a state of confusion and despair, the shadow of his former self. The poster makes it clear: his transformation is due to a new drug. "Only in the last decade has there been available—in Benzedrine sulphate—a therapeutic weapon capable of alleviating depression," it reads.

Amphetamines provided an alternative to psychoanalysis and electroshock therapy, a quick fix for "mild" depressions that could boost a person's energy and feeling of well-being. Amphetamine therapy, Abraham Myerson, a leading psychiatrist from Boston and the famed author of the 1925 book *When Life Loses Its Zest*, wrote, "is not in any sense curative and its effects are not permanent, but it helps to dissipate the morning apathy and depression. [I]ts ameliorative effect is sufficiently important to recommend it while the process of natural recovery is taking place." And for the restlessness, insomnia, and anxiety that often came from taking regular doses of a stimulant, sedatives were often prescribed in the evenings. "Clinically, the effects of the two kinds of drug when given in combination seem to be regarded as 'mutually corrective,'" the pharmacologist Hannah Steinberg wrote. While one drug lifted a person slightly closer to agitation and euphoria, the other brought them back down for a sleep.

Financially, these drugs were a huge success. In 1949 alone, Smith, Kline & French earned over $7 million from their sales in amphetamines. At their peak, pharmaceutical companies produced eight hundred metric tons of amphetamine pills every year, the equivalent of every person in the U.S. taking forty-three doses of amphetamine per year. Before the designer drugs like Valium, Prozac, and Xanax, there was Benzedrine sulfate. "It established the 'profile' that subsequent antidepressants would have to match," the historian Nicolas Rasmussen wrote, "the shoes that Prozac would ultimately have to fill." The difference is that modern antidepressants actually work, even in severe cases of depression.

My first week on citalopram was awful. It was spring and I sat on the sofa wrapped in a duvet feeling like I had the flu. Confused, nauseous, dipping in and out of a sleep, I swallowed my daily pill and waited for my brain to get used to its new injection of serotonin. The flu-like symptoms soon stopped, but my depression did not. In fact, it only got worse. The nausea was a daily occurrence, a constant reminder that I was a depressed person taking prescription drugs. This would have been a price I was willing to pay if the drugs worked, if the depression stopped. I could live with nausea. But the depression didn't abate and I tried to run from my demons.

I was twenty-five years old and thought that a freelancer could, and should, live anywhere he wanted. I had been in London for two years and wondered whether it was the reason for my anguish. After all, there was the extortionate cost of rent, the constant buzz of traffic, sirens, and the inability to escape to more tranquil places without an expensive train ticket. Although most of my friends were living in the city, it was rare that we met up and saw each other. I decided that I wasn't happy living in London and made plans to move to Berlin, a city that I had visited three times and thought would be a perfect place to start afresh. I had been dating Lucy for a few months and she had been thinking along similar lines. We moved in the summer of 2016, found a one-bedroom apartment that a friend was subletting for a few months, and thought that this might be a base for new beginnings.

There were good times, moments of a new relationship that I will forever cherish. We learned German together, walked through ancient woodlands filled with birdsong, and wrapped ourselves up against the dry continental winter that could dip to 5 degrees Fahrenheit. But, despite the distractions, it was soon evident that I was getting worse. My depressive episodes were becoming more frequent and more dangerous. I always made sure I had enough citalopram to last me until we returned to the UK to visit friends, but the drugs didn't seem to be doing anything. One day, I decided to stop taking them.

Although citalopram and other SSRIs are not addictive, they can come with some pretty horrendous withdrawal symptoms if you suddenly stop taking them. I wasn't told this. I quickly wished I had been. There's a German word for depression: *Weltschmerz*, or "world pain." And this withdrawal certainly felt apocalyptic. I locked myself in our bedroom and rocked myself into a stupor, sitting on the floor with my head in my hands. That weekend, one of my friends was visiting for a few days, but I barely saw him. He must have heard me crying and my screams. When he was leaving to catch his flight home, I hugged him and apologized. He left on a positive note: he admitted that he hadn't realized how bad depression could get. In that moment of departure, he became a closer friend.

"I feel trapped by a drug that I hate," I wrote at the time. If both group therapy and antidepressants hadn't made a dent in my depression, hadn't slowed the frequency or the severity of my episodes, then what point was there? Why go on when both fields of psychiatry—the psychological and the biological—had tried and failed to make a difference? As Lucy was working for a start-up that demanded eleven-hour days, I sat in our apartment and thought about an end. This world pain had to stop. I noted in my diary how Winston Churchill named his depression the "black dog," a pet name that helped separate his mental state from himself. As my first group therapy classes taught me, I'm not a depressed person, but a person with depression. But, as it took away more and more of my days, I couldn't see it like that anymore. I couldn't call my depression by a pet name. As I wrote in my diary, "Looking into the future—a difficulty at this time—I worry that others, my family, the media, Lucy, would know exactly what to call my depression. It would be my killer."

It was then, only six months after arriving in Berlin, that we decided to move back to the UK and reach out for the support of friends and family, and to look for the next stage of treatment. I was put on a waiting list for one-to-one cognitive behavioral therapy and my dose of citalopram was doubled to its maximum. Anything higher would lead to dangerous side effects such as the so-called serotonin syndrome. In mild cases, too much serotonin in the body can lead to vomiting, diarrhea, and muscle twitching. But it can also cause convulsions and, as one case report published in the *Journal of Medical Toxicology* in 2014 states, "fatalities may occur." "The clinician," the authors of this report added, "must be aware of the potential for large ingestions of citalopram to produce life-threatening effects and monitor closely for the neurologic, cardiovascular, and other manifestations that, in rare cases, can be fatal."

My doctor, a compassionate middle-aged woman who kept calling me Alexander, suggested that Lucy keep hold of my pills. Did we have a safe for which only she knows the code? Could she hide them somewhere I would never think to look? In that moment, we all agreed that I was a danger to myself. This prescription could be used for two purposes: for stability or for suicide. Unfortunately, it was the latter that seemed more likely.

For me, citalopram is a drug of side effects. After my dosage was doubled, my anxiety increased beyond what I knew was possible. I spent nights obsessing over thoughts that would never have troubled me in the past. My heart palpitated and a feeling of sickness extended from the back of my throat to the pit of my stomach. I knew that SSRIs work for many people. But, after two and a half years without relief, I also knew that citalopram wasn't for me.

In 2015, 7.2 percent of the U.S. adult population were taking antidepressants. In 2018, 16.6 percent of the UK population had received an antidepressant in the previous twelve months, half of which had taken these drugs for the entirety of that year. In Australia, these figures stand at 15 percent, a proportion that has doubled since 2000. In all countries, prescriptions are increasing year by year.

These data are striking. It could be argued that pharmaceutical companies have a grip on a large percentage of the population, and their pockets. But high uptake of prescription drugs could also be a sign that people are more willing to reach out for professional help, whether it is for their depression, anxiety, or any of the other illnesses for which these drugs are effective. These treatments have been stigmatized for a long time, and people have had difficulty accessing them. The only problem is that antidepressants are not always brilliant at treating depression. Some studies have found that they are only worth a patient's while if they are severely depressed—someone who experiences a broad range of symptoms for a long period of time. In such cases, antidepressants can do wonders, living up to the claim of miracle cures that made the headlines with the introduction of Prozac in the late 1980s. For milder depressions, the evidence for their use has led to heated debates and round after round of contradictory reviews.

In 2018, a meta-analysis—a kind of review that lumps similar studies together, filters through their differing methods, and takes an average of their results—published in the journal *The Lancet* seemed to provide some much-needed clarity. Reviewing studies on twenty-one different antidepressants, Toshi Furukawa, Georgia Salanti, and their colleagues found that they were all more effective than a placebo, but only just. In some cases, the gap between treatment and placebo was so small that the clinical benefit of prescribing these drugs remained inconclusive. Would patients actually *feel* such a slight reduction in their depression's severity? Would the side effects of these drugs make other parts of their life worse in the process? These are questions that continue to fuel heated arguments at international conferences and between psychiatrists in medical journals. I have sat and watched such debates and felt exasperated at the growing gulf between what psychiatrists study and the patients whom they aim to treat.

Clinical trials have a long history of cherry-picking patients who aren't representative of depression as a diverse collection of mental disorders. Usually more chronically depressed and not actively suicidal, they fit neatly into the *DSM*'s criteria of major depressive disorder. They have suffered with five or more symptoms for at least two weeks. Such patients are a specific subset of people with depression. But when someone reaches out for support, the doctor or psychiatrist doesn't get to decide whether they can choose another patient. They have to make a decision. Should this person receive an antidepressant or not? With limited time and resources, they have to make a call—and that call is usually an SSRI like citalopram.

I have often felt betrayed by this common practice. I thought that an SSRI was pushed into my life unnecessarily. But a study published in November 2019 provided a refreshing sense of understanding and clarity. Conducted across 179 general practices in four UK cities, this "pragmatic" randomized controlled trial followed the health of 640 people who weren't cherry-picked before the trial began. "Participants in our study ranged from those with very few depressive symptoms to those with severe depressive symptoms, therefore our results are more readily generalizable to the population currently receiving antidepressants in primary care," Gemma Lewis, a lecturer of psychiatric epidemiology at the University College London, and her coauthors wrote. For me, the results were reassuring. Even with such a broad sweep of depression, the SSRI sertraline outperformed the placebo on a range of measures. At six weeks, self-reported mental well-being and "quality of life" was significantly improved in the SSRI group, even at the six-week mark. Similarly, levels of anxiety—a common experience for people living with depression—were significantly reduced. At the twelve-week mark, a significant number of patients were in remission from their depression (51 percent), compared to the placebo (31 percent). Importantly, there was no correlation of clinical improvement with severity of depression; both mild and severe depressions responded similarly. The superiority of the SSRI over placebo, therefore, was unlikely to be simply a product of the more severely ill patients responding to antidepressants.

Reading through this study, I found comfort in knowing that antidepressants are able to help support such a diverse group of patients suffering from very different experiences of depression. Rather than being measured as a reduction of symptoms, this study provides an important reminder that an SSRI can make people feel better and reduce the anxiety that often can initiate and maintain a depressive episode, while also guiding half of these everyday patients into remission. Far from perfect and still prescribed through trial and error, this study shows that antidepressants are an effective weapon in our psychiatric arsenal.

G22355

In the middle of the 1950s, a meeting of psychopharmacologists was being planned for the Second International Congress for Psychiatry. Held in Zurich, a charming city of bridges, cobblestone paths, and church steeples that perch above the tranquil Limmat River, it was supposed to be a small gathering, not too different from a dinner with colleagues. But, as one of the organizers wrote, "the 'speck' of drugs to treat mental illness [had] developed into a small cyclone," and they had to make time and space for more than ninety participants. Over a dozen pharmaceutical firms provided the financial support for a three-day meeting, one that was separate from the general psychiatry conference. Nathan Kline was one of the introductory speakers. On the page, Kline was an eloquent and enthusiastic writer. In person, he was flamboyant and captivating, an orator who could cause waves of laughter in an otherwise sterile medical symposium. Some people loved him. Others detested him, thinking that he was narcissistic and would be more successful as a stand-up comedian than a serious academic.

Along with his growing art collection of Haitian sculptures, Kline amassed a vast archive of jokes over his lifetime. He often couldn't wait to return home from his private practice in Manhattan to test his latest quip on his daughter. At the international conference in Zurich, Kline started his plenary lecture—a talk that is scheduled for a time when the whole conference can attend—with his latest gag. He had isolated the cause of schizophrenia, he said. It was a bacterium all along—*the crafty Schizococcus*! A cure would soon be on the way. Whether anyone laughed was not recorded.

Over the next two days, Kline regaled his audience with his work into reserpine and iproniazid, two drugs that would lead him to the coveted Lasker Award in medicine, twice. (Kline is one of only two people who have received two Lasker Awards.) Although there had been reports of harmful side effects in patients taking iproniazid, such as dizziness, fainting, and a few cases of liver damage, Kline thought that these were unsubstantiated or, if they were real, part and parcel of this line of work. "It is almost a truism that a pharmaceutical [drug] that does not have side effects is a useless one," Kline told his audience. There is always a trade-off between cost and benefit, a middle ground between how much is taken away and what is restored. Every drug is a chemical compromise. For iproniazid, as it would later turn out, the costs were just a bit *too* costly.

Kline wasn't the only person in Zurich with a promising new drug for depression. After a decade without any advance on the existing amphetamines such as Benzedrine sulfate, two antidepressants were discovered within a few months of each other. Roland Kuhn, a psychiatrist from the bucolic town of Münsterlingen in Switzerland, presented his work on the last day of the general psychiatry conference. Largely unknown on the international scene, Kuhn hadn't been invited to the psychopharmacology section. Taciturn, sober, and with a whisper-soft voice, he was a very different character from Kline. He didn't have a plenary slot at the congress, nor any jokes for his small audience. All he had was an exclusive window into one of the greatest discoveries in the treatment of depression, one that would not only add to Kline's work but, in time, replace it.

But no one knew that in September 1957, not even those dozen people who stuck around and listened to his talk. Kline didn't even know who Kuhn was.

An unhurried town of a few thousand people, Münsterlingen sits on the southern shores of Lake Constance, the third-largest body of water in Central Europe that laps at the borders of Switzerland, Germany, and Austria. During the cooler months of the year, a thick fog blankets the streets at dawn, only lifting with the heat of the sun. By late morning, one can once again look out from the shore and see the sand-colored speck of Meersburg Castle in southern Germany, the invitingly blue and still lake in between. Working at Münsterlingen's psychiatric hospital, a trio of tall buildings that perch on the edge of the lake, it was this view that Roland Kuhn saw most of his working life. For a time, he lived on the hospital's grounds, opening the window of his apartment to the moist air and the din of waterfowl. It was like working where others might choose to holiday. It was so calm, picturesque, and peacefully quiet, that Kuhn rarely left.

He hoped that his research would put Münsterlingen firmly on the map. (Switzerland was a central hub of biological psychiatry and nearby Kreuzlingen was famous for its psychoanalytic institute, the Bellevue Sanatorium led by Ludwig Binswanger, but Kuhn and Münsterlingen were little known.) Since 1953, Kuhn had been working with a chemical called G22355, a direct descendent of chlorpromazine, the famed tranquilizer first discovered in Paris. These two chemicals shared the same three-ringed structure, a trio of honeycombs in a row. The only difference was in a few hydrogen and carbon atoms in place of an atom of sulphur. This slight change in structure, however, had dramatic effects in function. G22355 was useless, Kuhn found, for his hallucinating or manic patients; it either did nothing or made their symptoms worse. But it was brilliant for a subset of people who were depressed.

On January 21, 1956, Kuhn noted how Paula J., one of the first people to be given this chemical, had "undergone a transformation." "All of her restlessness and agitation has vanished," he wrote. "It is not entirely clear how the medication could have been responsible for such an abrupt change within a week's time, or whether this depressive phase abated spontaneously." This latter explanation seemed unlikely to Kuhn, however. She seemed like a completely different person after taking G22355. She slept better, was more sociable, friendly, energetic, and enjoyed reading books quietly.

Kuhn wanted to know more. Was Paula J.'s experience just a one-off? A placebo effect? After testing the drug on one hundred patients with schizophrenia who were also depressed (a diagnosis of schizoaffective disorder might be given to such patients today), Kuhn found that G22355 could elevate moods in forty of them. It wasn't a miracle by any means, but it was still a significant result in such a diverse mix of severely ill patients. On August 31, 1957, in his local bulletin, *Swiss Medical Weekly*, Kuhn wrote, "Feelings of guilt, delusions of impoverishment or culpability simply disappear or lose their affective importance, move into a distance, and the patient becomes indifferent and unconcerned with respect to these feelings." Previously bed-bound patients started getting up early in the mornings, initiated conversations for the first time in months, and wrote letters to people outside of the hospital's walls. In the wards, these formerly withdrawn individuals became social and even popular, elevating the mood of others. Just as the sun burned through the fog that blanketed Münsterlingen in the winter mornings, G22355 seemed to dissipate whatever had clouded his patients' minds.

A week after his publication in *Swiss Medical Weekly*, Kuhn traveled to the psychiatry conference in Zurich, a fifty-mile journey into the south of Switzerland. "The depression," he said in his native German, "which had manifested itself through sadness, irritation and a sensation of dissatisfaction, now gave way to friendly, joyous and accessible feelings." While Nathan Kline used terms like *psychic*, *ego*, and *id* to describe iproniazid, bridging the gap between psychoanalysis and drug therapy, Kuhn was fluent in medicine and scientific reasoning. Verbose and boringly thorough, he spoke with a philosophical monotone that made his once-in-a-lifetime discovery no more than a speculative whisper among the hubbub of other biological psychiatrists. When he had finished his presentation, the room remained quiet. The dozen or so people who turned up to this last-minute talk were wholly indifferent to the latest revolution in the treatment of depression.

Heinz Lehmann, a good friend of Nathan Kline, didn't make it to Kuhn's talk. He only became aware of the Swiss psychiatrist's work as he was flying back from the conference to his home in Canada. In 1937, Lehmann escaped from Nazi Germany by asking a friend to invite him on a skiing holiday in Quebec. Arriving at the airport with skis and a suitcase of clothes, he left his possessions behind and settled in Montreal, becoming the clinical director of Verdun Protestant Hospital in 1947. A decade later, he shared the Lasker Award with Nathan Kline (and others) for his research into chlorpromazine and its subsequent introduction into Canada and the United States. Returning from Zurich in September 1957, Lehmann read Roland Kuhn's article in its original German and couldn't wait to order some samples of this drug for his patients. Could he aid in this drug's dissemination, just as he had done with chlorpromazine? While Kuhn believed that qualitative observation was enough to determine whether a drug worked or not, Lehmann was a psychiatrist of the scientific method. He required quantitative evidence. He started a clinical trial, using a change in symptoms over time as a measure of recovery, improvement, or decline. Out of eighty-four patients with various forms of depression—endogenous or reactive, chronic or acute—he found that, roughly, two out of every three (60 percent) either fully recovered from their depressive symptoms or improved to such an extent that they could leave the hospital and return to their home or community.

Using different methods and styles of psychiatry, Kuhn and Lehmann both found that this drug was best suited to endogenous depression. In these cases, on average, three out of four people (75 percent) improved. "The effect is striking in patients with a deep depression . . . associated with fatigue, heaviness, feeling of oppression, and a melancholic or even despairing mood," Kuhn wrote, "all these symptoms being aggravated in the morning and tending to improve in the afternoon and evening." After a few days or weeks on G22355, Kuhn continued, "the patients express themselves as feeling much better, fatigue disappears, the feeling of heaviness in the limbs vanishes, and the sense of oppression in the chest gives way to a feeling of relief." Those who had been shrinking in size for months started enjoying their food and, over time, were able to maintain a healthy weight. Those that suffered from insomnia or woke up in the early hours of the morning and struggled to get back to sleep found that they not only slept, but they felt refreshed. They weren't "fatigued . . . as that so often produced by sleeping remedies," Kuhn wrote.

In 1958, two years after Kuhn's first trials, G22355 became a licensed drug. It lost its code name and became known as imipramine (marketed as Tofranil in the U.S.), the first member of a new family of drugs named tricyclic antidepressants after their three-ringed molecular structure. Prescribed in the U.S. and across Europe, imipramine represented another milestone in the treatment of depression: a drug that worked through a different chemical pathway from MAO inhibitors. "The pharmacological uniqueness of imipramine is of special significance," wrote the pharmacologist Fritz Freyhan in 1960, "since on the one hand its action cannot be explained in terms of enzyme inhibition, while on the other hand it appears to be the most effective antidepressant drug which is now available." Although imipramine also increases the concentration of monoamines in the synapse, it doesn't block MAO or any other enzymes. Instead, it wedges itself into the channels used by monoamines to return to the neuron that originally released them. These channels are like nanoscopic travelators that allow these chemicals to be reabsorbed, recycled, and reused. Its technical term in biochemistry is *reuptake*. In particular, later studies found, imipramine was particularly good at blocking the reuptake of norepinephrine, a neurotransmitter associated with sleep, appetite, and response to stressful situations. In our imaginary model of the synapse this process occurs along the sides of the clenched right hand, completing the circuitous flow of chemistry from the knuckles. Reuptake, rather than MAO inhibition, would prove of great importance in the future of drug therapy. It didn't lead to the same side reactions that would nearly doom many MAO inhibitors in the 1960s.

While imipramine was a great addition to the biological psychiatrist's medicine cabinet, there were doubts over whether it offered anything above existing treatments. "It should be noted," Heinz Lehmann wrote in 1958, "that the effects of the drug are much less spectacular than the therapeutic action of electroconvulsive treatment as regards both immediacy and intensity of its results. A deeply depressed patient who is suicidal might still require electroconvulsive therapy in order to control the situation rapidly, particularly if the patient is not hospitalized."

Although its use would sharply decline in the 1960s, ECT was still seen as the yardstick that any new therapy had to be measured against. "Electroconvulsive therapy (ECT) is considered by many clinicians to be the most effective treatment for depression," one study from 1966 states. "It is therefore important that any treatment suggested as a replacement for it be shown objectively to be at least equal if not superior in therapeutic efficacy." In the middle of the 1960s, two large-scale studies (incorporating over five hundred patients from four cities) in the U.S. and England put imipramine to the test, and ECT came out victorious in both. While the drug therapy led around 50 percent of patients into remission, ECT stood at 70 percent or even 80 percent. If delusions were present, as was the defining feature of melancholia, it was even more effective. More commonly known as "psychotic depression" today, this is a life-threatening illness that has a higher rate of suicide than depressions without delusions. But, amazingly, it is highly treatable. Studies in the 1970s and '80s that used simulated controls—a form of placebo where an anesthetic and muscle relaxant are used, but without an electric shock or convulsion—added further evidence that ECT is particularly effective in this type of depression. "Indeed, the significance of the original results, which show that real ECT is a more effective antidepressant than simulated ECT, depends very largely on the presence of 22 patients with delusions," a study of seventy patients concluded.

While these studies are now decades old and can't be repeated due to modern ethical standards (high-risk suicidal and psychotic patients can't be given simulated treatments), they all point in the same general direction: "[A] clear and significant advantage for ECT over medication," wrote Nancy Payne from New York University and Joan Prudic from the New York State Psychiatric Institute. "In fact, no study has found any treatment, including other forms of brain stimulation currently in development, to be superior to ECT in the treatment of major depression."

In the 1960s, however, imipramine was still seen as a viable replacement for ECT. Although it came with its own set of disabling side effects such as dizziness, tremors, and dry mouth, these were often temporary or reversible with a slightly lower dose. If the unquenchable, clacking dry mouth persisted, Nathan Kline recommended glycerine-based lozenges such as Pine Bros. cough drops. "[O]ther candies become cloyingly sweet after a short while," he wrote. Imipramine was also easier to administer than ECT. There was no need for a general anesthetic, injections of sux, and no risk of memory loss (a concerning side effect to most patients, even if it was usually temporary). But there were still concerns over whether imipramine was any safer.

The first reported case of an overdose on imipramine came less than a year after its release onto the pharmaceutical market. At 10:00 p.m. on February 19, 1959, a twenty-nine-year-old nurse who had previously been treated with ECT picked up her bottle of imipramine tablets and swallowed them all, a dose twenty times her daily norm. She felt nothing for three-quarters of an hour and then fell asleep. As she was nodding off, she hoped that she wouldn't wake up again. Two doctors from Barrow Hospital in Bristol, southwest England, investigated what happened next and published their report, "Suicidal Attempt by Imipramine Overdosage," in the *British Medical Journal*. "She has no recollection of any dreaming, and woke at 6:30 a.m. the next day in an agitated state, with gross involuntary movements of all her limbs, head and neck, which she described as being just like those of a fit and of a violent and terrifying nature," they wrote. "She remained in this state for some time, being unable to shout for help. Eventually her mother brought her a morning cup of tea, and, seeing the gross movements, asked what was wrong." Her daughter was unable to respond, but managed to gesture toward her empty bottle of pills.

This first overdose on imipramine didn't result in its first death. As later studies showed, the risk of overdose on these drugs is very small. One study found that, on average, four people died from an overdose for every million prescriptions, a percentage of 0.0004 percent. Still, these drugs had to be monitored and administered with care, especially when giving them to patients already at high risk of suicide. It was a paradox in a pill: imipramine could save lives from depression but could also be used for suicide, the most tragic consequence of this mental disorder.

Iproniazid wasn't as effective as Nathan Kline first thought. Although he originally reported that thirty out of thirty-one patients responded favorably to the drug, larger trials weren't so successful. While some studies found a response rate of 75 percent, others produced figures as low as 25 percent. This huge disparity demanded an explanation. Why would the same drug help three-quarters of depressed people in one study and then only a quarter in another? Was it the dose? Was it the length of time it was prescribed? As it turned out, the most likely explanation was that a group of depressed patients actually included people with a variety of different disorders, all lumped together as depression. Some may have endogenous depression, others might be more reactive. But iproniazid seemed to stretch the definition of what depression could be until it broke off into a new term altogether.

Atypical depression is, as the name suggests, an unusual form of illness. It is defined by its complete reversal of some of the defining features of endogenous depression. While this latter, more physical depression was characterized by weight loss, insomnia, and a cognitive dulling that negated any significant amount of anxiety, atypical depression was associated with eating too much (hyperphagia), sleeping more than usual (hypersomnia), and hyperactivity that manifested itself in tremors, restlessness, and crippling amounts of anxiety. Although these two conditions seemed complete opposites of each other, they were often lumped together under a single diagnosis: unipolar depression. (By 1959, bipolar disorder—a cyclical disease of mania and depression—had been separated into its own diagnosis and its own chemical treatment: lithium carbonate.)

For much of the twentieth century, doctors and psychiatrists weren't oblivious to the existence of an atypical subtype of depression. They met with countless patients who exhibited these symptoms. The problem was that they were often the most difficult to treat. Michael Shepherd, a psychiatrist from St. Thomas' Hospital in London, called atypical depression the bane of his psychiatric career. "For two years or more, [these patients] may have been complaining of vague depression, increased emotionality, diffuse anxiety, and sometimes of phobic fears of going out in the street, of traveling alone, etc.," hc said at a conference held in Cleveland, Ohio, in 1960. "They may also have become bad tempered, irritable, hyperreactive and aggressive; quite unlike so many of the more endogenously depressed patients. . . . In my twenty years' experience, this sort of [depressed] patient has been one of the most impossible and difficult to treat by any method."

Atypical depression usually responded to iproniazid within five to eight days. The only problem was when a patient came off the drugs, either by accident or to see if they had fully recovered, the depression returned like air being sucked into an area of low pressure. Relapse, as well as recovery, was a defining feature of MAO inhibitors such as iproniazid. People had to take these drugs for a long time to remain healthy. The same is true of both tricyclics and the third family of antidepressants, SSRIs, that were first used to treat depression in the 1980s.

Imipramine wasn't completely useless in atypical depression. It just took longer to have an effect. Rather than a few days, drugs like imipramine might take four or more weeks to lift someone out of this type of depression. ECT, meanwhile, made the anxieties that defined atypical depression worse and was best avoided for such patients.

More than a diagnosis, atypical depression was a lesson in how we study disease. It showed that clinical trials don't always fail because the drug doesn't work. Even if the large majority of patients didn't respond to a treatment, there might be a hidden minority that did. If the potential diversity of depression isn't taken into account, Peter Dally and Eric West, two doctors from St. Thomas' Hospital, wrote in 1959, "dramatic improvement in a few cases within the larger group may not reach statistical significance and may be ignored. Yet such improvements may be of greatest importance."

Just as Emil Kraepelin came to doubt his neat classification, these diagnoses weren't seen as a perfect guide to the diversity of depression. Atypical, endogenous, psychotic, reactive: they all had their own variations and overlapped with one another in their key features. Even if they were artificial constructs, they were also useful tools that helped sick people get the treatment most suited to their condition.

As it would turn out, this was particularly important for drugs like iproniazid, the chemical extracted from Nazi rocket fuel. This MAO inhibitor was soon found to kick-start a cascade of chemical explosions inside the human body. To prescribe this drug to someone who was unlikely to feel any benefit, therefore, could be catastrophic. Roughly one patient in every five hundred suffered from liver damage after taking the drug and became jaundiced—their skin turning a sickly yellow—from hepatitis. One in five of these patients died. Nathan Kline, who had been promoted to director of the recently created Rockland Research Institute (formerly Rockland State Hospital), argued that there was insufficient evidence to support these claims. "Among those first 400,000 Marselid [the brand name for iproniazid] patients there were some who got married, some who had ski accidents—and some who got jaundice," he wrote. "The marriages and ski accidents were not blamed on the drug, but the jaundice was. In fact, the case was never proved statistically, but the mere raising of the question created serious doubts." In 1961, a decade after it was first given to terminally ill tuberculosis patients on Staten Island, iproniazid was withdrawn from the medical cabinets of most countries around the world.

In one of his final pieces of writing, Kline recalled how it took six years for this antitubercular drug to be "discovered" by psychiatry. "The whole iproniazid story was one of these 'acres of diamonds' affairs," he wrote, "in which medicine kept stumbling over a treasure that had all along been lying there on its own neglected backyard." Although Kline had a personal attachment to this treasure, it wasn't a huge loss to psychiatry. By the early 1960s, there were already three new MAO inhibitors—phenelzine (Nardil), isocarboxazid (Marplan), and tranylcypromine (Parnate)—that could be prescribed. These were the drugs that would remain a part of the treatment of depression into the present day. As is the case for all drugs, they are not without risk. But Parnate was considered the safest because of its familiar chemistry. Although it functioned as an MAO inhibitor, it was actually a member of the amphetamine family and was therefore more recognizable, somehow safer; a drug that was unlikely to cause liver damage and could be prescribed with confidence.

The Mysterious Case of the Lethal Headaches

On June 23, 1962, a twenty-seven-year-old man walked into the Maudsley Hospital in London, slowly making his way past the Grecian columns that shadow its entrance. He felt giddy and off-balance, and he suffered from a painful headache at the back of his head, right at the point where the nape of the neck meets the underside of the skull. It pulsed with every heartbeat. Any movement made his head feel as rigid as wood. A nurse took his blood pressure. It was high, but nothing out of the ordinary. Soon after, he passed out. His blood pressure suddenly plummeted. A few hours later, he was dead. A postmortem examination revealed a burst artery, a hemorrhage, at the back of his head, right at the point where his headache had pulsed.

Though the consequences were less severe, three other patients at the Maudsley reported the same story: a "devastating headache" and, intriguingly, a prescription of Parnate. Each case was reported on and written up by J. L. McClure, the chief resident at the Maudsley and the supervisor of a twenty-nine-year-old intern named Barry Blackwell. A recent medical graduate, Blackwell listened to the reports and quickly lost interest. Three headaches and one death . . . So what? Thousands of people in the UK were taking Parnate to combat their depression. They weren't having any troubles. Further, there were no such reports from the U.S., a country that prescribed even more Parnate than Britain. If there was a problem with this drug, surely it would emerge there first.

As a medical student at the University of Cambridge in the early 1950s, Barry Blackwell spent more time at the local pubs and playing rugby than reading his textbooks. Having grown up fleeing from war in India and Nepal, and then being moved to a rural boarding school in the south of England to avoid German missile strikes aimed at urban centers like London and Bristol, his time at university was a welcome relief, a period of freedom and socializing. After he graduated—without distinction—he moved to Guy's Hospital in London, a soot-darkened set of buildings just south of the London Bridge. It was a leading institution for medical teaching, plus they had the oldest rugby team in the world. Over the next few months, he rotated as an intern in surgery, emergency medicine, and internal medicine. Although he published no case studies or reports during these early days, he did discover what he loved about being a doctor: chatting with patients, listening to their stories, and linking personal history to current health. "I was transformed by contact with patients," he would later say. "That unleashed both my energy and curiosity." Only when he discovered the people behind the abstract world of illness did he truly become a doctor.

Whether he liked it or not, Barry Blackwell seemed almost fated to unlock the medical mystery that was sweeping through the wards and forever tempting him to take a closer look. As he was eating in the hospital canteen, a long rectangular room with rows of pendant lights hanging from the ceiling and the smell of coffee infusing the air, a group of doctors were discussing a young female patient. She had recently suffered from a hemorrhage following a very painful headache. Blackwell was sitting within earshot, listening to their every word.

"Was she taking Parnate?" he asked, recalling his supervisor's report.

Yes, they said, she was.

From this chance encounter over lunch, Blackwell decided to spend some time investigating Parnate and these painful, potentially lethal, headaches. Was this drug definitely to blame? With so few cases, could it all just be a coincidence that people were taking Parnate? Was there a missing link between their prescription and these peculiar headaches? Blackwell sat down in the hospital's medical library and scoured through previous volumes of *The Lancet*, looking for any sign of headaches and Parnate from the last twenty months. He found six patient reports. Blackwell couldn't see any obvious pattern in the pages in front of him. The headaches affected both men and women, young and old, and didn't discriminate by job or location. The only whisper of a clue came from when these headaches occurred: they were most common in the evening. It wasn't enough to crack this mystery, but this temporal pattern would make sense when all the other pieces of the puzzle came together.

Without any firm conclusions, Blackwell published a short letter in *The Lancet* in which he proposed that these headaches might be far more common than previously thought. Were more people suffering—and potentially dying—from a drug that was supposed to help treat their depression? Although Blackwell had found a mere six cases, he was only looking in one journal and its recent publications. What about the other medical journals? More significant, were similar headaches and deaths going unreported? While he didn't have any answers to such questions, Blackwell was hoping that someone else might.

In the days before instant text messaging and email, he still didn't have to wait long for a response in the mail. Reading through the latest copy of *The Lancet* in his local library, a pharmacist working in Nottingham read Blackwell's article and thought he knew what the missing link was. Posting his own letter directly to Dr. Blackwell at the Maudsley, he explained how his wife had suffered the same back-of-the-head headache that the doctor described, not once but twice. In each case, he wrote, she had eaten a cheese sandwich for supper. "No effects are caused by butter or milk," he wrote. "If cheese is indeed the factor, it could perhaps explain the sporadic nature of the incidence of the side effect."

Could cheese react with Parnate to cause such blinding pain and, in rare cases, kill someone? Blackwell didn't believe it—it was, well, unbelievable. But as he was made aware of more cases and had time to reflect on the topic, the conversation he overheard in the canteen suddenly ballooned in significance, almost making him burst with the potential thrill of discovery. Perhaps that one case held the answer all along. Just as he had flicked through the pages of medical journals in the library, he rummaged through the kitchen's records (a sizable archive that was stored for monitoring patient allergies and other side effects), and found the menu from that day, the evening that the woman's headache started. He knew she was vegetarian and wouldn't have chosen the meat option. She had eaten a cheese pie.

The next step seemed obvious to Blackwell, a young intern with equal parts confidence and naivety. He prescribed himself twenty milligrams of Parnate from the hospital's pharmacy and took it daily for two weeks. Then, when he felt that he was sufficiently dosed up, he sat down to a breakfast of cheese.

Through their ability to elevate the amount of neurotransmitters in the brain's synapses, tricyclics and MAO inhibitors led to the most popular theory about depression: that it is an imbalance of brain chemicals. It was a logical conclusion. If these drugs worked—and they did for a large percentage of patients—then, by extrapolating backward, depression was ultimately a deficiency in brain neurotransmitters. First proposed in the mid-1960s, there were two imbalance theories for depression, one for norepinephrine and one for serotonin. Both were modern reversals of Galen's humoral theory. In place of too much black bile, a lack of neurotransmitters was considered the key determinant of low mood.

There were several observations that supported this idea. After taking drugs that reduce the amount of norepinephrine in the brain, for instance, people with no history of psychiatric illness developed a depressive-like condition that sometimes necessitated treatment with electroconvulsive therapy. (Subsequent studies would cast doubt on this association.) Research groups in Scotland and the Netherlands found that people with depression had lower levels of metabolic products of serotonin in their spinal fluid, an indication that serotonin itself was in short supply. And, finally, people who had died from suicides had less serotonin in the hind regions of their brain than people who had died in car accidents or from old age. Such patterns could be explained by many other mechanisms that weren't taken into account. Indeed, the authors of both imbalance theories were rightly cautionary in their conclusions. "It must be stressed that this hypothesis is undoubtedly, at best, a reductionist oversimplification of a very complex biological state," wrote Joseph Schildkraut in 1964, a then thirty-year-old research psychiatrist from Bethesda, Maryland, and the author of the norepinephrine imbalance theory.

Both theories of chemical imbalance, in other words, weren't seen as scientific truths. Not in the beginning, at least. They were put forward as guides or frameworks on which the research community could build more conclusive experiments and assessments. By the end of the twentieth century, however, this putative theory would be raised to the heights of a scientific fact, thanks to the third class of antidepressants that had few of the side effects of MAO inhibitors or tricyclics. Although they offered no improvement in efficacy, their main contribution to drug therapy was greater safety.

Even so, the first SSRI to reach public prescription was a dangerous failure. From its first studies in rat brains in 1978 to its eventual ban in 1983, zimelidine had a short life. Although early trials conducted in Sweden found no serious side effects other than nausea and headaches, more comprehensive studies found that the drug was associated with a twenty-five-fold increased risk of Guillain-Barré syndrome, a neurological disorder that leads to muscle weakness and, eventually, total paralysis. Over a year and a half of Swedish health records, there were thirteen reports of this otherwise rare disease.

Next came fluoxetine hydrochloride—Prozac. First created from an antihistamine-like molecule in the laboratory in 1972, it would take fifteen years for this drug to be released onto the market as a specific treatment for depression. As with zimelidine, there were safety concerns. There were doubts over whether it was actually effective. Plus, it takes time to fashion a popular brand that would change the course of history. Prozac, as it would turn out, isn't as effective in treating depression as most other antidepressants—MAO inhibitors, tricyclics, or even other SSRIs. Lacking in efficacy, it was pushed into the mainstream with the help of a shift in U.S. legislation. Previously limited to posting their leaflets and booklets to general practitioners and psychiatrists, pharmaceutical companies—such as Eli Lilly, the company behind Prozac—could create ads for their consumers, circumventing doctors. On radio, television screens, and on the nascent sites on the internet, SSRIs swept through the American populace along with ads for smoking cessation, lowering cholesterol, and hair-growth formulas for men. Eli Lilly hired a marketing mogul who had worked for McDonald's, Kellogg's, and Pillsbury.

Antidepressants were normalized with the introduction of Prozac and other SSRIs. As an article in *Newsweek* put it, "Prozac has attained the same familiarity of Kleenex and the social status of spring water."

SSRIs similarly popularized the monoamine hypothesis of depression. "These antidepressants will correct for a chemical imbalance in the brain" is a common mantra of doctors. While such a statement isn't categorically wrong, it is still a far cry from being right. This can be plainly seen in how long it takes for these drugs to have an antidepressant effect. SSRIs, tricyclics, and MAO inhibitors all increase the level of monoamines in the brain surprisingly quickly, sometimes peaking within just thirty minutes. But a person's depression might not budge for three weeks or over a month. The action of these drugs, in other words, doesn't match their effect. If there is an imbalance of serotonin in the brain, why don't patients respond to antidepressants in half an hour? Wouldn't that make more sense? Wouldn't that be fantastic?

This "lag phase" has been a glaring hole in psychiatry for decades, and many researchers have rushed to fill it with a suitably sized theory. Most persuasive, perhaps, is the idea that serotonin leads to the regrowth of brain regions that have been damaged from the stress of depression. Studies in mice have shown that parts of the hippocampus—a central part of the brain involved in memory—can regenerate new neurons after prolonged antidepressant treatment. But mice are not humans and there is still an active debate over whether neurogenesis, the cellular process of brain regrowth, is even possible once we reach adulthood. Further, recent work on fast-acting antidepressants such as ketamine has shown that drugs needn't take three weeks to have an effect but can lift someone from depression within hours.

Despite the name, antidepressants aren't the opposite of depression. Their mechanism doesn't necessarily offer an insight into the causes of this mental disorder. "People like this idea of whatever an antidepressant does then the converse of it must be what was wrong to cause depression," says Lisa Monteggia, a professor of pharmacology at Vanderbilt University in Nashville, Tennessee. "What I think we are studying is an antidepressant response." By understanding how current drugs work, Monteggia hopes to develop even more precise and effective treatments that come with fewer side effects. If there's a pathway in the brain that they are all using to lift people out of a depression, then perhaps a drug can be developed that targets that pathway specifically, while leaving the rest of the brain unaffected. "What we're currently doing, very well might not tell us much about depression and that's okay," she says.

Instead of precise mechanisms, we are left with the collective reports of what it feels like to be on antidepressants. "What antidepressants do is that they buffer you against the stresses of life," says David Nutt, a pharmacologist at Imperial College London. "So when you're on them, stress is less stressful. And therefore you can emerge from the depression slowly, as you're less pounded by waves of stress."

All SSRIs—Prozac, Zoloft, Paxil, Wellbutrin—are a blessing and a curse in pill form. The risk of overdose is so low that they have been handed out to almost everyone who had the slightest hint of depression. Who wants to talk about their problems and reveal their pathological thought processes when they could just take a pill every day and feel, in the famous words of a patient of the psychiatrist Peter Kramer, "better than well"?

• • •

In early 2018, I started taking a new antidepressant: sertraline, or Zoloft. While Prozac had risen to fame in *Newsweek* and had its catchy slogans—"Welcome Back"—sertraline came with an egg-shaped mascot and some lessons in basic chemistry. During a television commercial first aired in the early 2000s, a simple schematic of a synapse shows how monoamines travel from nerve A to nerve B. In depression, the soothing narrator explains, this flow is reduced. But sertraline corrects for this imbalance, he adds. "When you know more about what's wrong, you can help make it right."

I was unaware of such commercials (both because I would have been ten years old when they aired and I didn't live in the U.S.), but I still hoped that sertraline was an improvement over citalopram. Although they were similar in structure, I hoped there would be a dramatic shift in effect with this chemical SSRI. This can happen. One SSRI might not have any impact on someone's depression, while another can lead them into complete remission, and so I slowly reduced my dose of citalopram before starting my new prescription.

One hundred milligrams a day. One white pill. Then a fifty-milligram booster—two pills. The improvement was gradual, the pace of a glacier. Then, after weeks or months of feeling emotionally flattened, I started to notice things that I thought I had forever lost. I felt the basic desire to move again, the pleasure of activity and purpose. I felt moments of comfort and warmth inside my chest that I decided to call love. Such awakenings didn't happen all at once, and I barely noticed that I was changing. But I was slowly becoming stronger over time, more resistant to change. Lucy noticed my recovery before I did. But we were hesitant to celebrate. We both knew it wasn't over. Every day we remained vigilant, looking for any sign that I might be struggling.

I have spoken to many people who are also taking sertraline. Friends, family, the local tradesman who was incredibly open about his experience on the drug as he worked on our house. The side effects often circle around a single topic: sex. As well as reducing levels of anxiety and depression, sertraline—and other SSRIs—can make a person's libido shrink. In some cases, it not only makes sex unappealing, but uneventful. It can make orgasms stop entirely. For people who might already feel crippled by loneliness or their failures in life, this can put an enormous strain on relationships. One psychiatrist told me that one of his patients chose his depression over taking these drugs. The loss of intimacy was too much for him to bear. Unfortunately, he killed himself, as he thought that neither option was a life he wanted to live.

My side effects were minor. The sickness of citalopram disappeared altogether. But it still wasn't a perfect fit. Out of the blue, my depression would still return with a vengeance. These episodes were less frequent than they had been, and that was a relief. But it was still a return of my suicidal thoughts, the mental anguish that gripped my brain in a swirling vortex of activity, the lack of motivation that could leave me fatigued and listless for week after week after week. Were these drugs doing anything? Was I simply experiencing a placebo effect in the first few weeks of my treatment? Had I been seduced by a sertraline honeymoon? Such questions would follow me through the next two years of my life. Only when I decided to come off sertraline, in March 2020, would I realize the influence this drug had on my life.

To understand what something does, scientists often remove their topic of interest from their experiments. It is a common method of inquiry in genetics and ecology. What happens to the growth of a mouse without a particular gene? How does an ecosystem fare without a single species? In my personal experiment, removing sertraline from my body made me realize how much the drug supported me.

After eating cheese for breakfast, Barry Blackwell made sure to stay within the hospital's wards so he could be treated quickly in case of emergency. Half an hour passed without symptom, then hours. Nothing happened. Frustrated and confused, he returned to his patients.

With two young children at home and another on the way, Blackwell was moonlighting as a medical officer in an ambulance on weekends. He needed the extra money. One evening, a man called the emergency services and said that his wife was suffering from a severe headache after eating a cheese sandwich. The ambulance pulled up outside, and they entered the house to find that she was in the midst of a hypertensive crisis. For the next half hour, Blackwell watched as her blood pressure peaked and then fell back to a healthy state, her headache slowly subsiding. Just as his motivation was waning, this mystery seemed to offer another glimpse of discovery for Blackwell, pulling him back into the story.

He decided to try another test, not on himself, but on a willing patient at the Maudsley who was already taking Parnate. After she ate some cheese, Blackwell sat by her bedside, checking her blood pressure and pulse. After two hours, she showed no unusual activity and he left the room.

His buzzer beeped. A nurse was requesting a painkiller for his patient, the woman he had just left. She had developed a terribly painful headache and needed aspirin. Then, as he was walking through the Maudsley one night, a colleague rushed through the corridor to attend to two patients who were both suffering from headaches. Both were taking Parnate. And the cheese dish had made its weekly rounds in the cafeteria. It was the cheese! Blackwell wrote up these cases and published a paper in *The Lancet* in October 1963. Over nine months, he reported, there had been twelve cases of hypertension and headaches in patients who had taken Parnate at the Maudsley Hospital. Eight had definitely eaten cheese.

His paper was peer-reviewed and then ridiculed. His colleagues laughed at his ideas. "In a small way, the idea that a common dietary substance might kill someone was as ridiculous as it once was to consider the earth round or that man was descended from a monkey," Blackwell later wrote. One representative of a pharmaceutical company wrote that his findings were "unscientific and premature," which they definitely were. But Blackwell was no longer a lone investigator in this medical mystery. A few weeks after he published his first report in *The Lancet*, a research group at the Westminster Medical School in London provided a missing slice of the story: a molecule called tyramine, a product of tyrosine, an amino acid that takes its name from *tyros*, the ancient Greek word for "cheese."

Tyramine is a key monoamine, one that works alongside the better-known serotonin, dopamine, and norepinephrine. Its main function is to constrict blood vessels, a so-called vasoconstrictor, reducing the amount of space for blood to flow in our circulation systems. Like nipping the end of a hose pipe with your fingers, the pressure increases. When cheese is digested in the gut, this reaction is usually undetectable and reversible. But when taking an MAO inhibitor, a substance that blocks the enzyme that breaks down monoamines such as tyramine, the message to constrict the blood vessels isn't switched off. Homeostasis is knocked off-balance. The pressure can escalate until blood pulses through the body and bursts through the delicate vessels at the back of the head.

From his office at Rockland Research Institute, Nathan Kline read the worrying reports of the cheese reaction and soon realized that history was repeating itself. First iproniazid, now Parnate. Although the risk of a headache was low (0.0001 percent) and the risk of dying even lower (0.00001 percent), Parnate was withdrawn from U.S. markets by the Food and Drug Administration (FDA) on February 24, 1964, a decision backed by the evidence that Barry Blackwell had accrued. Out of three and a half million prescriptions, five hundred people suffered from the devastating headaches that were first recorded at the Maudsley. Forty people died from brain hemorrhages.

Unlike with iproniazid, however, Kline and other psychiatrists campaigned for its use in hospital settings and Parnate was reintroduced into the pharmaceutical canon later that same year, this time with strict dietary guidelines.

It wasn't just cheese. After Barry Blackwell's first reports from the Maudsley, a whole smorgasbord of foodstuffs were found to be high in tyramine and high-risk for people taking MAO inhibitors. There were cases of hypertensive crises after eating pickled herring and other tinned fish such as anchovies. A few headaches were thought to arise from poorly kept meat and game that had started to ferment. Yogurt, packet soups, yeast extract, beer, mushrooms, and even banana skins (when stewed, as in some Caribbean recipes) were added to the growing list of restricted substances.

When patients were warned of such dangers, the risk of headache, hemorrhage, or dying was negligible. MAO inhibitors are still used to treat depression today, albeit rarely and only in exceptional circumstances. "The people who really benefit [from MAO inhibitors]," says Ted Dinan, a professor of psychiatry at University College Cork, "who you can really pick up from day one, are the people who are hypersomnic (or oversleeping), hyperphagic with carbohydrate craving, or very anxious when they're depressed." In other words, people presenting the signs of atypical depression. Several clinical trials in the 1980s and '90s found that roughly three-quarters of patients with atypical features—such as gaining at least ten pounds in weight, sleeping over ten hours a day, and suffering from anxiety reactions such as panic attacks—responded to the MAO inhibitor phenelzine. Only half of such patients responded to a tricyclic antidepressant. Over thirty years of his career, Dinan says that he has had no problems with MAO inhibitors. Used for the right people with the necessary up-front precautions, he has only found success. When he meets someone with an atypical depression, he doesn't wait until all the other options have been exhausted. He uses them from day one. "[And] they respond,"—Dinan clicks his fingers—"like magic."

In 1970, eighteen leaders in biological psychiatry gathered at Taylor Manor Hospital in the city of Baltimore and discussed their life's work. Barry Blackwell began the proceedings with a speech on serendipity, scientific discovery, and cheese. Lothar Kalinowsky, now in his seventies and more skeletal than ever, provided a historical background of what came before both chemical convulsions and electroshock therapy. Nathan Kline and Roland Kuhn pontificated on their chosen family of antidepressants—MAO inhibitors and tricyclics—and made it quite clear that they could both claim the title of the first person to discover an antidepressant. Referring to iproniazid, Kline said, "[N]o drug in history was so widely used so soon after announcement of its application in the treatment of a specific disease." Given to four hundred thousand people in the U.S. in 1957, it was put into prescription a whole year before imipramine. But, Kuhn told his audience, his trials started in January 1956, nearly a year before Kline gave iproniazid to his hospital patients.

There was no doubt who was the face of antidepressant therapy, however. With two Lasker Awards, a best-selling book on depression (*From Sad to Glad*), and his obvious joie de vivre on radio and television, Nathan Kline embodied the excitement and discovery of drug therapy. In the 1960s, he was featured on the front cover of *Fortune* magazine and was referred to as one of the ten best-known men in the United States. During a time when the tensions of the Cold War fueled a global interest in scientific research, space exploration, and technological innovation, Kline convinced governments and NGOs that mental health was worth investing in. In 1955, Kline was one of three psychiatrists who persuaded Congress to donate $2 million to the study of new drug therapies in psychiatry every year. "The amount involved . . . in the 1950s was vast—so much so that those charged with administering it found it difficult to give it away," wrote the historian and pharmacologist David Healy in his book *The Antidepressant Era*. "By most accounts the figure who did most to extract this money was Kline . . . his efforts at lobbying were probably the single most important input to the establishment of a new psychiatry."

Throughout his career, Kline met with world leaders in medical research such as the deputy director general of the World Health Organization, Thomas Lambo, and the famed philanthropist and activist Mary Lasker. He was twice a guest on the popular *The Dick Cavett Show*, sitting in the same seat as Orson Welles, Janis Joplin, Truman Capote, and Stevie Wonder once had. He even met the Dalai Lama.

At the 1970 meeting in Baltimore, Kline gave his talk on MAO inhibitors an unusual title: "An Unfinished Picaresque Tale." Barry Blackwell, thirty-six years old and a recent immigrant to the U.S., didn't know what *picaresque* meant and looked it up in his *Oxford*

English Dictionary. Picaresque: "Relating to an episodic style of fiction dealing with the adventures of a rough and dishonest but appealing hero." Blackwell wondered whether Kline was referring to the drugs or to himself. Like iproniazid and other MAO inhibitors, Kline certainly had some toxic reactions to his personality, and his approach to science.

With MAO inhibitors often a second-choice treatment for depression, Kline became obsessed with his next project: endorphins. Naturally produced by the body, this family of "feel good" molecules were seen as the starting point for a new class of painkillers. Literally meaning "endogenous morphine," scientists hoped that endorphins—in particular beta-endorphins—could lead to the same level of relief as morphine without the risk of addiction. Nathan Kline and Heinz Lehmann, his old friend who would stay with Kline in his luxury apartment in Manhattan, wanted to test this substance on mentally ill patients. Could this be an alternative form of drug therapy, one that would replace imipramine just as it had replaced MAO inhibitors? It was novel, untested, and had no name to attach itself to; a scientific discovery up for grabs. Kline, as one author noted, "moved onto new ideas with a wheeler-dealer's arrogance and a cavalier manner." He invited the only supplier of beta-endorphin, C. H. Li, to his thirteenth-floor apartment, with its zebra skin rug on the wall and sliding glass doors that led onto a balcony, and "sweet-talked" his way to what he wanted. Li thought his host was "courageous" and agreed to provide him with some synthetic beta-endorphin, a hard-to-come-by commodity.

But psychopharmacology was no longer the Wild West it once was. In the 1950s, the main challenge for people like Kline had been getting their hands on new drugs such as reserpine and iproniazid. Once they had a sample, they could give it to anyone they wanted, both hospitalized and in the community (as Kline did in 1957). But since the thalidomide disaster that led to ten thousand babies being born with severe developmental defects (half of whom died within the first few months of life) in the 1960s, the laws surrounding drug testing had become much tighter. Although thalidomide wasn't certified by the Food and Drug Administration in the U.S. (thanks largely to FDA inspector Frances Kelsey), it nonetheless epitomized the damage experimental drugs could do if they were handed out like aspirin. New experimental drugs now required official certification from the FDA—usually after being tested for safety in at least two different laboratory mammals—and signed consent from the patient and their family. But, in his cavalier attempt to race ahead, Kline sometimes did neither. Li worried that he was moving too fast, later adding that Kline "liked the limelight too much. He wanted to make a big noise." According to a later lawsuit, Kline gave twenty-three hospital patients an unregulated, uncertified drug that, in the end, showed unremarkable effects.

In December 1977, Kline presented his early trials on six patients at an international conference at the Caribe Hilton in San Juan, Puerto Rico. As music and laughter from a wedding filtered into the room, Kline resolutely claimed that beta-endorphins were antipsychotic, anxiolytic, and antidepressant. Wearing a gray checkered blazer, he had curly white hair and a matching beard that, as one author later noted, gave him a "slight resemblance to Charlton Heston's Moses." But Kline was no longer a leader of his people. The man who had made jokes in Zurich and helped usher in a new era of biological psychiatry was gone. Standing at the podium, he was seen as a dangerous throwback, more Old Testament than New. One attendee, Avram Goldstein, a leader in endorphin research, stood up after Kline's presentation and spoke for ten minutes. He called Kline's work "a nonsense study" that could jeopardize trust if it were to continue. There was no control group, too few patients, and by claiming that this drug was a miracle cure before the trial began—which Kline did—any positive result was likely down to a placebo effect. "A considerable supply of the rare and expensive synthetic beta-endorphin was committed to an experiment that could not, in principle, have yielded positive results," he added. "[To] raise public hopes without sound scientific evidence . . . will serve science poorly in the long run."

Kline described the next five years as a kind of "slow agony." An audit of his files took three months. Representatives from the United States Department of Health and Human Services "hounded" him. He spent $100,000 on lawyer fees. Then, in March 1982, he sat in a courtroom in downtown Manhattan and signed a form that prevented him from studying "in any manner whatsoever any investigational new drug." His days as a scientist were over. In December, he retired from a thirty-year career at the Rockland Research Institute. But he still had his private practice in Manhattan. "With this unfortunate matter behind me," Kline said in a statement, "I will serve my patients and the medical community with a continued commitment to relieve the suffering from the debilitating effects of psychiatric disorders." A few months later, Nathan Kline was dead. Aged sixty-six, his aorta—the main artery that carries blood out of the heart—ruptured. He died on the operating table on February 11, 1983. After his death, the Rockland Research Institute was renamed the Nathan Kline Institute. Today, the golden statuettes and certificates from his two Lasker Awards are still found there, boxed away in a disused room next to a photocopier. Handwritten in crisp gold and blue calligraphy, his 1964 certificate looks like a royal script taken straight out of a children's fairy tale. "Dr. Kline, more than any other single psychiatrist, has been responsible for one of the greatest revolutions ever to occur in the care and treatment of the mentally ill," the award reads. "Literally hundreds of thousands of people are leading productive, normal lives who—but for Dr. Kline's work—would be leading lives of fruitless despair and frustration." Rough, dishonest, but appealing, Nathan Kline's life was itself an unfinished picaresque tale.

PART THREE

Getting Therapy

We live in secular and mobile societies . . . where the extended family and close communal ties of the immediate neighbourhood are little more than nostalgic images to be revered at Thanksgiving and Christmas . . . Thus, although the traditional forms of family and neighbourhood are less important, substitutes are being developed to provide the emotionally relevant equivalence in modern urban life. Psychological attention to interpersonal relations and the growth of professional psychotherapies offer secular, scientific, and rational responses to these needs.

Gerald Klerman, Myrna Weissman,
and colleagues (1984)

Unlike surgery, psychotherapy cannot be expected to have universally uniform techniques. This is because psychotherapeutic techniques are not pieces of hard technology but ways of soft communication. They have to be culturally relevant. Hence the idiom of any psychotherapeutic system has essentially to conform to the myths, legends, beliefs, and languages of the people with whom it is to be practiced.

J. S. Neki

[W]e ought to encourage the old to act as surrogate grandfathers and grandmothers to the community at large.

Nathan Kline

In Your Dreams, Freud

"He will *not* die!" Lizzie Temkin Beck yelled. "My son will *not* die!" Only seven years old, her youngest child, Aaron, had been playing in the playground when he was pushed from the slide, broke his arm, and within a few days had contracted a bacterial infection that multiplied through his bloodstream. He was put on the hospital's "danger list." Such cases of septicemia, the doctors said, had a 90 percent mortality rate. Put another, less pessimistic, way: Aaron Beck had a 10 percent chance of surviving.

And he did. Although he spent his eighth birthday in the hospital, his infection cleared and he was soon back at school. And though he was behind his year group, Beck soon caught up. In fact, he felt so determined to catch up that he actually overtook his classmates and skipped a year. From an early age, Beck faced disadvantage and flipped it on its head. "Psychologically," he later admitted, this experience showed "that if I got into a hole I could dig myself out, [and] I could do it on my own."

Artistic, tenacious, and quietly confident, Beck made a success of everything he put his mind to. He finished top of his class at high school, graduated with honors from Brown University in English and political science, and then decided that he wanted to study medicine. He was told that a Jewish boy wouldn't be able to get into a medical program without studying medicine or biology at university. After he completed a foundation year, however, he did precisely that. Between 1946 and 1948, he rotated between internships in surgery, dermatology, infectious disease, and neurology, the latter piquing his interest for its precision in diagnosis and the ability to see disease states under the microscope, just as Alois Alzheimer once did in Emil Kraepelin's clinic in Munich.

Beck thought that psychiatry was a bit of a joke. Hospitals and universities across the United States were dominated by psychoanalysis in all its different flavors, and Beck thought that this approach lacked any sort of accuracy or scientific rationale. "Everything that we would see in patients would be interpreted in terms of some deep, dark invisible forces," he said. When he did meet with mentally ill patients, he felt very uneasy providing any half-baked interpretation of their mental woes and what stage of their sexual development they might stem from. "I thought it was nonsense," he later admitted. Like it or not, Beck was forced into a six-month psychiatric rotation at Cushing Veterans Hospital in Framingham, Massachusetts. At this time, there were so few budding psychiatrists that this career move seemed like an opportunity too good to pass up. He could either stick to the overprescribed, competitive field of neurology or try his hand at psychiatry and have room to flourish. As months turned into years, he even started to put his faith in psychoanalytical theories and train as an analyst himself. His colleagues "had answers for everything," he said. "They could understand psychosis, schizophrenia, neuroses, every single condition that came in, you could get a good, sound—[or] apparently sound—psychoanalytic interpretation. And psychoanalysis also held out the promise that it could cure most people's conditions."

It wasn't long after he finished his psychoanalytical training that he realized that these promises were sometimes paper thin, that the theories that they were based on weren't always supported by scientific evidence.

From 1959 to 1961, Beck investigated one of the hallmarks of Freud's theory of depression: that it was a product of hatred turned inward, or "retroflected hostility." If this was the case, Beck reasoned, then his patients' dreams would be filled with themes of hate, masochism, or self-flagellation for their perceived failures in life. What he found was, in truth, quite banal. One patient dreamed of a pair of shoes that she adored but frustratingly found that they were both for the same foot. A man dreamed about a vending machine that accepted his coins but never released a drink or snack no matter how long he waited. Although the specifics of the dreams changed, themes of slight disappointment and mundane worries seemed to be common for many of Beck's depressed patients. They didn't dream of hating themselves.

For Beck, this was the first sign that Freud might have been wrong. The second came from a simple experiment. Beck started playing a rigged card game with his patients. He decided who won, himself or the patient. If people with depression were slouching torrents of self-hatred, Beck hypothesized that they wouldn't enjoy winning. They felt that they deserved to be punished. Losing was more in line with their worldview. Again, this isn't what Beck found. Compared to people who didn't suffer from depression, his depressed patients relished their wins and their self-esteem blossomed more than any other group of patients. To Beck, this meant one thing: his depressed patients had a negative worldview, an irrational idea that life was rigged against them. When this was shown not to be the case (as in the card game), they felt much better about themselves.

Just as the mythical figure Oedipus kills his father, Beck was providing a death knell for Freudian psychoanalysis. But a card game wasn't going to overturn decades of psychological dogma. A few dreams weren't going to nullify retroflected hostility. What Beck required was a theory of his own. Only then could his findings be tested, scrutinized, and compared to the Freudian era that he was hoping to leave behind.

As with his card game, he started playing more of an active role with his patients. Rather than sitting back and allowing their thoughts and memories to flow freely—as was the case for psychoanalysis—Beck asked them questions about what they were thinking of as they dipped into their emotional depths. As they did so, they came to contradictory, and often comical, conclusions. "A brilliant academician questioned his basic intelligence, an attractive society woman insisted she had become repulsive-looking, and a successful business man began to believe he had no real business acumen and was headed for bankruptcy," Beck wrote. These were extreme cases. But they exemplified a mold that most of his depressed patients could fit into. Whenever they thought of anything—whether it was about themselves, how others saw them, or predictions of the future—they were almost always negative and often irrational.

"In making these self-appraisals the depressed patient was prone to magnify any failures or defects and to minimize or ignore any favorable characteristics," Beck wrote. What's more, these thoughts (no matter how comical or paradoxical) were perceived to be as factual as the gravitational force that kept their feet on the ground. "Intrinsic to this type of thinking is the lack of consideration to the alternative explanations that are more plausible and more probable," Beck wrote. Through delicate questioning, Beck tried to reveal a depressed person's "distortions of reality," and tried to guide them back into a more rational view of themselves, others, and their future.

Beck called this technique "cognitive therapy," an offshoot of a field of cognitive psychology that was spearheaded by Albert Ellis. In the 1950s, Ellis posited that it was our thoughts and preconceptions of life events that were important, rather than the events themselves, a train of thought that reached back to the stoic philosophers of ancient Greece. In ninth-century Mesopotamia, Abu Zayd al-Balkhi fashioned his own cognitive theory of therapy in his book *Sustenance of the Soul.* Only in the 1970s, however, did the so-called cognitive revolution gather momentum.

Albert Ellis, Aaron Beck, and everyone else who stepped into this revolution were soon at the center of a lively debate that largely took place in medical journals. Since the 1950s, it was widely thought that behavior therapy would be the next psychotherapy to replace the outdated psychoanalysis, not cognitive therapy. Initiated by Ivan Pavlov's salivating dogs and then diversifying into pigeons pecking at levers and rats pressing pedals, behaviorism argued that a two-way system between a stimulus (a bell ringing for food) and a physical response (salivating) could explain our own lives. Although far more complex than the conditioning with food and a bell, human behavior was thought to be shaped by an endless series of interactions with our environment that determine our social connections, language, perception, memory, and our likes and dislikes.

Depression, for example, was seen as a learned trait in response to negative situations. One of the most famous depictions of this was called the "learned helplessness" model. In the late 1960s and early '70s, Martin Seligman, a psychologist from the University of Pennsylvania (the same institution where Beck worked), performed one of the most famous experiments into depression. It wasn't for the fainthearted. Medium-sized "mongrel dogs" were suspended in a harness with four holes for each of their legs. Then they were given a series of "severe, pulsating" electric shocks through a couple of copper plates touching their hind paws. No matter how much they wriggled or whimpered, these shocks didn't abate for fifty seconds. When they were then placed in a muslin-covered cubicle with a wire-mesh floor that could be electrified with the same intensity, they stood still, vibrating with the waves of the electricity that were out of their control. They didn't try to jump over a barrier that was easily jumpable. They were pictures of helplessness. They had learned that there was no point in trying to avoid their fate.

Dogs that hadn't been placed in the electric harness, who hadn't learned that electricity was unavoidable, were far more likely to try anything to escape from the electric shocks under their feet. They would crash into the sides, bark, urinate, defecate, and eventually find their way over the barrier and to safety. The only way to achieve this with the helpless dogs was to drag them across the barrier with a long lead. Seligman called this "directive therapy." "The initial problem seems to be one of 'getting going,'" he wrote.

The parallels with depression were obvious and hugely appealing. Seligman even found that rats that had been forced into this state of helplessness also suffered from weight loss, anorexia, and had lower levels of norepinephrine in their brain, the same monoamine that was the basis of the first chemical imbalance theory of depression in humans. As with his directive therapy in dogs, behavioral therapy for depression was based on countering negative stimuli in the environment with positive experiences. Rather than their patients sitting at home or lying in bed, it was the therapist's job to try and drag them back to former activities that they once enjoyed. "Although the extreme distress felt by the depressed person needs to be the first point of contact in therapy," one psychologist wrote in the *American Psychologist* in 1973, "the long-range objective needs to emphasize those positively reinforced behaviors that are missing." By countering helplessness with hope, some studies claimed that behavior therapy was successful in over 80 percent of depressed patients.

Behaviorists thought they had the psychological field at their fingertips. Then, in the 1970s, cognitive therapy was lauded as the next big thing. They weren't happy. "The cognitivists reject the conditioning theory of [mental illness] on the ground that they have a better theory which embodies an advance of such magnitude that they characterize it as 'the cognitive revolution,'" Joseph Wolpe, a behaviorist working in South Africa, wrote. "They see themselves as the bearers of a new paradigm that has the same relation to conditioning as [general] relativity has to Newtonian physics." Wolpe thought it was a sham. He thought that cognitive therapy was already a part of behavior therapy, a field that he had helped build since publishing his 1958 treatise, *Psychotherapy by Reciprocal Inhibition*. Highlighting irrational thoughts and countering negativity with positivity: to him, it still sounded like conditioning. As Wolpe once wrote, "Cognitive therapy is a subclass of behavior therapy . . . no matter how complex the contents may be."

Burrhus Frederic (or "B. F.") Skinner, a leading behaviorist in the United States who claimed that everything we do is a product of conditioned responses and, therefore, free will is an illusion, shared Wolpe's frustration. In 1977, three years after retiring from his professorial position at Harvard, Skinner wrote an article titled "Why I Am Not a Cognitive Psychologist." After providing a detailed insight into how behaviorism can explain everything from someone's sex to street crime, he added that the focus on internal thoughts and processing of information that defined cognitive psychotherapy was a dangerous game. To him, depression and other mental illnesses were a conditioned response to an environment. By ignoring this external element, Skinner feared, there would be no political pressure to change, say, unemployment, poverty, poor housing, or diseases that take their toll on the mental health of a population. "The appeal to cognitive states and processes is a diversion which could well be responsible for much of our failure to solve our problems," he wrote. "We need to change our behavior and we can do so only by changing our physical and social environments. We choose the wrong path at the very start when we suppose that our goal is to change the 'minds and hearts of men and women' rather than the world they live in."

For Beck and his students, it was as if they were stuck in the middle of two opposing forces, each unwilling to budge an inch. On one side were the biological psychiatrists with their ever-growing collection of tricyclic antidepressants and their startling statistics of remission and recovery. On the other: behaviorists who had their own association (the Association for Advancement of Behavior Therapy, or AABT), which was so inhospitable to cognitive psychologists such as Beck that he was forced to create his own journal. Some cognitive psychologists received "cease and desist" letters from their behavioral peers. "We felt very much on our own, that there was a need to kind of fight for our place in the department and fight for prerogatives in the department, fight for fair treatment," Ruth Greenberg, a psychologist and graduate of the University of Pennsylvania, recalled. "I don't have the sense that anybody necessarily discouraged [Beck], but I don't think there was a lot of support." Other students found these early years to be thrilling. "There was the excitement of being the underdog and coming up with something that was really challenging the establishment," one of Beck's students and colleagues, Brian Shaw, said.

History is sometimes primed for humor. Once rival factions, cognitive and behavior therapies would soon merge into one of the most popular treatments for depression: cognitive behavioral therapy, or CBT. While rationalizing negative thought patterns, depressed patients are also advised to fill in a diary of activities for each week and take note of positive experiences. "[B]ehavioral techniques are used to promote cognitive change," wrote cognitive therapist Marjorie Weishaar in 1993. "Increasing the daily activity of a depressed patient may directly contradict beliefs of ineffectiveness or predictions that he or she is more contented in bed." Internal thoughts and the external environment were seen as equally important in recovery.

There's an adage in cognitive therapy that Aaron Beck cured his mother's depression the day he was born. After his sister Beatrice died during the 1918 flu pandemic—a global tragedy that killed between 50 and 100 million people worldwide—his birth two years later put an end to his mother's agony and mental anguish, albeit temporarily. With three boys at home, his mother was protective of her youngest son. He wondered whether she would have preferred a girl instead of another boy, but she never voiced these thoughts or made him feel like a disappointment. While his father, a stoic shopkeeper, nurtured his interest in bird-watching and botany, his mother was a mercurial presence in the home, forever shaken by the experience of losing a child.

With such a beginning, it was almost as if Beck were destined to make a living from helping people who suffered from depression. It was in his blood. Unlike many other psychotherapists, however, he wanted to provide evidence that his cognitive approach was effective and to what extent. He sought confirmation that his theories weren't just logical on paper but actually worked in practice.

To begin, he needed a way to measure a person's depression. Rather than using qualitative metrics such as "I'm feeling better today and I am sleeping more," he wanted to be able to quantify its severity over time. Only then could he decide whether his psychotherapy was working or not. In 1961, he published his Beck Depression Inventory (BDI), a health questionnaire that has been used in thousands of studies over the decades since and, according to a review from 2011, is the third most cited paper in psychiatry. The BDI contained twenty-one facets of depression and, for each one, how they would best be described. Upon entering his outpatient clinic in Philadelphia, his depressed patients chose which best described their current state. The first topic was "Mood," and there were four options to choose from:

0. I do not feel sad.
1. I feel blue or sad.
2a. I am blue or sad all the time and I can't snap out of it.
2b. I am so sad or unhappy that it is very painful.
3. I am so sad or unhappy that I can't stand it.

After twenty more topics—such as pessimism, lack of satisfaction, crying spells, irritability, feelings of guilt, social withdrawal, sleep disturbance, and loss of appetite—Beck was handed a quick insight into his patient's depression. "As I look back on my life all I can see is a lot of failures," one statement reads. Another: "I used to be able to cry but now I can't cry at all, even though I want to." And, reaching into the very core of severe, endogenous depression, "I wake up early every day and can't get more than 5 hours sleep," and, "I have lost interest in sex completely." With these different levels of severity for each symptom, Beck could then track their depression over time, and see whether they were getting better (or worse) over the weeks and months of their therapy, whether it was psychotherapy, drug therapy, or otherwise.

As he continued to develop his theory, Beck trained himself to highlight his own cognitive distortions. Between 1962 and 1967, during a five-year "splendid isolation" from psychiatry, Beck spent each morning and afternoon at home writing down all his thoughts, their percentage of importance on his mind, and how each one could be rationalized or adjusted to better fit with reality. He used every piece of paper, notebook, and envelope he could find, scribbling down his thoughts in between playing tennis, napping, and going to the cinema with his wife, Phyllis. (His mother had also enjoyed visiting the cinema during times of stress and passed this on to her youngest son.) He toyed with different color schemes in his notes, constantly reorganizing and reevaluating his cognitions on paper. "It is difficult to describe in words the explosion after 1962 of spiral-bound notebooks, inside front and back covers of notebooks, pocket notebooks, steno pads, 3x5 notecards, and random pages on which line after line of his 'negative' thoughts are jotted down systematically," wrote the historian Rachael Rosner. "The template Beck used in his self-analysis became the foundation for the entire treatment protocol of cognitive therapy." He told Phyllis everything and, with her background in journalism, she was the perfect first editor to his ideas. "She was my reality tester," Beck said. "She went along with the newer ideas I had, and that gave me the idea that I wasn't in left field."

Nearly a century before, Sigmund Freud had similarly isolated himself to analyze his unconscious desires before treating patients in his apartment in Vienna. Beck even developed his own characteristic look. Just as Freud had his circular-framed glasses and thick cigar, Beck was rarely seen without a bow tie—a bright red one for special occasions—around his neck. His friendly demeanor and his bright, pale blue eyes were in stark contrast to the piercing gaze of Freud.

As he distanced himself from psychoanalysis, both in theory and appearance, Beck moved closer to Emil Kraepelin. By combining empirical diagnosis (Kraepelin) with talk therapy (Freud), Aaron Beck paved the way into a more collaborative future, one in which combinations of therapies—drugs and psychotherapy, for instance—would be used to help patients recover from depression. For Beck, however, there was only one treatment fit for the task. "We're not saying that patients can't respond to drugs," Beck told a journalist from the *New York Times* while sitting in his Cognitive Mood Clinic, a set of run-down offices with tattered chairs in Philadelphia. "We're simply saying that cognitive therapy is more effective."

More Than One Psychotherapy

In the late 1960s, Myrna Weissman, a soft-spoken but straight-talking social worker from Boston, found a new job working at Yale University with Gerry Klerman, one of the leading biological psychiatrists of the twentieth century. While he had spent his career studying pharmacological methods—becoming an expert on tricyclic antidepressants such as imipramine, for example—Klerman was hoping to branch out into psychological treatments. One question was dominating his mind, and he hoped Weissman could help provide an answer. Did talk therapy—whether it was marital therapy, group therapy, or psychoanalysis—actually make antidepressants less effective? Did it undo whatever good they were doing inside the brains of his patients? In a famous review of the scientific literature, the famed British behaviorist Hans Eysenck certainly thought so: there "appears to be an inverse correlation between recovery and psychotherapy," he wrote. "The more psychotherapy, the smaller the recovery rate."

Klerman wanted to test this idea in a clinical trial, just as he had tested tricyclic antidepressants in the past. For that, he needed a manual that could be used by different therapists to provide the same methods time and time again. Just as imipramine can be given in precise doses, he sought a talk therapy that could be replicated and compared to other treatments. Weissman worked two days a week and regularly met with Klerman to discuss and refine their growing manual.

This was the beginning of interpersonal therapy, a form of talk therapy that emerged at roughly the same time as cognitive therapy but was inspired by a more holistic view of depression. While Beck concerned himself with the internal workings of the mind, Weissman and Klerman focused on our ability to form strong social bonds, and what happens when they start to weaken and break. They were inspired by the work of Harry Stack Sullivan, the psychoanalyst who wrote, "The field of psychiatry is the field of interpersonal relations, under any and all circumstances in which these relations exist. . . . [A] personality can never be isolated from the complex interpersonal relations in which the person lives and has his being."

Today, as professor of psychiatry at Columbia University, Weissman has a framed photograph of Aaron Beck on her third-floor office's window ledge, a view of the George Washington Bridge and the Hudson River in the background. "We were friends," she says. "We were not competing." There were no grounds for competition. Neither Weissman nor Beck knew whether their theories were accurate, that their methods would be effective in the treatment of depression. Still, during her time at Yale University, Weissman was bolstered by some of the most celebrated minds to ever contemplate the origins of depression.

In the 1960s and '70s, any social theory of depression had to mention the work of John Bowlby, a British psychoanalyst who was still an active researcher when Weissman began her research. She remembers sitting in a lecture hall listening to Bowlby, a balding man with drooping jowls and a wheezing voice, and thinking that he was God. (He was also good friends with Gerry Klerman.) Taking inspiration from evolutionary theory and ethology (the study of animal behavior), Bowlby was a new breed of psychoanalyst, one that was guided by science and its pursuit of objective truth. Instead of focusing on the Freudian dogma of erotogenic zones (oral, anal, genital), he proposed that the most important development in a child's early life was one of basic attachment. Babies cling, grasp, and like to be rocked. They cry for attention and smile when smiled at. They pine for their mother and follow her every move. From birth, Bowlby reasoned, we are primed to make strong attachments. If we weren't, we wouldn't survive. Born helpless and with underdeveloped brains compared to other primates, humans require years of dedication and maternal care. "It is fortunate for their survival that babies are so designed by Nature that they beguile and enslave their mothers," Bowlby wrote in his 1958 paper "The Nature of the Child's Tie to Its Mother." This behavior, he added, is "as much the hallmark of the [human] species as the red breast of the robin or the stripes of a tiger."

Mental disorders, Bowlby wrote in his 1969 book, *Attachment and Loss*, are the cost of our biological need for attachment. It needn't be a loss of a mother. It could be any of the strong social bonds that we make throughout our life, all of which follow in the footsteps of our initial maternal bond. As this book was published, Gerry Klerman and his colleagues at Yale University were providing some of the best evidence to support Bowlby's theories. Through several surveys and questionnaires, they had found that, in the six months prior to the onset of symptoms, "departures from the social field" were three times as common in people with depression as those who were not depressed. Divorce, disease, or death of loved ones. Unemployment, a change of job circumstance, or a new job. Family members leaving the home. Stillbirths and pregnancy. All these events were seen as breaks in attachment, confirmation of Bowlby's theories into the social origins of depression.

Support also came from London, the city where Bowlby had trained and worked for much of his professional career. In the 1970s, the sociologist George Brown and his colleague Tirril Harris were finding similar ties between loss and depression in Camberwell, South London, a catchment area that included rich women with titles to their name and families living in dank Dickensian conditions. This diversity in setting was matched by a surprising variety of life events. "Some [events] were unexpected but others had been eagerly sought; some were major catastrophes but others involved no more than a change in routine," Brown and Harris wrote in their book, *Social Origins of Depression*. "The message is again that it is loss and disappointment rather than change as such that is important."

Whether it was called loss, disappointment, or exits from the social field, depression was being seen as a product of our social connections, an imbalance within our communities rather than just chemical communication across a synapse.

Interpersonal therapy started with common sense. As with drug therapy, a diagnosis seemed like a logical place to begin. Did the patient meet the criteria for depression? Were there symptoms numerous and frequent enough to warrant treatment at all? Second, as guided by the work from New Haven and London, could Myrna Weissman and Gerald Klerman pinpoint what might have triggered the episode? "It made sense that when people come in [for therapy], we find out what is going on in their life," Weissman recalls. Whom are they dependent upon? Do they have someone who is giving them trouble or causing stress? Has anything happened recently that was unexpected or disruptive, even if it was a positive change? "You find out the problems that are associated with the onset of symptoms," Weissman says. "Although it may be apparent to the world what's going on, it may not be apparent to the person. So it's first clarifying that." Only by understanding the problem areas could they then be worked on and remedied. Over twelve to sixteen weeks, interpersonal therapy provided the stage for an intimate conversation about present stresses of life, and how they might be resolved or accepted.

Most of the manual was Weissman's work, but Klerman added an important piece that he thought was missing. Every patient, he said, should be told that they were ill: that they weren't a depressed person, but a person with depression. They weren't defined by their illness. As with someone with a virus or a bacterial infection, they could get better. This so-called sick role was first theorized in the 1950s and granted ill people a greater amount of freedom to focus on their well-being, even at the expense of long-held plans. No one would expect someone who had pneumonia (a noncontagious bacterial infection) to attend a wedding or have friends over for dinner. The same should be true for people with depression.

• • •

In 1979, the National Institute of Mental Health (NIMH) decided that the time was right for a large-scale trial into psychotherapy. As part of the Treatment of Depression Collaborative Research Program, cognitive therapy and interpersonal therapy were chosen as the two leading therapies used in the United States and would be compared to imipramine, the gold-standard drug therapy at the time. All three were then compared to a more basic level of care or, in the drug trial, a placebo pill. Undertaken at three sites (George Washington University, the University of Pittsburgh, and Oklahoma University) and using the same methods and health questionnaires, the only real difference was which treatment the 260 patients received. How would they compare after sixteen weeks? What about eighteen months after the trial had finished? Taking nearly two years to plan and peer-review, this multisite trial was the best way of finding out.

Aaron Beck believed that this was the opportunity for cognitive therapy to show its true value. In 1977, he and his colleagues had published their first clinical trial and found that, compared to twelve weeks of imipramine treatment, cognitive therapy helped three times as many patients into full remission from depression (79 percent vs. imipramine's 24 percent). What's more, these benefits were maintained over the next three to six months. By this point, thirteen out of nineteen patients (68 percent) given imipramine had returned to drug therapy, while only three out of twenty-two patients (14 percent) given cognitive therapy required further treatment. Beck hypothesized that these long-term benefits were due to the patient learning about their disruptive thought patterns and being able to apply new techniques to their management. Just as someone might be made aware of a glaring grammatical mistake they had been making for years, they would learn from their therapeutic lessons and rarely repeat the same errors in the future. Beck called it a "sense of mastery" over one's emotions. No drug could offer such mastery.

There were problems with this study. It was a small sample of people with depression: only forty-one patients (twenty-two for cognitive therapy and nineteen for imipramine). Perhaps people who were somehow primed to respond to cognitive therapy just happened to be in that group. Only when greater numbers were included could these random biases be reduced. However, the most problematic element of the paper wasn't its content, or that Aaron Beck was a coauthor, but where it was published: *Cognitive Therapy and Research*, a journal he had just created. The centerpiece of its inaugural volume, this paper kick-started Beck's new journal with a bang. Everyone in that trial wanted to believe that this therapy would work. And such excitement can be infectious.

The Treatment of Depression Collaborative Research Program was more neutral. It took place away from the buildings affiliated with either cognitive therapy or interpersonal therapy and the therapists were trained as part of the trial. This took three months. Beck complained that this wasn't sufficient. He needed a year to train anyone up to a level required for cognitive therapy. This would allow them to learn not only the theory but also how to practice it in a way that was human, empathetic, supportive, and, if necessary, jovial. "Some of the beauty," said Jeffrey Young, a close colleague of Beck's and one of the trainers in the trial, "got lost in the translation into a manualized therapy." With therapists receiving only three months of training, Beck worried that this trial would be a test of the individual therapist rather than his cherished therapy. Frustrated, he compared the therapists in this trial to a set of junior doctors performing delicate heart surgery.

The NIMH had their reasons for this brief opportunity for training. They wanted to test this therapy as it would be performed in the real world, often by people who were recent psychology graduates and who didn't have a complete grasp of all of Beck's methods. If this therapy was to be adopted nationally—or even internationally—it had to be scalable. "The immediate goal of the training programs was to efficiently and effectively train therapists at the different research sites to attain specified levels of mastery in their respective treatment approaches," the coordinators of the trial, Irene Elkin and Tracie Shea, wrote. Training every single therapist up to Beck's preference might have resulted in more effective practitioners, but it wasn't seen as an efficient approach.

Beck resigned from the study. "I think his underlying feeling was 'this won't [turn] out that well,'" Young recalled, "which is what did happen." While all three forms of therapy were superior to a placebo, only interpersonal therapy and imipramine excelled in the more severe cases of depression. Cognitive therapy was significantly less effective in reducing depressive symptoms in this group of patients.

Why might this have been the case? Cognitive therapists argued that because interpersonal therapy was rooted in psychoanalytical theories of loss, these therapists were better placed to deal with these more severely ill patients. They had decades of practice in a very similar form of talk therapy. The freshly trained cognitive therapists, meanwhile, didn't have the same depth of experience that was crucial when dealing with the most challenging patients. Elkin and Shea acknowledged this discrepancy: "[T]he most intensive training program is that for CB therapy, in which a new (or relatively new) framework is taught for conceptualizing and treating the problems of depressed patients. Somewhat less intensive is the IPT training program, in which the selected therapists already share a fairly common framework, and in which the training, although providing them with some new procedures and strategies for treatment, mainly serves to focus, channel, and sometimes restrict their usual treatment approach."

And yet, for Beck and his colleagues, there was still one positive outcome of this trial. A year after treatment had ended, the number of patients who relapsed back into depression or who returned to treatment was lowest for those who received cognitive therapy. As with their first study in 1977, Beck and his colleagues put this down to the lessons that the patients had learned about how their thoughts work, and how they can control them. A drug might make someone feel better through increases in monoamines, but cognitive therapy was like a tutorial in mental health care, a set of tools that could be used long after the last therapy session had finished. "I think cognitive therapy is likely to compare quite nicely with other existing interventions like pharmacotherapy in terms of short-term effects," said Steven Hollon, a close colleague of Beck's and a lead researcher in cognitive therapy since the 1970s. "But I think it's in the long-term effects—the prevention of relapse, the prevention of recurrence, and possibly the prevention of the initial onset in the first place—that's cognitive therapy's major contribution."

Cognitive behavioral therapy is the most frequently prescribed psychotherapy for depression for good reason. "Problem-solving does work and interpersonal does work," says Helen Christensen, the director of the Black Dog Institute in Sydney, Australia, "but there's no doubt the strongest evidence is for CBT."

In the autumn of 2017, two years after my group therapy sessions, I caught the bus from our rented apartment in Bristol, a city in the southwest of England, to a mental health center on the outskirts of the city. I filled in my Patient Health Questionnaire (PHQ-9) as I waited for my therapist to collect me from the waiting room. In a tiny room with little more than a desk, a computer chair, and a padded chair for me to sit on, we went through the familiar choreography of cognitive behavioral therapy.

I wrote down my "personal summary," a list of thoughts or events that were making me feel low. We would then spend the next six weeks working through these key areas. I filled a "behavioral activation" diary every week, detailing what I was going to do each day and making sure I made time for pleasurable activities as well as chores. If anything made me feel good about myself or my current situation, I jotted it down in my "positive data log." With homework and photocopied printouts, this was a far cry from lying down on a sofa and speaking about any dreams or anything else that came into my mind. Unlike psychoanalysis, CBT is rigidly focused on the realities of the present. Through discussions and homework, negative thought patterns are tested against reality and, hopefully, retrained toward a viewpoint that is more healthful. I doubted my abilities as a writer, for example, even though my pitches were still being accepted by science magazines. Making this latter point clear in my head, I felt more confident that I had a future in this career of my choosing. When I thought that my family was in tatters, I was reminded that I lived with Lucy and that we were happy together.

As I left each session, I felt a slight buzz that we were making progress. I felt that I could train myself away from depression. As I caught the bus back to our apartment, I watched as the trees were shedding their yellowed leaves for the coming winter. They were changing, ensuring that they could make it through the coming winter.

The sixth and final session was a recap. What had I learned over the last few weeks? What can I still work on? Can I implement the behavioral activation diary and the positive data log into my everyday life? I felt like I needed less reviewing and more training. With the first session having been spent introducing my problems, I only had four fifty-minute therapy sessions to work through a mental illness that I had struggled with for over two years. It was as if a doctor had given someone a cast for their broken leg and then, before the bone had healed, taken their support away again. The bone might hold, or it might break and lead to an even longer rehabilitation.

My PHQ-9 scores were still showing that I was moderately depressed at the end of this "low-intensity CBT." I was honest in my review. I wanted more time, the option of high-intensity CBT in order to continue with any progress we had made. Before we said goodbye, my therapist told me that she would refer me for a phone review. I should expect the call in two weeks or so and it shouldn't last more than forty-five minutes. If they decided that prolonged treatment was necessary, then the waiting list was two months at least.

The next week I received a letter from my therapist. It was sent with good intentions but made me feel much worse about my future. "I would encourage you to continue practicing the techniques that you have learnt throughout your treatment," she wrote. "Should your situation worsen, or if you feel like you are in a crisis you should seek support from your GP or the emergency services. It has been good to meet you and I wish you every success in your ongoing recovery." In my head, it was a confirmation that I was being abandoned. I wasn't important enough to continue with the therapy that I had found useful. When it was invented, CBT was meant to last between twelve and sixteen weeks, and I thought that my "low-intensity" course was just a means of ticking boxes, of covering the health system's back if I were to kill myself.

My disappointment is reflected in the scientific literature; low-intensity CBT doesn't fare well in the few clinical trials that have assessed its efficacy. Although it has similar effects as SSRIs or twelve weeks of CBT in the short term, these benefits aren't maintained over the subsequent months. In 2017, for example, a study of four hundred people living in northeast England found that over half (54 percent) of patients who had responded to low-intensity CBT had returned to their depressed state twelve months after the trial had finished. They had relapsed. In comparison, the rate of relapse for normal CBT practice—as taken from the average of seven clinical trials—is only 29 percent, nearly half the number of people. In fact, high-intensity CBT has some of the longest-lasting effects of any treatment for depression. In 2016, a study encompassing 469 patients from seventy-three general practices in the UK found that the benefits of twelve to eighteen sessions of CBT are maintained for at least four years, when compared to a "treatment as usual" group. What's more, the majority of these patients had suffered from severe depression that had been unremitting for at least two years. They were terribly unwell, suffered from various long-term health conditions, and had failed to respond to a course of antidepressants. And yet, with a full course of CBT, they were able to stay healthy long after the treatment had ended.

The long-term benefits of CBT have been known for decades. It is this longevity that separates this psychological therapy from antidepressants. Drugs such as SSRIs only work for as long as a person takes them. Take them away, even slowly, and the depression is likely to return. Although the reasons are still unclear, this difference has been explained by the educational element of psychological therapies that Aaron Beck first noted in the 1970s. A patient learns how to deal with some of the causal mechanisms of their depression.

They notice their negative thoughts that might spiral into despair and a feeling of helplessness. Just by noticing and rationalizing such thoughts—"I'm not a failure at everything"—they gain some element of self-sufficiency, a means of controlling their mental health that an elevation of serotonin cannot provide. If this is the case, perhaps the five or six sessions of low-intensity CBT don't provide enough time for such lessons to be learned.

Low-intensity CBT isn't pointless. For some people, it can have lasting benefits. But six weeks of therapy shouldn't end abruptly, especially if someone shows signs of improvement in their mental health. Rather than being told to expect a further phone review a few weeks in the future, an immediate continuation of the therapy should be offered. After bravely reaching out for professional help, a person with depression shouldn't have to go through that process again. In my case, I skipped my phone interview for further therapy, as I thought I was wasting everyone's time by even asking for more CBT. Someone else might need it. Plus, who knows what state I might be in after two months? I had heard that some people were waiting for eight months. I wondered how bad things would have to get before someone took my case seriously. When Lucy booked me in for another emergency doctor's appointment—calling my GP as I was advised to do—the doctor recommended a type of antidepressant, known as venlafaxine, that doesn't just change the serotonin levels in my brain but also norepinephrine, another neurotransmitter. I had heard of this drug and had read reports that its withdrawal effects were notoriously severe. After my experience in Berlin with citalopram, I felt terrified about becoming stuck on this drug, too. Lucy and I agreed that this was a bad idea, a medical version of trial and error that we didn't want to get involved in. After hearing that I was suicidal, the doctor finished our conversation by saying, "I hope I don't have to worry about you tonight."

I ignored him and walked out. Lucy and I made the slow journey home together, arm in arm. She wanted to scream as much as I did.

Health care systems aren't built for depression, or mental illness, more generally. This is a problem of funding rather than a lack of knowledge. As to avoid unnecessary side effects and becoming stuck on a prescription, the sensible option would be to treat depression with CBT first and, only if that fails, try antidepressants such as SSRIs. But this doesn't reflect the clinical reality. A prescription of antidepressants is cheaper, and easier, than a full twelve to sixteen weeks of patient-focused CBT.

Around the world, mental health care is criminally underfunded. In the UK, for example, only 1 percent of the £3.5 billion health care budget is allocated for adults with depression or anxiety disorders. That's £500, roughly, per person. In comparison, the investment for physical rehabilitation in a person's last year of life amounts to £60,000. As the psychiatrists Richard Layard and David Clark write in their 2014 book, *Thrive*, "We spend so much on trying to extend the *length* of life (often by very small amounts), and yet so little on improving its *quality*."

There are many ways to argue that mental health is worth investing in. There's the economic argument (it is cost-effective to treat depression so that people are more efficient at work), the scientific argument (these treatments work when given to the right patients), and the global argument (depression is a leading cause of disability around the world and requires significant action). But even the best statistical models amount to nothing compared to the moral argument: people with depression are suffering, at high risk of suicide, and families are affected over generations. It is a tragedy of incalculable proportions that they don't receive the treatment they need.

"If Mom Ain't Happy Ain't Nobody Happy"

CBT is often what people mean when they say they are "getting therapy." But it isn't the most suitable option for everyone. Interpersonal therapy, for example, can work for those people who require guidance through problematic life events. Whether someone is trapped by their own thoughts or stuck in a tricky social situation, psychotherapy should match the concern. "There shouldn't be one psychotherapy," says Myrna Weissman. "I think that people who say, 'This is *the* psychotherapy,' are doing a disservice to patients. It's like saying there should only be Prozac."

Interpersonal therapy was just one element of Weissman's work. Since completing her PhD in 1974, she was, first and foremost, an epidemiologist, an academic who tracks diseases through time and populations, just as cartographers keep up to date with the growth of cities. Weissman turned to academia at the right time and in the right place. She studied at Yale University when a vocal civil rights movement was campaigning for equal opportunities at workplaces for everyone, regardless of their race, sexual orientation, or gender. "I applied for the PhD program, in the 1970s, as a young, married mother of four children under the age of seven," Weissman later wrote. "I did not fit the historical picture of a Yalie." Without the constant push for equality, a mother of four was unlikely to be seen as a potential candidate for any PhD program.

Her doctoral research focused on chronic diseases such as cancer and cardiovascular disease, some of the best-known and most-studied epidemiological subjects. Depression and other mental illnesses weren't studied in such detail. The few surveys that had been conducted were based on abstract terms such as *well-being* and *social functioning*. With the arrival of the third edition of the *Diagnostic and Statistical Manual for Mental Disorders* (*DSM-III*) and Beck's inventory for depression, symptom clusters and strict diagnosis were becoming staples of psychiatry, replacing the Freudian notions of depressive neurosis or anxiety reactions. It was a watershed moment for depression. In one go, a medley of different diagnoses that biological psychiatrists had used were all lumped together underneath the term *major depressive disorder*, or *MDD*. Involutional, atypical, reactive, or endogenous—it was all the same: MDD. The only detail was whether there were melancholic features present. Was the depression significantly worse in the morning than the evening? Did the patient show excessive guilt or marked slowing in thought and movement? Had they lost a large amount of weight recently? This wasn't a separate disease entity, not according to the *DSM-III*, anyway. It was a more endogenous *subtype* of major depression.

Although major depression was no doubt an oversimplification of the true nature of depression, the *DSM-III* was an excellent tool to find people who might require psychiatric care. As it was the same diagnosis no matter where it was used, populations across the United States could be studied and, importantly, compared. Already in contact with the psychiatrists—known as neo-Kraepelinians—working on the *DSM-III*, Weissman saw an opportunity to branch out from physical diseases and study the patterns of mental illnesses in the American population.

Psychotherapy and epidemiology, the two silos of Weissman's work, were deeply connected. How can you treat depression if you don't know how common it is? How can you deliver therapy of any kind, chemical or psychological, to the people who need it if you don't know who they are? In her first survey, encompassing 511 people living in the New Haven area, Weissman suggested that psychiatrists had been focusing on the wrong demographic. At this time, in the 1970s, there was a widely held theory that depression was predominantly a disorder of menopausal women. From her survey, Weissman didn't find any evidence to support this claim. Yes, women were twice as likely to suffer depression as compared to men. But they certainly weren't menopausal. They were more likely to be in their twenties or thirties—the prime of their reproductive lives. Even adolescents, a demographic who were thought to have insufficient "ego" development to suffer from depression, were found to show the hallmark signs and symptoms of major depression.

This was a small study. Just over five hundred people in a single community might not be relevant to other communities, both in the U.S. and around the world. Perhaps there was another reason that the first episode was occurring at younger ages in this sample. To support (or refute) these preliminary results, Weissman needed larger samples from multiple sites across the United States.

As with her academic induction at Yale, Weissman found a political atmosphere that was highly conducive to her ongoing research. Jimmy Carter, the then president of the United States, and his wife, Rosalynn, were investing millions of dollars in mental health research and the dissemination of evidence-based therapies. Weissman sent the handwritten drafts of her small New Haven survey to the secretary of health, education, and welfare, Joseph Califano, highlighting the early onset of depression. If confirmed, could adolescents and young women in particular be targets of either psychotherapy or antidepressants? The funding for what would become the largest survey of depression in the United States came shortly thereafter.

The Mental Health Epidemiologic Catchment Area (ECA) survey, published in October 1984, included seventeen thousand people from five different study sites across the U.S. (St. Louis; Baltimore; Los Angeles; Durham, North Carolina; and New Haven). Weissman, the then director of the Depression Research Unit at Yale University, and her colleagues focused on the local region of New Haven, a leafy city in Connecticut where one hundred twenty-six thousand people lived. Familiar patterns emerged from the data. With help from Priya Wickramaratne, a statistician who could make sense of the large stacks of raw data, Weissman found that, on average, women were indeed twice as likely to have depression than men. The first onset was in the young, on average at twenty-seven years old, and these rates gradually declined after middle age and menopause (in contrast to the previous theories). On average, nearly 5 percent of American people would experience, or were currently experiencing, a depressive episode. In other words, as Weissman walked down the tree-lined streets of New Haven, one in every twenty people she passed were, on average, either suffering from a current depressive episode, were in recovery from a previous episode, or were likely to become depressed later in life.

Weissman's surveys soon spanned state lines and crossed national borders. Hundreds of individuals became thousands. Although the prevalence differed from country to country (the rates of depression in Taiwan and Korea were significantly lower than in New Zealand and France, for example), the same patterns were found no matter where in the world she looked. Depression was roughly twice as common in women than in men. It had a first onset in adolescence to early adulthood. And the risk of depression was two to four times higher in people who had been divorced or separated.

In October 1985, Richard Glass and Daniel Freedman, two leading psychiatrists in the U.S., found such epidemiological work both exceptional and deeply worrying. Not only were the rates of depression and other mental illnesses higher than expected, the majority of sufferers weren't receiving support or psychiatric care of any kind. "[D]ata from the ECA, as well as previous epidemiologic surveys, indicate that most persons with psychiatric disorders in the community are untreated," they wrote in the *Journal of the American Medical Association* (*JAMA*). "The costs of untreated illness are unknown, but may be astronomical in terms of lost productivity alone." Absenteeism, lack of interest at work, and fruitless visits to the general practitioner who thought aches and pains had nothing to do with the patient's mental state—it all added up. In the U.S. alone, Glass and Freedman wrote, the cost of untreated mental health conditions could be as high as $185 billion per year.

Merging her expertise in interpersonal therapy and epidemiology, Weissman saw an opportunity to do something about this mental health gap. In addition to the NIMH Treatment of Depression Collaborative Research Program, this form of talk therapy had been tested alongside antidepressants and found to provide a significant reduction in symptoms, compared to just using the drugs on their own. In particular, while tricyclics seemed to target the sluggish cognition, sleeplessness, and bodily complaints, such as constipation, interpersonal therapy provided a boost to a person's self-esteem, while reducing thoughts of guilt, suicide, and helplessness. "Our data," Weissman told the *New York Times* in 1978, "show that the combination of drugs and psychotherapy has an additive effect. And we have seen that at the end of the treatment, usually four to 12 weeks, there is a noticeable difference. People function better. They feel better. They eat better. They can sleep. And most important, when the crisis is over, they can begin to work at identifying the problem that led them into the depression in the first place."

Today, interpersonal therapy is a first-line talk therapy for depression. Depending on which clinical trials are used, it is generally seen as similarly effective as cognitive behavioral therapy and antidepressants. On average, roughly half of the patients who receive this form of psychotherapy improve to such an extent that their depressive symptoms fall below a healthy threshold. Due to its focus on social ties and understanding of life transitions, IPT has been shown to be particularly useful for depression in mothers.

Myrna Weissman has experienced many of the departures from the social field that define her work on interpersonal therapy—divorce, transitioning into a new career and city—but it was only in 1992 that she had to stop what she was doing and focus on her own health. On April 3 of that year, after a long battle with diabetes, Gerry Klerman died of kidney disease at the age of sixty-three. After both of their first marriages had ended in divorce, Weissman and Klerman had been married for seven years.

It felt as if an entire field of academia was in mourning. "The world of medicine has lost a distinguished, energetic, and prolific member of its community," Martin Keller, a psychiatrist from Brown University, wrote. "[Klerman's] ability to challenge and stimulate trainees was extraordinary and although this often created some performance anxieties in his students, his style was greatly admired and emulated." At his funeral in New York, Daniel Freedman, the editor of the journal *Archives of General Psychiatry*, said that "Gerry always brought a kind of cleansing clarity to anything he touched. We all grieve him, but the things that he made happen really are happening now and will be in the future." His influence reached out from biological psychiatry and stimulated the field of psychotherapy.

Although it wasn't unforeseen (Klerman had been on dialysis for years), his death turned Weissman's life upside down. They shared much of their social space. He was a mentor and a close colleague, a friend and the stepfather of her four kids, a loving husband. "I put [interpersonal therapy] aside," she says. "I had to." After publishing the first edition of *Interpersonal Psychotherapy of Depression* in 1984, and finding this treatment's efficacy replicated by their own research group and elsewhere, this was no longer a professional hobby that she could devote herself to. John Markowitz and Lena Verdeli, two colleagues from Columbia University, took the reins. Since then, interpersonal therapy has been adopted and adapted by psychiatrists from around the world. Her manual has been refined through four editions and translated into seven languages. The latest edition was published in 2018. Gerald Klerman is credited as an author on all of them.

As is the case in interpersonal therapy, Weissman's approach to life is to focus on the problems of the present and not to dwell so heavily on the past. Divorce and widowhood are labels that no longer define her. Today, she is a married woman with four adult children and seven grandchildren. She has grieved two husbands and a first marriage that ended in divorce. But these were natural responses, and only if they continued for many months or even years would such pain become something that might require psychiatric treatment. The line between grief and depression is diffuse and impossible to define. But Weissman knew that an incomplete grieving process was a dangerous precursor to mental health problems. "Depression [can be] associated with abnormal grief reactions that result from the failure to go through the various phases of the normal mourning process," the first *Interpersonal Psychotherapy of Depression* textbook states. "A delayed or unresolved grief reaction may be precipitated by a more recent, less important loss. In other cases delayed reactions may be precipitated when the patient achieves the age of death of the unmourned loved one."

In order to feel stronger and return to work Weissman knew she had to accept the rumination, guilt, anger, and obsession that followed Klerman's death. As she stepped aside and allowed others to take interpersonal therapy into the future, Weissman moved from studying how depression spans continents to how it gets passed on in the home.

For decades, Weissman has been investigating one of the most intriguing patterns in psychiatry: the children of depressed parents are at particularly high risk of depression. Although the reasons are still foggy, the most obvious explanation for this pattern is that a depressed parent is less able to nurture, care for, and generally take interest in their child as they grow. "That's not rocket science," says Weissman. "That was obvious. If you have children and realize how much energy it took, how could you be depressed and take care of kids?" Using her New Haven surveys as controls, Weissman quantified this risk over the decades, as the cohort of children became adults and parents themselves. At the ten-year mark, the risk of depression was three times as high for people who had lived with depressed parents compared to those who hadn't. (This figure has remained relatively constant at the fifteen-, twenty-, twenty-five-, and, most recently, thirty-year mark.)

The next step screamed from her data sets. "In epidemiology

there's the concept of the 'modifiable risk factor,'" Weissman says. "And the modifiable risk factor was the depressed parent: you can modify that, by treating them."

As she was waiting for a flight back to New York from a conference held in Hawaii, Weissman started chatting to a psychiatrist, who told her about the results of an impressive trial about to be published. Known as STAR*D, it was a test of modern drug therapy and how quickly it could lead to remission (a near-total absence of symptoms of depression) in over four thousand people. If one drug failed, another was immediately tried. If that also failed, a third option was prescribed. Weissman asked whether she could study the children of these patients. By treating the parent, would the depressive symptoms of the child also improve? With the help of several research groups around the United States, this is exactly what Weissman found. Just as a keystone species might allow neighboring life to flourish, the health benefits of being free from depression seemed to trickle down from parent to child. The crucial element was remission, not just treatment. Only those parents who were rid of their depression seemed to benefit their children (as judged by questionnaires that assessed symptoms of depression, anxiety, and disruptive behavior disorder, as well as their social functioning).

This study was covered in the *New York Times*. To celebrate this publication, Weissman's children—all adults at this point—bought her a customized T-shirt. Across its front, it read: "If mom ain't happy ain't nobody happy."

Published in 2006, it was a foundational study that provided a parental focus for the prevention of depression. But it also created some confusion in the psychiatric community, even among those who were part of the trial. "We do not know from this study that treating parents causes the children to do better," says Judy Garber, a clinical child psychologist at Vanderbilt University, Tennessee, who led one of the sites for the STAR*D trial. "It is simply that the parents are doing better, and so are the children. And it can be for a massive number of reasons." The parents of the children that got better could have been less depressed to begin with, Garber says. It could be due to shared life events or common genetic risk that weren't taken into account in the sample. Perhaps the correlation was the other way around: the children getting better facilitated the parents' response to treatment.

Although Weissman could see that the sequence of events was from parent to child (and not the reverse), she knew that the study needed refining and replicating. "So we did another study, which was a clinical trial," Weissman says, a laconic statement that makes the process of scientific publication—planning the study, applying for funding, selecting the patients, conducting the trial, analyzing the results, and editing the manuscript following to peer review—sound quick and easy. Published in the *Journal of American Psychiatry* in 2015, the study found that children of depressed mothers treated with the antidepressant escitalopram (Lexapro) showed significant improvements in their social functioning and mental health. "We showed the same thing [as before]," Weissman says, "if the mother got better, with treatment, the kids got better." A few hundred miles west of Weissman's office in Manhattan, at the University of Pittsburgh, some of her former colleagues at Columbia had continued to study interpersonal therapy. In 2008, they replicated Weissman's STAR*D results with the therapy that she helped create. Treating depressed mothers with interpersonal therapy led to improvements in the mental health and social functioning of their children.

As with vaccinations and dental care, the responsibility of mental health care in children shouldn't be wholly on their parents. In an ideal world, national governments should provide evidence-based therapies for families. These can either be universal (given to everyone regardless of risk) or targeted (given to those in high-risk demographics), the two main forms of preventative medicine. According to Helen Christensen, universal prevention is the only way forward. "You don't know who's going to get depressed," she says. "We might have risk factors and so on, but that's not actually going to tell us which particular person on the ground is likely to get depressed. . . . So you've got to actually provide these sorts of prevention interventions for everybody."

For children and adolescents, the classroom has been seen as the most logical place for universal psychotherapy, delivered by visiting therapists or using internet-based programs. Currently, however, the evidence supporting such initiatives is weak at best. Large school-based clinical trials in Chile and the UK have failed to find significant reductions in depressive symptoms compared with other approaches such as attention control or Personal, Social, Health and Economic education (PSHE) lessons. In fact, the latter study, led by researchers at the University of Bath in southwest England, found that children given cognitive behavioral therapy in schools were *more* likely to show signs of low mood twelve months after the trial had finished. As Robyn Whittaker, an associate professor at the National Institute for Health Innovation, University of Auckland, and her colleagues succinctly wrote in 2017, "Universal depression prevention remains a challenge."

The bulk of the evidence currently supports the use of targeted prevention programs. In 2009, for instance, Judy Garber and her colleagues provided a cognitive-behavioral program for adolescents who either had symptoms of anxiety or hopelessness (known antecedents of depression) or their parents had a history of depression. "What we found was what we call a moderator effect," she says. "Those teens whose parents weren't currently depressed benefited from the intervention. With the program, they had fewer depressive episodes. But when we looked at the teens whose parents were currently depressed, the intervention didn't have a significant effect." The reasons why are still unknown. "It could be genetics; it could be that [those] parents were more ill," Garber says. "It might also be that the currently depressed parents were having trouble getting the youth to the intervention." Whatever the case is, it's clear that depression is a disorder that runs in families. Whether it's cognitive behavioral or interpersonal, psychological intervention is a powerful tool in the long-term treatment and, potentially, prevention of depression. The main challenge for psychiatry is getting these services to people who need them, no matter where they live.

"Happier Than We Europeans"

During her psychiatric training at the Maudsley Hospital in London in the late 1980s, Melanie Abas was faced with some of the most severe forms of depression known. "They were hardly eating, hardly moving, hardly speaking," Abas, now a professor of global mental health at King's College London, says of her patients. "[They] could see no point in life," she says. "Absolutely, completely flat and hopeless." Any treatment that might lift this form of the disease would be lifesaving. By visiting their homes and their general practitioners, Abas made sure that such patients were taking their prescription of antidepressants for long enough for them to take effect. Working with Raymond Levy, a specialist in late-life depression at the Maudsley Hospital, Abas found that even the most resistant cases could respond if people were given the right medication, at the correct dose, for a longer duration. When all else failed, she turned to ECT. "That gave me a lot of early confidence," says Abas. "Depression was something that could be treated as long as you persisted."

In 1990, curious to expand her horizons, Abas accepted a research position at the University of Zimbabwe's medical school and moved to Harare. Unlike today, the country had its own currency, the Zimbabwean dollar. The economy was stable. Hyperinflation, and the suitcases of cash that it necessitated, was over a decade away. Harare was nicknamed the "Sunshine City," a bright centerpiece of a country only liberated from British rule a decade before. Called Southern Rhodesia (after Cecil Rhodes, a British businessman who lived in South Africa) until 1980, Zimbabwe had taken hold of its own future. Yet, although there was sunshine in Harare, the country was descending into a painful period marked by decolonization. After instilling the people of Zimbabwe with a sense of national pride and individual freedom, the political activist Robert Mugabe soon became a political tyrant who was closer in kinship to the cruel colonial elites that he had replaced than a homegrown savior of his people. From 1983 to 1987, an estimated twenty thousand Ndebele citizens were slaughtered in the western region of Matabeleland North, a political stronghold for the rival ZAPU party. Carried out by the government's North Korean–trained military, this became known as Gukurahundi, a name that translates into "the early rain that washes away the chaff before the spring rains." In the parlance of international law, it was genocide.

Execution was only one of the fatal products of Mugabe's regime. While he remained wealthy and well-fed, the majority of people in the country struggled to get by. What was described as "the worst drought in living memory" struck the country in 1992, drying up riverbeds, killing over a million cattle, and leaving cupboards empty. Malnutrition, and a lack of medicine against infectious diseases such as tuberculosis, cholera, and malaria, meant that children—the most vulnerable demographic, along with the elderly—struggled to survive. For every one thousand live births in Zimbabwe, around eighty-seven children died before the age of five, a mortality rate eleven times higher than that of the UK. The death of a child left behind grief, trauma, and, as Abas and her team found, a father who might abuse his wife for her "failure" as a mother. Then there was HIV, a virus that would infect a quarter of the population by the mid-1990s.

With such loss, poverty, and disease, it was surprising that depression, a mental disorder that thrives on such breaks in the social sphere, was rarely found in Zimbabwe. A survey from Harare reported that fewer than one in every four thousand patients (0.001 percent) who visited the outpatient department had depression. "In rural clinics, the numbers diagnosed as depressed are smaller still," Abas wrote in 1994. In comparison, around 9 percent of women in Camberwell, the southern region of London where Abas was working at the Maudsley Hospital, were depressed. Essentially, she had moved from a city where depression was prevalent to one in which it was so rare it was barely even noticed. It was a complete reversal of the research she was accustomed to. From busily treating the most severe cases of depression with drugs or ECT, her research in Harare had to start at the very beginning: finding people who might need her help.

This low rate of depression fit snugly with a racist dogma that had first taken root during the Enlightenment era of the eighteenth century. At this time of eureka moments and insightful musings, the German philosopher and conductor Jean-Jacques Rousseau eloquently proposed that civilization was the scourge of our species. He romanticized the notion that, historically, humans lived in peaceful union with nature. "[N]othing is so gentle as man in his primitive state," he wrote. "[T]his period in the development of human faculties . . . must have been the happiest and most durable epoch." As such, people who lived outside of the so-called civilized Western world were viewed as cheerful, peaceful, and, in a term wrongly credited to Rousseau, "noble savages." This Western worldview soon encircled the globe. Upon meeting Aboriginal people of Australia, for example, James Cook, explorer and captain in the British Royal Navy, wrote that they "live in a Tranquillity which is not disturb'd by the Inequality of Condition: The Earth and sea of their own accord furnishes them with all things necessary for life, they covet not Magnificent Houses, Household-stuff . . . they live in a warm and fine Climate and enjoy a very wholesome Air, so that they have very little need of Clothing. . . ." They were, in short, "happier than we Europeans."

Australian Aboriginals, African bushmen, Native Americans, and the Inuit were all thought to resemble the pure spirit of humankind. Untouched by the ways and words of civilized society, they were free of the stresses and strains of modern life. Unhappiness, never mind depression, was as alien to these cultures as the woolen stockings, petticoats, and ruffles that men of the Enlightenment found so fashionable. Rousseau believed that our species was originally destined to remain in this state of peaceful union, but had been enchanted by the promise of individual expression, industrial revolution, scientific breakthroughs, and the overall prosperity that civilization would bring. This "progress," Rousseau wrote, "has been in appearance so many steps toward the perfection of the individual, and in fact toward the decay of the species." Civilization was a poisoned apple that Western society had consumed with relish. There was no cure, no turning back. Other, less developed cultures were in danger of being similarly corrupted and pushed out of their peaceful ignorance.

This view was reversed during the nineteenth century's "Scramble for Africa." As European nations sought ever-greater geopolitical power, Africans were recast as savages who were in desperate need of civilization and, most commonly, the guiding faith of Christianity. It was an enforced alignment with the values of their colonial masters. There was nothing peaceful about living in a colony. Natural resources such as gold, diamonds, rare dinosaur fossils, and sometimes entire villages were stolen and transported back to Europe for sale or exhibition. Tens of millions of Europeans traveled to see the mock mud huts, ceremonial dresses, drums, dances, black skin, and curly hair of these so-called savages. Any rebellion or violent uprising was quelled with an inequality in weaponry; spears and bows and arrows were no match for rifles and bullets. The cost of colonialism is incalculable.

Whether a colonialist spoke French or English, German or Italian, Spanish or Portuguese, the same rhetoric was taken as fact. Before the colonization of African countries, they said, there was no civilization there. It was a continent of sparse populations, rudimentary dwellings, and illiteracy. Small villages, tribal leaders, and subsistence farming were the rule through centuries of time and over thousands of miles of coastline, savanna, desert, and forest.

Psychiatrists—the Western equivalent of traditional healers and shamans—adopted this worldview into the middle of the twentieth century. Depression, it was said, was a Westernized disease, a product of civilization. It wasn't found on, say, the plateau of eastern Africa, the shores of Lake Victoria, or anywhere else south of the Sahara Desert. "In general, it seems," the psychiatrist Geoffrey Tooth wrote in 1950, "that classical depressive syndromes are seldom seen, at least in Africans untouched by alien influences." These aliens—Western people—might not have been from outer space, but they were seen as products of other worlds: urban life and the stress that modernity brings. "[T]he very notion that people in a developing black African nation could either be in need of, or would benefit from, western-style psychiatry seriously unsettled most of my English colleagues: their skepticism was rampant," a psychiatrist working in Botswana in the 1980s wrote. "They kept saying, or implying, 'But surely they are not like us. It is the rush of modern life, the noise, the bustle, the chaos, the tension, the speed, the stress, that drives us all crazy: without them life would be wonderful.'"

In Zimbabwe, the grip of colonial psychiatry had to be particularly tight. While still Southern Rhodesia, the country provided indisputable evidence that, if made public, could overturn the Rousseauesque notion of happy people untouched by modern life who lived peacefully with nature. Since the first Portuguese explorers journeyed through the region in the sixteenth century, there had been reports of huge cities of granite in southern Africa, local populations that reached into the thousands, and evidence of former trade networks with China, India, and the Middle East. Nowhere was this more obvious than in Southern Rhodesia, a place where hundreds of ruins speckled the landscape in every direction, none more impressive or controversial than the drystone walls of Great Zimbabwe. Before the colonial missionaries, explorers, and psychiatrists of Europe arrived, civilization had thrived here. Built in the eleventh century and inhabited for the next four or five hundred years, Great Zimbabwe comprises a trio of fortresslike structures that are spectacular in both their scale and architectural fluidity. Constructed atop a hill and snaking into the valley below, a series of impressive drystone walls—eleven meters tall and five meters wide in places—were built entirely from granite mined from the local bedrock. The Hill Complex blends into a natural crag of granite rocks that perch atop a raised mound of earth. A stairway in this African acropolis leads to a natural balcony among the boulders. Below, an oval structure known as the Great Enclosure stands as the centerpiece of a former city. At its height, archaeologists and mathematicians have estimated that between ten and eighteen thousand people lived at Great Zimbabwe—a few hundred behind its curved walls, while the rest eked out an existence in wooden huts outside.

But no colonial country wanted to admit to the true origins of such a place. Reuniting Southern Rhodesia with its history could lead to a resurgence in pride and individuality, revolution, and, ultimately, independence. And so, since the 1870s and into the twentieth century, colonial journalists and archaeologists had argued that the local Bantu people were incapable of building such walled fortresses and, instead, proposed that Great Zimbabwe was the work of the mythical Queen of Sheba, the Erythraean people of southeast Asia, travelers from southern India, or Arabs from the Middle East. "Censorship of guidebooks, museum displays, school textbooks, radio programs, newspapers and films was a daily occurrence," a former curator at the Museum of Great Zimbabwe recalled. "Once a member of the Museum Board of Trustees threatened me with losing my job if I said publicly that blacks had built Zimbabwe. . . . It was the first time since Germany in the thirties that archaeology has been so directly censored."

As with architectural competency, presumed levels of intelligence have long been tied to the predisposition to mental illnesses such as depression. In 1953, the recently created World Health Organization, based in Geneva, Switzerland, published a report titled "The African Mind in Health and Disease." Written by John Carothers, a British doctor who had worked at Mathari Mental Hospital in Nairobi, Kenya, in the 1940s, this report set the tone for other psychiatrists working across sub-Saharan Africa. He quoted several authors who compared African mental states to that of Western children, to immaturity. A black person, Carothers wrote, "has but few gifts for work which aims at a distant goal and requires tenacity, independence, and foresight." In an earlier paper he compared the "African mind" to a European person's brain that had undergone a lobotomy. The lack of depressive cases at his mental hospital seemed to confirm these views. How could someone be depressed if they lacked foresight and the pursuit of distant goals? How could someone worry or feel guilty over past deeds when they are so embedded in the present? Out of the 1,508 patients Carothers saw between 1939 and 1948, there were only twenty-four cases of depression, a mere 1.6 percent of total admissions. "Even allowing for the fact that other cases have doubtless been missed," Carothers wrote, "the condition as a whole is relatively rare."

Carothers's views were becoming outdated soon after he had written these words. Thomas Adeoye Lambo, a psychiatrist, author, and member of the Yoruba people of southern Nigeria, wrote that Carothers's conclusions were nothing but "glorified pseudo-scientific novels or anecdotes with a subtle racial bias." Refraining from calling out blatant racism, Lambo was a calm and impressive character in everything he did. "Nobody could meet him without coming away with a powerful impression of being in the penumbra of a force-field that radiated power and presence," the psychiatrist and novelist Femi Olugbile wrote. Lambo wore his traditional blue robes to otherwise besuited global health conferences and would later become the deputy director general at the World Health Organization in Geneva. Carothers's studies, Lambo concluded, contained so many gaps and inconsistencies "that they can no longer be seriously presented as valuable observations of scientific merit."

The tide was turning fast. From the late 1950s, colonial states in Africa were giving way to independent nations. With this shift in cultural identity and freedom from foreign emancipation, it was as if a veil were slowly removed from an entire continent. This was even seen in the shift in psychiatric epidemiology: rates of depression soared as decolonization spread. Studies in two Ugandan villages—

places of rolling hills, fields of crops, and quietude—found that the rates were twice as high as in people who lived in South London. Among the Yoruba tribe, the rates were still four times as high as a population in Stirling County, Canada. This didn't mean that decolonization had caused the increase in depression. It was simply that representatives from Western countries began communicating with people from other cultures, not seeing them as underlings but as equals. The same phenomenon was being recognized the world over. "For one trained in one's own culture and by those familiar with the patients' culture . . . it becomes evident that the mood of sadness in oriental depressives is no less frequent than in the western co-sufferers," Venkoba Rao, a professor of psychiatry at Madurai Medical College, India, wrote in 1984. Speaking the local language, living in communities rather than in mental hospitals—the hallmark signs of mental suffering revealed themselves in ways that were unmistakably familiar to the Western term *depression*.

Kufungisisa

Following in the footsteps of researchers in Uganda and Nigeria, the reality of depression in Zimbabwe was revealed a decade after it gained independence in 1980. Between 1991 and 1992, Melanie Abas, her husband and colleague, Jeremy Broadhead, and a team of local nurses and social workers visited two hundred households in Glen Norah, a low-income, high-density district in southern Harare. They contacted church leaders, housing officials, traditional healers, and other local organizations, gaining their trust and their permission to interview a large number of residents. They wanted to know whether *depression* even translated into the local Shona language. What if there was no word for *depression*? Could this be another reason why this so-called Western disease had been so neglected in sub-Saharan Africa? Through discussions with traditional healers and local health workers, her team found that *kufungisisa*, or "thinking too much," was the most common descriptor for emotional distress. This is very similar to the English word *rumination*, which describes the negative thought patterns that often lie at the core of depression and anxiety. "Although all of the [socioeconomic] conditions were different," Abas says, "I was seeing what I recognized as pretty classical depression."

It wasn't just a case of thinking too much, however. There was also the lack of sleep and loss of appetite. A loss of interest in once enjoyable activities. And a deep sadness (*kusuwisisa*) that is different from a normal state of sadness (*suwa*). Using culturally appropriate terms such as *kufungisisa* to help identify mental suffering, Abas and her team found that depression was nearly twice as common in Harare as in a similar community in Camberwell, South London. In direct contrast to the previous surveys that found a rate of one in every four thousand (0.001 percent), Abas's community surveys found that it was closer to one in every five people (20 percent).

As with the epidemiological studies in the U.S. and England, Abas wondered whether stressful life events were associated with depression in Harare. Even if the words were different, did the same social triggers transcend cultural and national boundaries? Adopting the methods of George Brown and Tirril Harris, famed sociologists who studied the relationship between life events and depression in Camberwell, she found a strong pattern emerge from her surveys. "[We found] that, actually, events of the same severity will produce the same rate of depression, whether you live in London or whether you live in Zimbabwe," she says. "It was just that, in Zimbabwe, there were a lot more of these events." Poverty, malnutrition, unemployment, cholera, HIV: all took their toll. In particular, HIV and depression fit together into a tight spiral that slowly twists toward mortality. Not only are people who are HIV positive twice as likely to be depressed than an HIV-negative person, but the depression then makes the virus more lethal. The same is true for poverty: it is both a trigger for depression and made worse by being depressed.

These cyclic relationships were also an opportunity. Would treating the depression, Abas wondered, have knock-on effects for other facets of a person's life? Would they take their antiretroviral medication and feel less stigmatized by their HIV status? Would they find employment and earn a living for themselves? At the time, in the 1990s, over one hundred million people around the world were said to suffer from depression, but only a select few were receiving treatment. Even today, 90 percent of people living in low-income countries don't have access to evidence-based therapies like antidepressants and psychotherapy. In high-income countries, this figure hovers around 60 percent. As Shekhar Saxena, the former director of the Department of Mental Health and Substance Abuse at the World Health Organization, once put it, "[W]hen it comes to mental health, we are all developing countries." Little did Abas know that her work wouldn't just help treat depression in Harare or even Zimbabwe. It would eventually lay the foundation for a revolution in mental health care that would reach out of sub-Saharan Africa—a place previously thought to be absent of depression—and across the world.

In 1992, inspired by the work of David Ben-Tovim, a doctor working in Botswana, Abas helped introduce a form of psychotherapy that nurses and doctors could provide to their patients in health centers across Harare. Known as the "7-step plan," it was a mixture of Myrna Weissman's interpersonal therapy and Aaron Beck's cognitive therapy, and could also include antidepressant treatment. Starting with questions about patients' personal lives and what they might be thinking too much about, nurses were advised to listen and empathize: "[T]ry and share their sadness," Abas wrote. In the first meeting, they would also evaluate the risk of suicide, and after a few days they'd repeat the same steps and discuss potential support networks—friends, family, church, or a welfare agency. If symptoms persisted for three or more weeks, a prescription of amitriptyline, a tricyclic antidepressant, was advisable. The patient should be informed of the potential for dry mouth, sedation, constipation, and the lag time between taking the pills and the relief of the depression (or *kufungisisa*). Three months later, any persistent or severe depression should be referred to one of the ten psychiatrists serving a country of ten million.

The most revelatory element of this seven-step plan was its most basic: the knowledge that *kufungisisa* was a common and treatable illness opened up the potential for a healthy future. Brochures and posters were put up in health centers across the city to reduce stigma and enhance understanding of this common disorder:

> Thinking too much makes people ill.
> We call this illness kufungisisa-depression.
> It is a mental illness but it is different from madness
> or kupenga and it is not caused by Ngozi [evil spirits]

But who takes notice of posters and pamphlets, especially when they are busied by their own thoughts? To really drive home the message, a drama group transformed these words into a song and dance that, they hoped, would resonate with a culture more attuned to live performances, music, and rhythm than an audience trying to avoid eye contact with strangers in the stuffy waiting rooms of the Western world.

Over her two and a half years in Harare, Abas had developed a soft spot for the Sunshine City. It was a place of intriguing contrasts. In areas of great poverty, there was a shared feeling of camaraderie and joy. The roads were filled with potholes, but they were lined with flowers and, in the summer months, the violet blossoms of jacaranda trees. It was welcoming. A second home. But Abas had to complete her training as a psychiatric epidemiologist and return to England, accepting a permanent position at King's College London. Over the next decade, as the political and economic situation in Zimbabwe continued to spiral into crisis, the number of psychiatrists in Zimbabwe dropped to just one or two. And with a health service already stretched to breaking point, the nurses and doctors more focused on HIV, cholera, and tuberculosis than *kufungisisa*, Abas worried that her work would be forgotten, dismissed as just another casualty of Robert Mugabe's authoritarian regime.

The study of mental health—its prevalence, diagnosis, and treatment—around the world has many names and different origins. John Carothers, for example, was one vocal member of the field of transcultural psychiatry, a colonial practice built upon ethnocentric ideologies, racism, and the construction of Western-style mental institutions in foreign lands. In 1995, this antiquated field underwent a much-needed rebrand with the publication of a book. Coordinated and cowritten by Arthur Kleinman, a professor of anthropology and psychiatry at Harvard University, *World Mental Health* put forward an agenda that could tackle the high prevalence of depression and other psychiatric disorders around the globe.

It was a time of fresh insight and hope, an era in which common mental disorders were understood as the real threats we now know them to be. At one time, the metric used to gauge whether a disease required action and attention was its overall mortality. How many people did it kill and how quickly? By this count, cancer, heart disease, and infectious diseases took center stage for any policy decisions for the allocation of funding for health services. Even when a patient dies from depression it is often labeled as by "suicide," a blurred view of the true cause of death.

But diseases aren't solely fixated on the line between life and death. They influence our lives in myriad ways and aren't always fatal. Just before the publication of *World Mental Health*, this biased conception of diseases changed with the introduction of the disability-adjusted life year (DALY) in the early 1990s, a metric that took into account not only death but also the number of lost healthy years. "One DALY can be thought of as one lost year of 'healthy' life," the World Health Organization states. Based on this more accurate model of the way disease encroaches upon health, depression, and other mental disorders immediately became some of the most costly diseases. In 1992, for example, a study from the World Bank estimated that mental health problems accounted for over 8 percent of all healthy years lost. This was more than cancer (6 percent), heart diseases (3 percent), and malaria (3 percent). "Yet despite the importance of these problems, they have received scant attention outside the wealthier, industrialized nations," Kleinman and his colleagues wrote in *World Mental Health*. "[A]llocations in national health budgets for preventing and dealing with these problems are disproportionately small in relation to the hazards to human health that they represent."

In his mid-seventies, goateed and focused more on writing than research, Kleinman says that his initial optimism after the publication of this book was quickly replaced with frustration and the dreadful feeling of being pushed to one side. As the nineties turned into the early aughts, the concerns over world mental health were replaced by another health crisis: HIV. "I watched us being knocked off the agenda by the AIDS movement," Kleinman says.

This isn't to say that the AIDS movement didn't require urgent action and funding. It did. The production of antiretroviral drugs that kept HIV loads down and their dissemination around the world is one of the greatest success stories of medicine, one that continues today. It was once thought to be impossible to disseminate antiretroviral drugs to people living in sub-Saharan Africa in order to stem the HIV crisis. Although HIV is still an international problem, viral loads are decreasing, and with the correct treatment and advice, people can live happy, productive lives while being HIV positive.

It is time for the same success for the treatment of depression, Kleinman says, an illness that is far more widespread and common than HIV and has a suite of effective treatments available. Psychotherapy, antidepressants, electroconvulsive therapy: they all work for different severities or subtypes of the disease. But even if modern, evidence-based therapies were made universally available, they still wouldn't reach everyone. In some parts of the world, psychiatrists are highly stigmatized and few people might want to see them, or be able to afford them. Many might prefer to see a local religious leader or a traditional healer. Many more wouldn't want to speak about their problems outside of their home for fear of shaming or discrimination from their friends or family. An effective therapy is often a compromise, a blend of the scientific model of medicine and the local customs and culture that allow such a model to be accepted if and when it is made available. Lastly, in order to tackle a mental health crisis that numbers in the hundreds of millions of sufferers, treatments must be scalable. If they are too costly, require too much time for training, or are so specific that they can't be implemented both in sprawling cities and isolated villages, then they won't work. In order to ensure sustainability, the allocation of treatment has to be a part of the public health system, and initiatives must not be governed by short-term contracts from universities or NGOs, but instead require a regular slice of the national health budget year after year. "That's the only way to go," Vikram Patel, a professor of global mental health at Harvard University and co-creator of Sangath, a community health service in Goa, India, said in June 2018 at a global mental health workshop held in Dubai.

Sitting in his third-floor office at Harvard University in Boston, Arthur Kleinman doesn't know when this shift will happen. "I'm seventy-seven years of age," he says. "Let's say that I live to ninety-five, that would be optimistic. So that's eighteen years. In eighteen years will we see substantial monies come into this field? I hope so, I sincerely hope so. I have no idea if it will or not. So I'm not planning, necessarily, to see the victory. But will there be a victory in this area? Absolutely. Will it happen in the twenty-first century? Absolutely."

Helen Verdeli, a clinical psychologist and a colleague of Myrna Weissman at Columbia University, traveled to southwest Uganda in February 2002. It was her first visit to Africa and she traveled away from the capital city of Kampala and began her work in dozens of villages in the southwest of the country. Surrounded by fields of crops, she met with village elders and the next counselors in interpersonal therapy for depression. Without trained psychiatrists or doctors in this area, her plan was to train employees from World Vision International, an NGO that worked in the area, providing aid and education to the local communities. But, at the last minute, they told Verdeli and her colleagues that they were too busy. They couldn't take on any more responsibilities. As an alternative, they asked their relatives if they could help instead.

Initially a problem, this change of plan became the beginning of the latest approach to global mental health: training the local community in psychological therapy. With zero experience and no medical degrees, could laypersons be effective therapists? A few little-known studies from the 1970s and '80s suggested that this might be an understatement. Although not limited to depression, one review found that, out of forty-two published studies, thirteen showed that community health workers (or "paraprofessionals") were more effective than trained professionals. Only one study found professionals more effective, while the remaining twenty-eight studies found no difference between the two. "The clinical outcomes that paraprofessionals achieve are equal to or significantly better than those obtained by professionals," the author, Joseph Durlak from Southern Illinois University in Carbondale, wrote in 1979. "The provocative conclusions from these comparative investigations is that professionals do not possess demonstrably superior therapeutic skills, compared with paraprofessionals. Moreover, professional mental health education, training, and experience are not necessary prerequisites for an effective helping person."

Nine people living in the Rakai and Masaka provinces of southwestern Uganda fit the criteria Verdeli, Weissman, and their colleagues were looking for. They had grown up in the local area and had at least a college-level education. Plus, they all spoke English and Luganda (the most common indigenous language in Uganda), and could communicate with both the American psychiatrists and their neighbors. Over an intensive two weeks of training, Verdeli and her colleague from Columbia University, Kathleen Clougherty, taught them the ins and outs of interpersonal therapy.

The identification of causal life events and their solution isn't an easy task in any therapy session. But these difficulties were raised to frightening extremes in a low-income country like Uganda. As Abas found in Zimbabwe, the same life events were found here as anywhere else on earth. Death, divorce, disease, unemployment, natural disasters, poverty. It is their frequency that is unavoidable. Throughout the 1990s, nearly a quarter of the people living in Uganda were infected with HIV. Not only did the virus usually come with a death sentence, it also led to social isolation, stigmatization, and the death of loved ones. The loss of breadwinners was particularly devastating, leaving families without the financial support they were once dependent on. Widowed, the wife or wives (polygamy is common in Uganda) of the deceased had few skills to earn a living and fell further into poverty. Civil wars, political repression, and widespread malnutrition stirred depression into every corner of the country. A survey from Paul Bolton, a cross-cultural psychiatrist from Johns Hopkins University in Baltimore and a colleague of Verdeli, found that 21 percent of people in southwest Uganda met the criteria for major depression. A previous study had found a similar rate. Roughly one in four people required psychiatric care, whether it was psychotherapy or antidepressants. But few received any. Even traditional healers in Uganda expressed their inability to help people with depression-like illnesses. For people with depression, the isolation and helplessness were crushing.

As was the case in Shona, there wasn't a word for *depression* in Luganda. But there were close equivalents, Bolton found. *Yo'kwekyawa* and *okwekubazida*, roughly translating as "self-loathing" or "self-pity," overlapped with the Western diagnosis of depression. Since it came with suicidal ideation, *yo'kwekyawa* was seen as the more severe form of the two depression-like illnesses. As well as the usual symptoms of fatigue, lack of interest, and feelings of hopelessness, *yo'kwekyawa* and *okwekubazida* also included a few nuances of their own: People with these illnesses tended to not respond when someone greeted them. They hated the world and people in it. And even when they were offered generosity or assistance they were unappreciative.

As an illness of thoughts, emotions, and memories that encompasses a diverse range of symptoms, depression is shaped by a person's cultural traditions and language. The symptom or feeling that becomes definitive of the condition—low mood, tension, fatigue—depends on how someone expresses themselves, communicates, and connects their inner world to the environment that surrounds them. "The experience of this condition is so deeply embedded in one's social world," says Vikram Patel. In China, depression was more commonly referred to as *neurasthenia*, a term referring to nerve weakness that was popularized in the nineteenth century. Under the dictatorship of Chairman Mao, it was more socially acceptable for Chinese citizens to suffer from such a physical illness akin to fatigue. A diagnosis of low mood or mental depression may have been taken as an insult to the socialist state, or even to Mao himself. Only in recent decades, as the country has become more Westernized, has the symptom of low mood become a definitive descriptor of mental illness in China. In India, meanwhile, depression is felt as "tension." The muscles seize up and everyday movements and walking feel like the air has turned into thick, gloopy syrup. "We never use the word 'depression' because no one understands it," says Patel. "The literal translation will become something like 'Are you sad?' And they'll say, 'Of course I'm sad, my life is miserable.'"

The most common descriptor of the Western diagnosis of depression, however, is "thinking too much," a phrase used across swathes of Africa, Southeast Asia, and South America.

These subtle variants don't mean that depression is different depending on where someone was raised. The underlying biology and symptoms are far more similar than they are dissimilar. Still, unlike cancer or heart disease, depression can't be reduced to a failing organ or an uncontrollably dividing cell. Without a universal marker for depression, its social context has to be taken into account before it can be identified and treated.

Given the cross-cultural differences between the United States and Uganda, interpersonal therapy couldn't simply be translated into Luganda, word for word. The manual had to be made culturally appropriate. Although the key concepts were the same, there were a few important adaptations that Verdeli and Clougherty had to adopt. Most obvious was the therapy format. Sessions weren't one-on-one discussions between a lay counselor and a patient. Instead, a group of five to eight people met together, all of the same sex. "In these cultures," Verdeli and her colleagues wrote, "people tend to see themselves as part of a family or community unit before they see themselves as individuals." Even marriages and funerals are attended by the whole community, regardless of relation. Using the standard format of interpersonal therapy was, therefore, thought to be unsuitable. Led by a trained community member, people with *yo'kwekyawa* or *okwekubazida* were given the opportunity to support and advise their neighbors during each session, offering solutions to life events that they themselves might have faced in the past.

There were more subtle adaptations. Speaking negatively of the deceased—even those that had been abusive or unfaithful—was strictly prohibited in Ugandan culture. "The dead are living among us" was a common phrase heard in the villages. To respectfully approach any turbulent relationships, lay counselors were taught to ask questions such as "Were there times in your life together when you felt disappointed by the dead?" As polygamy was common in these villages, a common life event for women was how to deal with their husband's new wife, a shift in their household that often changed the role they once had in the family. Most important, however, was the conflict arising from HIV: How can widowed women earn money for their children or find someone else to support them? Simply knowing that HIV wasn't a form of punishment but a common infection that could be treated with the right medication could help allay some of the guilt and self-pity characteristic of *yo'kwekyawa* and *okwekubazida*.

With these modifications, group interpersonal therapy was an effective means of reducing depressive symptoms, especially in women. Compared to a control group that was offered a more basic form of care, the reduction in symptoms was three times as great, as judged by a standardized health questionnaire. Not only were the results highly significant, but they were also stable. During a six-month follow-up, the prevalence of depression had barely increased. There was an unexpected explanation for these long-term effects. Many of the groups continued to meet up without their lay counselor, supporting each other through shared hardships. Group interpersonal therapy wasn't just a sixteen-week trial, in other words. It was a process that built long-lasting social connections. The one group that didn't meet up after the trial showed significantly higher rates of depression than the others.

This informal "social support system," as Verdeli and her colleagues wrote, provided a vital component of any treatment used in low-resource settings. With no option for frequent or prolonged treatment, any beneficial effects need to stand the test of time. Published in the *Journal of the American Medical Association* in the summer of 2003, Verdeli and her colleagues had shown that laypersons—people with no former training in psychiatry or medicine—

were a valuable resource in the treatment of depression. That same year, two randomized controlled trials were published in *The Lancet* from similar projects in low-income regions in Santiago, Chile, and Goa, India. Although they trained nurses or social workers (and not laypersons), all three studies showed that a psychology degree isn't a prerequisite to be an effective therapist.

Care by the Community

Dixon Chibanda spent more time with Erica than most of his other patients. It wasn't that her problems were more serious than others'—she was just one of thousands of women in their mid-twenties with depression in Zimbabwe. It was because she had traveled over 160 miles to meet him.

Erica lived in a remote village nestled in the highlands of eastern Zimbabwe, next to the border with Mozambique. Her family's thatch-roofed hut was surrounded by mountains. They tended to staples such as maize, and kept chickens, goats, and cattle, selling surplus milk and eggs at the local market. Erica had passed her exams at school but was unable to find a job. Her family, she thought, wanted her only to find a husband. To them, the role of a woman was to be a wife and a mother. She wondered what her bride price might be. A cow? A few goats? As it turned out, the man she hoped to marry chose another woman. Erica felt totally worthless. She started thinking too much about her problems. Over and over, thoughts swirled through her head and began to cloud the world around her. She couldn't see any positivity in the future.

Given the importance that Erica would hold in Chibanda's future, it could be said that their meeting was fated. In truth, it was just the product of extremely high odds. At the time, in 2004, there were only two psychiatrists working in public health care in the whole of Zimbabwe, a country of over 12.5 million people. Both were based in Harare, the capital city. Unlike his besuited colleagues at Harare Central Hospital, Chibanda dressed casually in a T-shirt, jeans, and running shoes. After completing his psychiatric training at the University of Zimbabwe, he had found work as a traveling consultant for the World Health Organization. As he introduced new mental health legislation across sub-Saharan Africa, he dreamed about settling down in Harare and opening a private practice—the goal, he says, for most Zimbabwean doctors when they specialize.

Erica and Chibanda met every month for a year or so, sitting opposite each other in a small office in the one-story hospital building in central Harare. He prescribed Erica amitriptyline. Although it came with a suite of side effects—dry mouth, constipation, dizziness—they would, he hoped, fade over time. After a month or so, Erica might be better able to cope with the difficulties back home in the highlands.

Isolated and unable to afford the bus journey to Harare, Erica took her own life in 2005.

Over a decade after her death, Erica was still at the front of Chibanda's mind. "I've lost quite a number of patients through suicide—it's normal," he says. "But with Erica, I felt like I didn't do everything that I could." Soon after her death, Chibanda's plans were flipped on their head. Instead of opening his own private practice—a role that would, to an extent, limit his services to the wealthy and provide ample financial support for his growing family—he founded a project that aimed to provide mental health care to the most disadvantaged communities in Harare.

Chibanda's roots are in Mbare, a southern district of Harare. Its low concrete buildings, wooden shacks, and dusty roads buzzing with traffic fail to capture the importance of this place. Even though it is half an hour from the city center by road, Mbare is widely considered the heart of Harare. As one waiter working in a barbecue joint puts it: "If you come to Harare and don't visit Mbare, then you haven't been to Harare." Mbare has long been a political stronghold for the Movement for Democratic Change (MDC), a party that has fought long and hard, sacrificed tears and lives, against the ruling ZANU-PF that was led by Robert Mugabe since independence in 1980. Back then, when Zimbabwe was called Southern Rhodesia, Harare, the capital city, was called Salisbury. Mbare, the southern district, was called Harare. As the country would later take this name for its capital (replacing Salisbury in 1982), it was a prescient reminder of the influence this southern district has on its country's future.

Chibanda's grandmother lived here for many years, seeing firsthand its passing from Harare into Mbare. Whatever name it has been given, it has attracted people from all over the country to buy or sell groceries, electricals, and retro, often counterfeit, clothing. The line of wooden shacks that make up the daily market is a lifeline for thousands, an opportunity in the face of inescapable adversity.

In May 2005, the ruling ZANU-PF party initiated Operation Murambatsvina, or "Clear Out the Rubbish." It was a nationwide, military-enforced clampdown on the people whose livelihoods were deemed to be either illegal or informal. An estimated seven hundred thousand people across the country, the majority already in disadvantaged situations, lost their jobs, their homes, or both. Over eighty-three thousand children under the age of four were directly affected. Those places where resistance might have emerged, such as the MDC's stronghold of Mbare, were hit the hardest. The destruction also took its toll on people's mental health. With unemployment, homelessness, and hunger taking hold, depression found a place to germinate, like weeds among the rubble. And with fewer resources to deal with the consequences of the destruction, people were wrapped up in a vicious cycle of poverty and mental illness.

Chibanda was among the first people to measure the psychological toll of Operation Murambatsvina. After surveying twelve health clinics in Harare, he found that over 40 percent of the people scored highly on psychiatric health questionnaires, a large majority of whom met the clinical threshold for depression. Chibanda presented these findings at a meeting with people from the Ministry of Health and Child Care and the University of Zimbabwe. "It was then decided that something needed to be done," Chibanda says. "And everyone sort of agreed. But no one knew what we could do." There was no money for mental health services in Mbare. There was no option to bring therapists in from abroad. And the nurses already there were far too busy dealing with infectious diseases, including cholera, tuberculosis, and HIV. Whatever the solution—if one actually existed—it had to be founded on the scant resources the country already had.

Chibanda returned to the Mbare clinic. This time, it was to shake hands with his new colleagues: a group of fourteen elderly women. In their role as community health workers, grandmothers have been working for health clinics across Zimbabwe since the 1980s. Their work is as diverse as the thousands of families they visit, and includes supporting people with infectious diseases, teaching people how to dig latrines, and offering community health education. "They are the custodians of health," says Nigel James, the health promotion officer at the Mbare clinic. "These women are highly respected. So much so that if we try to do anything without them, it is bound to fail."

In 2006, they were asked to add depression to their list of responsibilities. Could they provide basic psychological therapies for the people of Mbare? Chibanda was skeptical. "Initially, I thought: How could this possibly work, with these grandmothers?" he says. "They are not educated. I was thinking, in a very Western, biomedical kind of sense: you need psychologists, you need psychiatrists." This view was, and still is, common. But Chibanda soon discovered what a resource the grandmothers were. Not only were they trusted members of the community, people who rarely left their local district of Harare, but they could also translate medical terms into everyday words that would resonate culturally. With the buildings of the clinic already full of patients with infectious diseases, Chibanda and the grandmothers decided that a wooden bench, built by local craftsmen, would provide a suitable platform for their new project. Placed under the dappled shade of a tree, the locals could sit and talk at any time of day.

At first, Chibanda called it the Mental Health Bench. The grandmothers told him that this sounded overly medicalized and were worried that no one would want to sit on such a bench. They were right—no one did. Through their discussions, Chibanda and the grandmothers came up with another name: Chigaro Chekupanamazano, or, as it became known, the "Friendship Bench." Its beauty was in its simplicity. There was no need for a formal appointment. There were no doctors or medical prescriptions. But, like the surrounding district of Mbare, the Friendship Bench was greater than the sum of its parts. This group of grandmothers, their benches, and the words that they use to express and soothe distress would soon be at the forefront of a global revolution in how we treat depression, one that would reach out from southern Harare into the towering cityscapes of New York and London. It was the Western world learning from sub-Saharan Africa. Although depression was once thought to be absent in countries like Zimbabwe, lessons in its successful treatment were soon to be passed from Africa to the United States—a reversal of damaging colonialist practices.

This startling reversal had been foreseen. "Scientific knowledge tends to travel more easily from the developed world to the developing world," wrote Ricardo Araya, a psychiatrist from King's College London, while working in Santiago, Chile. "However, it is possible that something could be learned from studies using simple interventions with intensively trained personnel."

Late in 2009, Melanie Abas was working at King's College London when she received a call. "You don't know me," she remembers a man saying. He told her he'd been using her work in Mbare and that it seemed to be working. Dixon Chibanda told her about his project, his team of fourteen grandmothers, and their use of the seven-step plan for depression at three clinics. To identify people who might benefit from this project, Chibanda used the work of Vikram Patel, who, also working in Harare in the mid-1990s, had adopted Abas's research into the local idioms of distress to create a screening tool for depression and other common mental disorders. He called it the Shona Symptom Questionnaire, or SSQ-14. It was a mixture of the local and the universal, of *kufungisisa* and depression. As with the PHQ-9 used in English-speaking countries, it was incredibly straightforward. With just a pen and paper, patients answer fourteen questions and their health worker determines if they are in need of psychological treatment. In the last week, have they been thinking too much? Have they thought of killing themselves? If someone answered yes to eight or more of the questions, they were considered to be in need of psychiatric help. Fewer than eight and they weren't.

Patel acknowledges that this is an arbitrary cut-off point. It makes the best of a bad situation. In a country with few health services, the SSQ-14 is a quick and cost-effective way to allocate scant resources. Once patients were identified, Chibanda and his grandmothers could then lead them through the seven-step plan. With its strong focus on social relationships and enforcing support networks, he thought that it would be pertinent to Mbare, a place where disputes over money or domestic violence are found in abundance. As with interpersonal therapy, patients are guided toward their own solutions.

Although the results were promising—in 320 patients, there was a significant reduction in depressive symptoms after three or more sessions on the bench—Chibanda was still apprehensive about telling Abas. He thought his data wasn't good enough for publication. Each patient had only received six sessions on the bench and there was no follow-up. What if they just relapsed a month after the trial? And there was no control group, an essential counterbalance to rule out whether a patient wasn't just benefiting from meeting with a trusted health worker and spending time away from their problems.

Abas hadn't visited Zimbabwe since 1999, but still felt a deep connection to the country where she had lived and worked for two and a half years. She was thrilled to hear that her work had continued after she had left. Straightaway, she decided to help.

Chibanda traveled to London to meet Abas in 2010. She introduced him to people working on the IAPT (Improving Access to Psychological Therapies) program at the Maudsley Hospital, a nationwide project that had started a couple of years earlier. Abas, meanwhile, pored over the data he had sent her. Together with Ricardo Araya, she found it to be worthy of publication. In October 2011, the first study from the Friendship Bench was published. The next step was to fill in the gaps—adding a follow-up and including a control group. Together with his colleagues from the University of Zimbabwe, Chibanda applied for funding to conduct a randomized controlled trial, one that would split patients across Harare into two groups. One would meet with the grandmothers and receive a form of therapy that would help patients develop their own solutions to whatever social problems they brought to the bench. This was called problem-solving therapy, a descendent of cognitive behavioral therapy that aims to reduce depressive symptoms by guiding patients through potential solutions to any difficulties in their life.

One of the major life events that can be tied to the onset of depression is domestic violence. "We have an epidemic of that," Chibanda says. "A lot of these women are in abusive relationships because they have absolutely nothing. They depend on this man for everything—literally, everything." From the beginning, therefore, it made sense for the grandmothers to help their clients earn some money of their own. Some thought it was a good idea to ask relatives for a small kick-starter to buy and sell their chosen wares. Others crocheted handbags, known as Zee Bags, from colorful strips of recycled plastic (originally an idea from Chibanda's actual grandmother). With greater independence, the grandmothers have found, marital disputes and domestic violence decreased. As Juliet Kusikwenyu, one of the grandmothers in Mbare says, "[C]lients normally come back and report that, 'Ah, I actually have some capital now. I've even been able to pay school fees for my child. No longer are we fighting about money.'"

Other problems are harder to solve. Tanya, a forty-two-year-old woman living in eastern Harare with her husband and two children, was diagnosed as HIV positive in the early 1990s at the height of the epidemic. At this time, antiretroviral drugs were in short supply. HIV was seen as a demonic curse rather than an infectious disease. But the most pressing problem for Tanya was that her husband was found to be free of HIV. After his blood tests came back negative, he was convinced that she had been unfaithful and had caught this sexually transmitted disease from another man. Tanya was thrown out of her house, separated from her two children, and made to think that her HIV status made her guilty, promiscuous, unworthy of life.

Serodiscordant relationships—in which one partner is HIV positive and the other isn't—are quite common. It can be the case that one person has had sex with someone who has HIV. But that isn't the only explanation. Blood transfusions can transfer the virus from one person to another. Sharing needles to inject drugs is another common pathway. There is also the possibility that the infection was present before the relationship and lay dormant for a decade or more. In this scenario, the virus may not have passed on to the partner by chance or if they had a natural resistance to that strain of the virus. Tanya claims that she was faithful to her husband, but nothing would change his mind. She was never allowed back into the home they once shared.

Tanya spent years on the streets begging for food and money, and thinking about what happened to her over and over again. She felt trapped by her own thoughts. How were her children? When would this virus kill her? How was she HIV positive when her husband wasn't? When she wasn't sleeping on the streets, Tanya spent time in a mental institution. With beds and drugs in short supply, she was soon forced out, just as she had been evicted from her own home. The church was not an option, either. Being HIV positive was seen as the mark of the devil or a punishment from God. Was she being punished for a sin she didn't think she had committed? Tanya became delusional. She was convinced that she had to kill herself and her two children, saving them from this cruel world. Nothing made more sense.

One day she walked into her local health clinic, the place where she collected her HIV medication, and heard about a new project that was available for people who were thinking too much. It was a form of talk therapy that was outside of the clinic walls, completely confidential, and unrelated to the church. After years of toil and suicidal thoughts, she sat down next to an elderly woman wearing brown overalls and started to talk. She was told that HIV was nothing to be ashamed of. She said that *kufungisisa* is normal, especially for people with an infectious disease. But it doesn't have to make you feel miserable. Over the coming weeks and months, Tanya got back in touch with some family members and was reunited with her children. Although her husband's views were unchanged, Tanya held her head up high. At a "holding hands together" meeting in 2018 in Harare, her eyes shaded by a wicker hat with the logo "Ending TB" on the front, she told a small gathering of Friendship Bench patients that she was HIV positive, had *kufungisisa*, and was alive.

In 2016, a decade after Operation Murambatsvina had cleared out much of Mbare, Chibanda, Abas, Araya, and their colleagues published the results from the multicenter clinical trial, incorporating 521 people from across Harare. Roughly half sat on the Friendship Bench and spoke to the grandmothers. The other half met with doctors or other health workers, receiving support but no problem-solving therapy. Although they averaged the same score of depression on Patel's SSQ-14 before treatment, only the group from the Friendship Bench showed a significant decrease in symptoms, falling well below the threshold of depression. When tested against the gold standard of medical science (the randomized placebo-controlled clinical trial), Chibanda and his group of grandmothers had passed with flying colors. "They didn't have an intervention for depression before, so this was completely new in primary health care," says Tarisai Bere, a clinical psychologist who trained 150 grandmothers across ten clinics. "I didn't think they would understand it the way they did. They surprised me in so many ways. . . . They are superstars."

Rudo Chinhoyi, a grandmother with faded roses on her bandana, says that she carries her printed manual—the screening tools and the therapy—in her brown canvas bag everywhere she goes. Anyone might need her services, not just the people who visit local health clinics. "The Friendship Bench has become imaginary," Chinhoyi says.

"I Live and Breathe Peer"

When she was a student at the New York Institute of Technology, Helen Skipper discovered the extracurricular activities that took place beneath the glitzy skyscrapers of Manhattan. Skipper had an analytical mind and an obsessive appetite for new information, but she suffered from the crushing side effect of boredom. Nothing could satisfy her for long. Textbooks for an entire semester were read—and understood—in an evening. Playing the class clown soon got old. Comic books were repetitive. After her first toke on a crack pipe, however, she felt like she had found what she had always been looking for. Her boredom dissipated in an instant. College seemed so mundane, meaningless, in comparison. Pleasure, success, friends, and life itself: it could all be bought for a few dollars a day.

Skipper was quickly pulled into the 1980s crack cocaine epidemic. While the wealthy sniffed expensive lines of cocaine in the tower blocks above, those people living in lower-income neighborhoods, often African American and Hispanic, were more likely to be addicted to crack. The punishment for mere possession of the latter was severe, a political sleight of hand that allowed presidents such as Richard Nixon and Ronald Reagan to punish people of color, lock them up in prison, and take away their housing, employment, and suffrage. The local devastation of this so-called Crack Era would later be compared to the aftermath of war or economic depression. "The subcultural behaviors associated with crack use also led to much interpersonal violence, duplicity in relationships, increased prostitution, child neglect and abuse, and family dissolution," wrote Eloise

Dunlap, a sociologist from the University of California, Berkeley, in 2006. "Crack users placed a heavy burden on families . . . extended kin, and community members who sought to support these persons. Crack users also greatly disappointed their offspring who might otherwise have depended upon them, thereby placing additional burdens on family, kin, and community."

Skipper, who was raised in a middle-class African American household, was first arrested when she was seventeen years old. She would spend the next twenty years trapped in a cycle of imprisonment, rehab, homelessness, and crack cocaine. It was the drug that offered her work, money, and friendship. It was the New York State judiciary system that failed to help her get off the streets.

Years passed her by. Her mother kicked her out of the family home, a refuge with three bedrooms upstairs and two cars in the driveway. Skipper gave birth to two sons but barely saw them. She missed their lives as they unfolded without her. There were no first steps, no first words. Separated from her family, ousted from her home, there was no one in the world to talk to but herself.

Skipper inherited a deep mistrust of professionals. The police. Lawyers and judges. People with "MD" after their name. "We never wanted to see a doctor, much less a therapist, or a psychiatrist, or a psychologist, or anything like that," Skipper says. "It's multigenerational. You know, like Grandma never went for help, and Grandma was sitting in the attic talking to herself at the time, and then my dad, he used to go into the backyard and talk to himself. . . . Me, I'm in the closet talking to myself. We all do it, it's normal." The compulsory stints in rehab facilities that followed her imprisonment only stopped her from taking drugs. It didn't do anything to ameliorate her craving for crack, her psychological need for satisfaction and pain relief.

When all else seemed lost, she still had her smarts. After her stints in prison, she would write a résumé and land a menial admin job at one of New York's many financial institutions. She worked at Chase. She made it onto Wall Street. When she had earned enough money, it was a smooth slide back into old habits, like putting on some comfy slippers. She smoked, sold, and delivered crack cocaine up and down the Upper West Side. When she inevitably lost her job, she begged for money. "If you're working that area, you know you don't say anything to the black people, because they won't give you anything," she says. White tourists with cash, their eyes transfixed by the bright lights and advertising boards, were her prime target.

Over two decades, Skipper slept on rooftops, in homeless shelters, and in mental institutions. Diagnosed with schizoaffective disorder, a dizzying mix of schizophrenia and depression, which she thought were "just pretty words," she was given pills to swallow that made her passive and groggy, nothing like the crack she was used to. One day, she watched a fellow patient sit in the corner of the room with his tongue lolling out, a living picture of suppression and enforced passivity. It was then that she decided enough was enough. "I didn't want that for me, and I don't care how much you say it helps me," she says. "You're not going to do me like that."

In her forties, she reached out to a therapist as part of a drug and mental health rehabilitation program. She told the therapist that she didn't want to be on medication for the rest of her life. She didn't want to continually relapse back into addiction. She wanted a life free of drugs, both medical and recreational. "As luck would have it, I had a good therapist," Skipper says, "someone who really talked to me like she really *got* me. By talking to her and doing what I'm supposed to be doing with the rest of my life—like addressing my drug addiction, repairing the damage I did to my family, building up myself, my psyche, my self-esteem—I was finally able to come off it." For those sleepless nights, she took sleeping pills to ease her racing mind but, on the whole, she was free of chemical dependency. Off the streets, out of jail, with no threat of institutionalization, she decided to help others as her therapist had helped her. Hired by the the New York City Department of Health and Mental Hygiene (DOHMH) in 2007, she became a "peer," someone who can reach into underserved areas of the city and talk to people about their problems. Skipper had no medical degrees or psychiatric qualifications of any kind. She was as far removed from the health system as one can get. But what she had was "lived experience." "I've been in every situation, so there's not a lot that I cannot talk about from personal experiences," she says. Addiction, poor mental health, homelessness, unemployment, dropping out of education, she has been there and done it all. "Basically," Skipper says, "I live and breathe peer."

In 2017, Skipper, in her early fifties with short tan-colored dreadlocks, half-rim glasses, and a voice that quavers with the ups and downs of her past, applied for a new program as part of the DOHMH. It was called Friendship Benches and they needed a peer supervisor, someone to advise and train other peers across New York in mental health care for the community. Her application was a textbook example of what other people might try and hide from potential employers. A criminal record. Mental health problems. Institutionalization. Drug addiction. But, for this particular role, they were requirements for the job. A doctor requires years of education, exams, training, and rotation. A peer supervisor needs to have lived in the underbelly of New York City as if it were a second home.

Skipper got the job, becoming the only peer supervisor in the DOHMH. By this time, she had been clean for over a decade and hadn't taken any medication for seven years. Her eldest son had a child of his own. Although Skipper was still estranged from her family, she was, by definition, a grandmother. Like the community health workers in Harare, she was a respected elder of the community, someone who broke down the barriers between clinical psychiatry and the people who might benefit from its care the most. As with *kufungisisa* in Zimbabwe, the Friendship Benches of New York had to be made culturally appropriate. The benches were made out of bright orange plastic and looked like giant Lego bricks. Peers were trained to advise people on opioids such as heroin or fentanyl, drugs that, like crack cocaine, were fueling an epidemic of addiction across the United States. Often with lived experience themselves, Skipper's peers often knew where the best rehab centers were and were walking examples of recovery.

Move a few miles away from the manicured gardens of Central

Park, and the unmistakable calling cards of drugs like heroin and fentanyl blend in with the Yellow Cabs and traffic lights. Empty needles litter the floor. Fast-food joints have their own security guards on the doorstep. Huge billboards advertise naloxone, an injectable drug that can reverse the effects of opioids in the event of overdose. "People come to New York and they don't see this; they see the flashy lights," says Beth Rodriguez, the coordinator of New York's Friendship Benches, as she drives from East Harlem up into the low-income neighborhoods of the Bronx, a borough that has the highest rate of lethal overdoses in the country (34.1 per 100,000 residents). "This is darkness."

Helen Skipper's ups and downs aren't uncommon experiences for people across the United States. An estimated one in ten people with disorders such as schizophrenia, depression, or bipolar disorder encounter the police before a psychiatrist or therapist. Imprisonment has, in part, replaced institutionalization. Care homes have become the latest mental hospitals for chronically ill elderly persons, those with depressions that were once called involutional melancholia. As the number of psychiatric beds in American mental hospitals declined from 558,922 in 1955 to 159,405 in 1977 (largely a result of the mass rollout of medications), community health services were stretched to the breaking point, leaving those who might require long-term care on the streets, forced into crime and at risk of imprisonment. A classic study from a Los Angeles County jail in 1983 found that out of those inmates with psychiatric disorders, "only 14 percent were receiving mental health treatment (primarily medications only) at the time of arrest, and only 25 percent altogether had ever received any form of outpatient mental health treatment at any time in their lives."

Peers are just one way of filling this gap. Skipper doesn't use a diagnosis or medical reports in her work. Depression is just one of the many problems that she sees every day, from drug abuse to unemployment. "Our discussion might not have anything to do with mental health," she says. "But it could well be about something that, if left unchecked, will lead to poor mental health. We don't discriminate: we don't say, okay, come back and see us when you have a mental illness. We want to stop it before it gets hold o' you.

"If we've gotta sit on the curb and have a conversation with you, we'll do that," she adds. "Because of my history I'm not afraid to go into a shooting den or something like that to talk to someone. I'm not afraid of that. I've been there. I've done that."

In January 2018, Chibanda traveled from the summer of Harare into a freezing East Coast winter of the U.S. He met with his new colleagues and the First Lady of New York City, Chirlane McCray. He was blown away by the support from New York's mayor, Bill de Blasio, the number of people the project had reached (over forty thousand), and by Skipper and her team. But it wasn't to last. In 2019, Friendship Benches in New York hired a clinical director who tried to medicalize peer support. Skipper felt so disheartened and nearly left her profession altogether. Instead, she returned to college, studying for a degree in criminal justice with minors in psychology and sociology.

As expected, her grades weren't an issue (she achieved 3.9 out of 4 on her grade point average, or GPA). The major challenge was finding the time to pack in all her ideas and interests. After her freshman year, she applied for a new job as peer coordinator at the Criminal Justice Agency (CJA). After nearly leaving peer support altogether, she decided to use her lived experience to add a third tier to peer training: criminal justice. Currently, there are only two certifications for peers in New York: mental health and substance abuse. "My job is to take elements of both those certifications and apply them to Criminal Justice," Skipper says. "I had complete control of the program from hiring to formulating policy and procedure." Six months after she started, Skipper was given a raise and a budget to hire twelve peers. "I am now the manager of Peer Services at CJA and I do this job with the lens of creating spaces for those of us with lived experiences in an agency that is reforming the criminal justice system in New York. Not only that—I am completely validated in my role. The agency fully understands and sees the importance of peer support." With training in mental health, substance abuse, and criminal justice, Skipper is creating a community of care that, crucially, is becoming culturally relevant to American society. Mental health care in the United States needs to encompass the epidemics of drug addiction from crack cocaine in the 1980s to opioids today. It must educate people who are affected by a legal system that targets black Americans and inner-city neighborhoods, and strips away basic rights such as suffrage, affordable housing, and employment. It must, in other words, channel Helen Skipper, a former drug addict, prison inmate, inpatient, and homeless person who is on a mission to reshape New York City from the ground up.

PART FOUR

The Universe Within

Depression is emotional pain without context.

Helen Mayberg

Many forms of insanity are unquestionably the external manifestations of the effects upon the brain substance of poisons fermented within the body. . . .

Johann Ludwig Wilhelm Thudichum
(1884)

. . . either LSD is the most phenomenal drug ever introduced into treatment in psychiatry, or else the results were evaluated by criteria imposed by enthusiastic, if not positively prejudiced people.

Louis Jolyon West

It Feels Like Spring

On May 23, 2003, Helen Mayberg watched as a thin wire was guided into the middle of her patient's brain. Pushed through a guide tube, it reminded her of a cooked piece of spaghetti. At its end was an electrode composed of four rounded contacts that looked like the metallic ends of a battery. Each just over a millimeter in diameter, they were arranged vertically, like traffic lights, with tiny spaces in between. It was delicate work, requiring steady hands and years of medical experience. Although the neurosurgeon, Andres Lozano, was highly skilled and had performed this procedure many times before on patients with various movement disorders such as Parkinson's, Mayberg, a neurologist raised in California and an unmistakable maverick of her profession, was nervous. What if something went wrong? Even with thousands of successful operations in the past, the surgeon's experience, and the knowledge that the wire could just as easily be taken out as put in, accidents are always possible.

Lozano wasn't blind during this operation. After studying the brains of people with depression since the 1980s, Mayberg told him exactly where to put the electrode. Unlike antidepressants that were discovered by chance in a sanatorium for people with tuberculosis, this was a test of a hypothesis, a trial that asked a specific question—does this part of the brain control depression?—and sought answers based on whether critically ill patients got better or worse.

Once everything was in position, the lowest of the four contacts was switched on, creating a gentle buzz of five milliamps into the surrounding tissue. Nothing. It was switched off. What about the next electrode up, second from the bottom? Still nothing. After the third electrode was switched on, however, the patient asked, "What are you doing?"

"Why do you ask?" Mayberg said.

"I suddenly feel relaxed." Trying to further describe the feeling, she struggled to come up with a suitable metaphor.

"It's like the difference between a laugh and a smile," she said.

Okay . . . What the hell does that mean? Mayberg thought to herself. She watched as her patient tried to explain emotions that she obviously hadn't felt in a long time. "She's struggling, it's clearly profound," Mayberg recalls. "She's getting kind of pissed off. We aren't understanding what she is trying to describe."

Then she had it. It's like looking out the window in winter and seeing a crocus poking out of the snow, she said. "It feels like spring."

With a signal to her colleagues, Mayberg switched the electrode off. As though a light had been switched off, the feeling of spring disappeared; the crocus died. Perhaps it was nothing.

This patient was one of six people with depression to receive deep brain stimulation between 2003 and 2004. In case reports, they were all classed as "treatment-resistant." Their depression seemed intractable. A diversity of drugs—antidepressants, antipsychotics, tranquilizers, mood stabilizers—couldn't budge their mental anguish. Psychotherapy had been a disappointment. For five out of six patients, even ECT had failed. On social benefits, living between the home and the hospital, these were some of the sickest people seen by a psychiatrist. "These people are morbidly ill," Mayberg said, "and the goal is to help them."

Although she studies psychiatric patients, Mayberg isn't a psychiatrist. She trained in neurology, a medical specialty that focuses on diseases of the brain and spinal cord. "Psychiatry is interesting to a neurologist, but their language is just ridiculous," she says. Strict classifications like major depressive disorder work pragmatically but don't make much sense biologically. Despite a general agreement among researchers and clinicians that depression has a biological basis that involves the brain, there are no definitive biological markers to identify such a diagnosis. From a young age, Mayberg pined for acuity, precision, something to anchor a messy field of classification into its discrete parts. Her father was a physician and her uncle, whom she greatly admired, was a biochemist and nuclear medicine physician. Mayberg would take elements from all of these disciplines—chemistry, nuclear science, medicine—to understand the biological diversity of depression. Instead of opening up the skull and slicing through the brain as a neurosurgeon, she turned to the burgeoning field of brain scanning. After training in neurology in New York, she moved to Johns Hopkins University in Baltimore to study positron emission tomography (PET) scanning, a method that used radioactive tracers to visualize molecules of interest as they moved and settled within the receptors of the brain. It was nuclear science (radioactive particles) meeting neurosurgery (seeing inside the brain), but without the explosive or bloody manifestations of either.

Parkinson's, Huntington's, stroke: Mayberg studied various neurological disorders in the late 1980s and '90s as she moved from Baltimore to south-central Texas. She found that these neurological diseases, interesting in their own right, often came with depression. She was hooked by the opportunity to understand this mental disorder as a facet of other, well-studied diseases. Movement disorders and stroke had been mapped to regions of the brain and understood in significant detail. There were regular patterns in activity and atrophy. By comparing those who had depression to those who didn't, patients with these diseases were the perfect blueprint to understand how depression reshapes and distorts the brain. "I like controlling as many variables as possible," Mayberg says. By studying the brains of well-known neurological diseases, she adds, "we can match patients who are alike in all aspects of their illness except for being depressed or not." To her, stroke, Parkinson's, and Huntington's were anchors to which she could test her burgeoning theory of how depression takes control of the brain, regardless of the cause.

Depression is often associated with other diseases. When it

tags along with something like cancer or cardiovascular disease, it is termed "secondary depression," a mental consequence of an underlying (primary) disease. Take Huntington's disease. Roughly 40 percent of patients with this degenerative disorder show the hallmarks of depression, which often precede the more typical symptoms of speech impediments, muscle twitching, and paralysis. Depression can be the first indicator of the future disease trajectory. This association had been noted for as long as Huntington's had a name in medical science; in 1872, George Huntington, the discoverer of this disease, wrote, "[T]he tendency to insanity and sometimes to that form of insanity which leads to suicide is marked."

Although secondary, these depressions still respond to antidepressants, ECT, and psychotherapy. They aren't inevitable ramifications of a body in decline. "In the last thirty years or so, depression in its various forms has come to be recognized as not an acceptable or normal part of growing old," says Charles Reynolds, a professor of geriatric psychiatry at the University of Pittsburgh.

From her early research into depression in patients with neurological diseases, Mayberg found a surprisingly regular pattern of brain activity. Wherever she looked, whether it was Huntington's, stroke, or Parkinson's disease, there was a biological blueprint that separated anyone who had depression from those who did not. Depression came with a significant reduction in activity of the frontal cortex (specifically the prelimbic frontal and temporal cortex), the part of the brain that folds around the more central regions like the fingers of a clenched fist. Mayberg's later move from PET scans to fMRI (from radioactive tracers to powerful magnets) found similar patterns in activity, each pixelated blob in the front of the brain visualized in a cool blue color, a sign of decrease in blood flow. Then she looked at the brains of patients with primary depression, people who didn't suffer from neurological disorders, and found exactly the same thing. The same blue dots in the prelimbic frontal and temporal cortex appeared from her data sets. As Andres Lozano, the neurosurgeon and colleague of Mayberg, would later say, these brain regions were literal markers of "the blues."

What was going on? Why were these parts of the frontal cortex—previously linked to rational thought, attention, and decision making—showing up as blue in their scans? Why were they hypoactive (the opposite of hyperactive)? The answer, Mayberg proposed, was to be found in a region of high activity toward the center of the brain, a blob of tissue that sat in the middle of the curled fingers of the frontal lobes. This was Brodmann area 25, a nexus of the brain that modulates memory, mood, appetite, and sleep, and the future target of her deep brain stimulation trials. Showing up as red (for increased blood flow or glucose metabolism) in these early brain scans, area 25 was a scream of activity that, she thought, caused other regions of the brain to shut down. Just as a booming voice might dominate a debate, not allowing the quieter, and perhaps more rational, views to be heard, was this part of the brain leading the whole system to break down from within?

This idea closely matched the reports from the depressed patients in her brain scanning studies. They often complained of feeling trapped within themselves, unable to reach out to others, deprived of any feeling of connection with those people closest to them—even their own children.

In 1999, Mayberg published one of her most famous studies in *The American Journal of Psychiatry*. It was a short paper, only six and a half pages long and with two figures of blurry brain scans, providing a concise insight into area 25 and how everyone—not only those who are depressed—can experience its fury. When asked to think about an extremely sad experience from their past, area 25 lit up like a beacon in the center of the healthy volunteers' brains, while parts of the frontal cortex switched off. When they stopped thinking about this sad experience, the brain returned to its previous state. Area 25 quietened. The rational frontal cortex jumped back to the fore. In people with depression, however, this same pattern was only achieved with antidepressant treatment. Perhaps, Mayberg reasoned, people who were depressed were stuck, they couldn't switch off this "sadness center." It just kept firing, and firing, and firing. With such a constant barrage of painful signals, the circuitry within the brain starts to take on a different shape. Pieces start to malfunction. Parts of the cortex are silenced and pushed out of practice. Even after slight disappointments, full-blown depression becomes the norm. The brain settles into a disordered state that medication can often help disentangle.

This was the idea, anyway, a theory that went beyond the influence of monoamines such as serotonin. Depression was a dysfunction in the circuitry of the brain, the neurological highways that carry the traffic of signals from one region of the brain to another. Mayberg was particularly interested in the limbic system—of which area 25 was a key component—that connects regions of the frontal cortex involved in motivation, drive, and rational thought with those more central parts of the brain that are crucial to memory and emotional regulation. Through the sadness center, could this circuitry be recalibrated or rewired? Such a question echoes the "fixed thoughts" theory that led Egas Moniz to perform the first prefrontal leucotomies in 1935. But where Moniz was reckless, impulsive, and impatient, Mayberg had spent thirty years judiciously understanding how a depressed brain works before any operation began. During the session in 2002, it was as reversible as turning a light off.

With a bob of brown hair and thick-framed glasses, Mayberg is unapologetically straight-talking. She speaks rapidly, and often jumps from one topic to another as if her words are chasing thoughts through her brain's networks. "I talk too much," she admits. "I like action." In the same breath she adds how she tried her hand at neurosurgery as a student (following in the footsteps of her father), but found herself unsuited to its demands. "I probably had the hands. I didn't have the patience." When discussing her career in neurology and nuclear imaging, fields dominated by men, she says that she didn't feel any issue or discrimination over her sex. "When people got blocked or undermined I always considered it the competition of science. It's not a gentlemanly sport," she says. "Even when only men are involved. It's high stakes and low money. All you have is ego and data. And good ideas are not common. I was brought up to stay out of trouble, be sincere, and fight for what's yours; but work twice as hard as everybody else and take half the credit if it gets you where you want. I'm accustomed to unfairness. As Dad always told me: What made you think that life was fair?"

After studying in Baltimore and Texas, Mayberg packed up her things in 1999 and moved to Toronto, Canada, a city where she would eventually meet the treatment-resistant patients for the first DBS trial. There, she was also introduced to patients who contradicted all her previous work. Similarly depressed and in need of treatment, their brain scans revealed the reverse of what she had thought to be the common blueprint of depression. Parts of their frontal cortex were overactive, seething with red blobs where, she thought, there should only be blue. Their area 25 was quiet—hypoactive and not hyperactive. It was as if the brain scans she had been accustomed to had been reversed by image-editing software. What was usually reduced was increased. This was her first hint of an exciting project that would run parallel to her trials in deep brain stimulation: Are there distinct forms of depression? If so, might this explain why some people respond to certain treatments while others don't? Might data from brain scans be used to guide treatment?

For as long as there have been antidepressants and evidence-based psychotherapies, they have been handed out through trial and error. Often, the only metric used to evaluate whether a patient was prescribed an antidepressant was the number of "endogenous" or "atypical" features that they showed. Would they benefit from imipramine or an MAO inhibitor? But Mayberg's patients in Toronto provided another window into the diversity of depression. Although their brains were showing different blueprints of activity, their symptom profiles were of similar severity to those she had seen in Texas. They were no less depressed. But, as shown in trials that began in Toronto but continued in 2005 when she moved to Emory University in Atlanta, they responded to CBT far better than they did to antidepressants. With therapy, their overactive frontal cortex quietened—the reverse of medication. This, Mayberg, along with Boadie Dunlop, Edward Craighead, and her PhD student Callie McGrath, wrote in 2014, was a "psychotherapy-responsive" form of depression.

The other kind of depression—the one with reduced activity in the frontal cortex, area 25 screaming in a flash of red pixels—is a little less straightforward to define. While some patients in this group responded well to antidepressants, there were some who didn't. They weren't the embodiment of an "antidepressant-responsive" depression. Mayberg and her colleagues had seen this pattern before, however. It was the same brain type of those patients with treatment-resistant depression that were put forward for deep brain stimulation. This begged the question of whether it was a more stubborn form of depression, one that required a more aggressive approach to treatment. In such medication- and therapy-resistant patients, one of the oldest treatments for depression can be used with remarkable effects. For severe depression that often comes with delusions, electroconvulsive therapy is still the most effective treatment available. Over eighty years since it was invented in Rome, it is also now one of the safest procedures in all of medicine.

Rebirth

In a brightly lit operating room in Brooklyn, New York, Alice lies on a padded gurney, feeling very anxious. As requested, the computer monitors that display her blood pressure, heart rate, and oxygen levels have been put on silent. If they weren't, the incessant beeping that they make—and her body's rhythms that they monitor—would only add to her nervousness. The silence in the room is broken only by the general chitchat with the lead nurse and a routine set of questions that Alice answers quietly and without hesitation. What's your name? Date of birth? And what procedure are you having today? "ECT," Alice says.

She is then told that she will feel a cool sensation, as if her circulatory system is being fed by icy glacial streams, due to the anesthetic that is administered through a cannula in her right arm. Alice falls into a deep sleep within seconds. She doesn't respond to any more questions. Since she recently had hip surgery, the anesthesiologist increased Alice's dose of sux, the muscle relaxant that is kept cool in a small fridge in a corner of the room. Her eyebrows and facial muscles twitch as the drug starts to take effect. The anesthesiologist switches her monitor's sound back on.

As the regular beeping returns to the room, the most important number on the anesthesiologist's screen is the concentration of oxygen in Alice's blood. Recorded by a peg-like instrument attached to her index finger and displayed in bright blue numbers, it hovers around 98 percent and is kept near to 100 percent throughout the procedure. Early practitioners of electroshock therapy didn't always use artificial respiration, and cellular damage could occur as the brain was deprived of oxygen. In 2018, however, the year that ECT turned eighty years old, Alice doesn't have to worry about this. Her anesthesiologist is equipped with an artificial respirator (a plastic rugby ball–shaped instrument that connects to an oxygen mask) that she squeezes and expands throughout the session. "It's my job to keep the patient breathing," she says. Since all muscle relaxants stop the diaphragm from moving, artificial respiration is a necessity in modern ECT.

When all is ready and the numbers are looking stable, Tricia Paperone, a thirty-year-old psychiatric resident, pushes the red button on the front of the ECT machine with her thumb. The word *TREAT* flashes on the red LED display. It beeps for a few seconds, then the room is still and silent but for Alice's regular heart rate. She hardly moves a muscle. But inside her brain, an electrical storm is circulating. Invisible and roughly recorded by scribbles on a piece of paper, her brain swirls with activity. Her synapses are flooded with neurotransmitters and growth-inducing molecules that, together, trigger cascade after cascade of signals that will continue long after Alice wakes from her sleep and returns to her apartment on the Upper West Side of Manhattan. The convulsion is over in less than a minute but has long-lasting effects. Although it is still an active area of research, it is this burst of activity—and perhaps the regrowth of connections in the brain—that is thought to be at the core of ECT's efficacy in depression. Prolonged stress and severe depression are known to prune the brain's delicate wiring. Parts of the brain start to noticeably shrink. ECT is thought to undo this damage. Rather than damaging the brain, as many psychiatrists and activists have argued over the years, modern ECT is more akin to a fertilizer of new connections, a regenerative procedure that can reverse the destructive combination of stress, depression, and time. An alternative theory is that ECT acts as a reset button on regions that are hyperactive in depression, including Mayberg's area 25.

Whatever the true mechanism, Alice is simply happy that she has been given access to this life-saving treatment. (She was shocked to find how vehemently some people stigmatize this treatment, and writes a blog to explain how boring and everyday ECT actually is.) Her own depression first arose in her mid-teens and resulted in her admission to a mental hospital, an ever-changing pick and mix of pills, and ten years of recurrent depression. Nothing worked. "I've always been a bit of a difficult case because people could not understand why I wasn't responding to medication," she says. Alice lost count of how many psychiatrists she has seen, how many drugs she has been given. In 2014, just before she turned thirty-five years old, her latest psychiatrist finally admitted that this "shotgun" approach wasn't working. She needed a different type of treatment. She needed to meet Dr. Kellner.

As of 2018, Charlie Kellner had performed thirty-five thousand sessions of ECT and literally wrote the book on how it should be given and to whom. His first session was in 1978, the year he became a certified doctor in the United States. It was an unpopular and precarious career choice. After the 1960s, ECT had fallen into disrepute. The reports from across the Atlantic were deeply concerning for psychiatrists around the world, and their patients. In the UK, the procedure was performed by untrained medical students in open wards where privacy was impossible. The machines themselves were often outdated and didn't conform to the latest medical safety regulations (some didn't have an automated pulse of electricity and the duration of the shock was dependent on the practitioner's finger). "If ECT is ever legislated against or falls into disuse it will not be because it is an ineffective or dangerous treatment; it will be because psychiatrists have failed to supervise and monitor its use adequately," an anonymous letter published in *The Lancet* in 1981 stated. "It is not ECT which has brought psychiatry into disrepute. Psychiatry has done just that for ECT." Most worryingly, perhaps, was that ECT was given to people who would never benefit from its effects, while suffering from its very real side effects. Short-term memory loss, headaches, and confusion are common risks with this procedure, even today.

A change was badly needed. Like the patients it treated, ECT required a transformation of its own.

From his base at Mount Sinai Hospital in Manhattan, Charlie Kellner has spent his career teaching the next generation of psychiatrists—such as Tricia Paperone—how to do ECT properly. It starts with selecting the right patient. "For the right kind of illness, [ECT] is almost as specific as penicillin [in the treatment of] pneumococcal pneumonia," Kellner says. People with psychosis, mania, catatonia, and even a form of self-harming autism have all been shown to respond to ECT. But severe depression is still the primary indication. *Endogenous*, *delusional*, and *psychotic* are terms still used today, but for Kellner, this disease doesn't require a name. He knows it when he sees it. These people are often suicidal, wake early in the morning, and suffer from a sluggishness that is more often associated with Parkinson's or Huntington's disease. Their symptoms improve as the day progresses, but the next morning, when they wake at three o'clock, they will be back in the same dangerous position. Their minds are fixated on a single delusion: their bodies are empty; their tissues are rotten; they are dying from a cancer that no doctor can detect.

Psychotic depression might be unfamiliar, even alien, to the general public, but it isn't rare. In 2002, a large survey of 18,980 people living in Europe found that nearly a fifth of people who fit the diagnosis of major depression also fulfilled the criteria for psychotic features. For people over sixty years of age who have been hospitalized with depression, delusions are found in up to half of all cases. At high risk of suicide, such patients have often failed to respond to lists of medications—both antidepressants and antipsychotics—and feel like they are falling through the gaps, like grains of sand through an upturned hourglass.

Alice responded after the first session, a rare but not unheard-of occurrence. As she woke up from the general anesthetic, she was met with two familiar faces: Dr. Kellner on her left and her mom on her right. Both have been supportive throughout her recovery, remissions, and recurrences of depression. Almost immediately, Alice says, "I felt like I had some sort of life, and some sort of chance."

As with any medical treatment, the side effects are sometimes painful and disorienting. Headaches are the most common problem after receiving ECT. Painkillers and coffee are known to help soothe the pain and both are provided in the waiting area of Kellner's clinic. Then there are the memory disturbances. "[They] are very real," says Alice. "Anyone would admit that. There are memory problems. What I would say to that is that they do come back." To help coax them back into her consciousness, Alice has started taking photos of her activities in the days and weeks preceding her appointments. "I can look at the picture and be like, 'Oh yeah! I remember doing that.'" Although some people do lose personal moments forever, Alice's main concern is how long it takes her to remember them.

Once a treatment for anyone and everyone, ECT is now mostly limited to the wealthy and the insured. It costs a lot of money just for an anesthesiologist. Then there are the nurses, the trained psychiatrist, and the receptionists who organize the busy schedules and follow-up appointments. And then there are the modern machines, sedatives, and sux, which needs to be refrigerated during both transportation and storage. Added up, the price of just one session of ECT varies from $300 to $1,000. But the potential cost of not using ECT is far higher. As it can start to work after the very first treatment, ECT is one of the safest and most effective methods to prevent suicides. It can literally be a lifesaver. But it is rarely used as such. People who have not responded to several lines of antidepressants, such as Alice, might never be told about ECT. Psychiatrists are still hesitant to offer it, perhaps because they don't know much about it or they can't perform it themselves. "People are allowed to remain sick for years," Kellner says, "and, after fifteen to twenty medications, finally they get ECT and they end up saying, 'Why didn't anybody tell me about this?'"

The moral argument of using ECT is its ability to make critically ill people better. But there is an economic argument here, too. Even with its high up-front cost, the fact that it works and it works quickly can help people return to their jobs and be more productive when they do. Plus, all those failed medications add up. Why not use something that works in a few weeks rather than throwing different varieties of pills and hoping one sticks? Plus, longitudinal studies show that depression is harder to treat with each successive episode. This makes it even more important to find a treatment that works early in its course. For people with delusions, ECT should be a first-line treatment, Kellner says. While only 34 percent of such patients respond to antidepressants, 82 percent respond to ECT. Even when antidepressants and antipsychotics are combined—along with side effects such as weight gain and Parkinson's-like symptoms such as tremor, muscle stiffness, sluggish speech and movement—they still don't reach the same efficacy as ECT on its own.

A study from 2018 put clinical outcomes to one side and focused on the economic argument for ECT. Daniel Maixner, Eric Ross, and Kara Zivin at the University of Michigan in Ann Arbor found that ECT became cost-effective after two failed medications for severe depression. "Two failed medication trials, not twenty-two," Kellner says. "We should get away from this 'last resort' business."

Morally, economically, and scientifically, ECT shouldn't be pushed into the fringes of psychiatry. It should be offered to those people who need it. It should be used as part of a standardized regimen of treatment, one that includes regular psychotherapy and antidepressants. Doing otherwise, Kellner says, is tantamount to medical malpractice. Lothar Kalinowsky said exactly the same thing half a century ago. The effectiveness of ECT is the same no matter where in the world it is performed. For psychotic depressions, recoveries in 70 to 95 percent of patients can be expected. It is one of the most effective treatments in the whole of medicine. And it is also one of the safest. The mortality rate is estimated to be 0.2 to 0.4 per 10,000 patients, a risk no higher than that of the general anesthetic itself.

Away from sensational Hollywood movies and the misplaced opprobrium of the Church of Scientology, ECT has always had a few vocal advocates of its benefits. In her 1994 memoir, *Undercurrents*, the psychiatrist Martha Manning writes that people are so shocked to hear that she received ECT for her depression that they think she must have been abused. "People say, 'You let them do that to you?!' I didn't let them," she writes, "I asked them to do it." In *Shock: The Healing Power of Electroconvulsive Therapy*, Kitty Dukakis wrote, "Feeling this good is truly amazing given where I am coming from, which is a very dark place that has lasted a very long time. It is not an exaggeration to say that electroconvulsive therapy has opened a new reality for me." Even in Sylvia Plath's *The Bell Jar*, while the protagonist's first experience with unmodified electroshock therapy is a horrific one—"a great jolt drubbed me till I thought my bones would break and the sap fly out of me like a split plant"—the second was given with the correct safety precautions and was effective in treating her depression. "All the heat and fear purged itself," the narrator explains, mirroring Plath's experience with ECT in her own life. "I felt surprisingly at peace."

Alice, a concert violinist, didn't have the platform of a famous poet or a psychiatrist. She couldn't express her views through a protagonist or count on the support of a publisher. But, in a world where anyone can create a website for free, she did have her blog. Over four years, she wrote about her experience with ECT, how it saved her life, how to manage the side effects, and what she has to say to anyone who wants to ban its use. "If you consider yourself a 'victim,'" she wrote in 2018, "what gives you the right to want it to be BANNED for the many, many people who have benefitted? Are OUR lives less important? That's like the equivalent of wanting open heart surgery banned because your husband happened to die on the table."

Alice wrote her blog anonymously. (Her name and identifying details have been changed for this story.) One day, however, she hopes to be able to use her real name. She wants to celebrate her recovery and, she hopes, give other people the confidence to speak up about their own treatment. It isn't an experience limited to people who write memoirs or novels. ECT should be a part of public discourse, like chemotherapy or surgery. "Amazingly, the people who claim to have been harmed are OK with giving THEIR names," Alice wrote in November 2018, "because they are supported by the general [view] of the public. So WHEN can we give ours?"

• • •

Italy, the birthplace of ECT, has nearly banned its use entirely. It is only employed by a select few psychiatrists and only in the most life-threatening situations. This decline in popularity has an explanation outside of medicine. Ugo Cerletti and Lucio Bini created electroshock therapy at a time when fascism was at its height in Europe. As with most academics living in Italy and Germany at the time, neither Cerletti nor Bini spoke out against the despot that ruled their country. Whether they actively supported Mussolini or not isn't clear, but their creation was still tainted by the political environment in which it was invented. Six years later, in 1944, its reputation took an insurmountable hit when Emil Gelny modified an electroshock machine and killed at least 149 patients. "[T]he unconscionable misuse of those treatments in former mental institutions," a group of Italian doctors from the University of Naples wrote in 2016, "still profoundly overshadows the very true nature of today's psychiatric practice." Even today, ECT is widely thought to be a fascist treatment, a barbaric throwback to a time when millions of people were imprisoned and murdered for their race, religion, gender, or disability. In truth, electroshock therapy was rarely used in Nazi Germany; eugenics and the prevention of hereditary diseases—through sterilization or euthanasia—was more important than treatment. While an electroshock machine was built at Auschwitz in 1944, it was used to "make emotionally disturbed people fit for work again," and not as a form of punishment. Elsewhere in the country, the public believed that "psychiatry [was] becoming more and more superfluous since mentally ill people would soon become extinct due to racial hygiene laws," wrote Ernst Rüdin, Emil Kraepelin's colleague at Munich. Despite the rapid decline in the popularity of eugenics in the twentieth century, electroshock still didn't become a regular form of psychiatric treatment. For decades, it was rarely performed in Germany. And yet neighboring countries that are similar socially, politically, and culturally became global leaders in ECT research and its clinical practice. The progressive Scandinavian countries have some of the highest utilization rates of ECT in the world. "Tiny little Denmark has been a world leader for decades," Kellner says. The division between one country and another is clear: history and politics—not science—dictate whether this treatment is available for people with depression and other mental disorders. Evidence is weighed down by the baggage of the past.

One of the most common arguments against the use of ECT is that we don't know its long-term impacts. In response to this question, Martin Jørgensen, a psychiatrist from the University of Copenhagen, says, "Well, that's bullshit." In fact, ECT has some of the best longitudinal studies in psychiatry. Making use of the Danish cohort study, for example, Jørgensen and his colleagues assessed whether there were any differences between people with severe depression who had received ECT and those who hadn't. Over a twenty-year period, did they show any notable differences in their health? They did. For patients over seventy years of age, there was a significant decrease in dementia for those who received ECT compared to those who didn't. One explanation for this pattern is that the most severely ill people might also suffer from physical diseases that could make them unsuitable for ECT. But Jørgensen argues that it is exactly these patients whom ECT is suitable for. Elderly patients with age-related diseases often can't take antidepressants because they simply don't work, or they interact with their list of other medications. For them, ECT is the safest option.

The most likely explanation for this pattern—better long-term health outcomes in patients who receive ECT—is that the treatment reduces the risk of dementia by successfully ameliorating the depression, a known risk factor for other age-related diseases. "[We know] there is an increased risk of dementia in people with depression," says Poul Videbech, a psychiatrist-turned-neuroscientist who leads the Center for Neuropsychiatric Depression Research in Glostrup, a town a few miles west of Copenhagen. By removing one disease, another is also halted or significantly delayed. When taken over decades, in other words, ECT actually reduces the chances of significant memory loss.

A tall, stubbled, stylish man in a black turtleneck, Videbech likes to tell the story of another neuroscientist, Gabriele Ende, who investigated whether ECT led to permanent brain damage. As a student at University of California at San Francisco, Ende had studied the hippocampus—the brain region devoted to memory—of epileptic patients and found the classic markers of brain damage. A physicist and not a physician, she had never heard of ECT until she moved to the Central Institute of Mental Health in Mannheim, Germany. There, she learned of its remarkable efficacy in depressions but, given her previous research, hypothesized that it damaged the hippocampus in these patients. "She did all these very good studies and she had to admit that she couldn't show any signs of damage," Videbech says. Quite the opposite, in fact: she found very early signs of neural regeneration, possibly neurogenesis—that is, the growth of new connections in the brain. Since her first paper on this topic was published in 2000, Ende's research has been replicated by many others.

In 2019, Videbech and his student Krzysztof Gbyl provided the latest study to investigate the rejuvenating potential of ECT. While other studies had found that the hippocampus—the memory center of the brain that is often shrunken in people with depression—increased shortly after the ECT treatment, this study found that a similar effect occurred in the frontal cortex, the part of the brain that sits behind the forehead and is associated with the higher functions of emotion, intelligence, and memory retrieval. Importantly, this neural growth was only found in patients who responded to the treatment. Out of eighteen patients with endogenous or psychotic depression, fourteen responded to ECT treatment. The brain regrowth was not found in the four who didn't respond.

Such studies are small in size and need to be replicated by other research groups. But the story so far is not one of damage. ECT, whether assessed through a brain scanner, animal studies, or blood tests, is a catalyst of regeneration.

Outside of Videbech's office in Glostrup is a scientific poster of their latest research. It shows a scan of Gbyl's own hippocampus from when he volunteered as a healthy, nondepressed model. Gbyl, a clinical psychiatrist who is studying for a PhD in neuroscience, shows it to visitors like a proud father flicking through photos of his children. "Do you want to see my hippocampus?" he asks. Videbech, in his cool and measured voice, likes to make the joke that, with this brain scan, "at least we know he has a brain."

With such evidence for safety as well as efficacy, the rates of ECT are starting to increase in countries such as Britain, the United States, and even Germany. But there is a long way to go. The stigma surrounding the treatment can make some psychiatrists refuse to prescribe ECT, even to those who might be most likely to respond. "Psychiatry has to come back into the medical fold and ECT has to become a part of the treatment of severe psychiatric illnesses," Kellner says. "It's not on the fringe. It's not a religion, I don't believe or not believe in it. It is what it is: demonstrably the most effective treatment for the most severely ill people."

The effects of even the best treatments for depression don't last forever. In the majority of cases, patients require regular checkups, fine-tuning, and repeat prescriptions. It is rare for depression to be a one-off event. It requires constant vigilance and combinations of approaches. Alongside her ECT sessions, Alice continues to take antidepressants and regularly visits her therapist for CBT-based psychotherapy. They aren't in competition with each other. Psychotherapy, drugs, ECT: they are collaborators.

In late 2018, Alice performed at a large concert theater in midtown Manhattan. Along with the strings, winds, and percussion, Alice's violin is just one sound among many, a sound wave that harmonizes the orchestra, just as a pulse of electricity helps recalibrate her brain waves. She has played this gig many times before. But, for Alice, this one was particularly special. Although she couldn't see him among the bright sea of faces, she knew that Dr. Kellner was somewhere in the audience. "For someone who I really consider has saved my life to be there, with his wife, watching me do something that I can do, in part, because of him . . . it was just so significant for me," she says. "[Plus], there's pretty much nothing cooler than standing on the stage, playing Handel's *Messiah* to a full house." In three parts, the symphony progresses through the birth, sacrifice, and resurrection of Jesus Christ, a powerful celebration of a miraculous return to life.

The Epitome of Hopelessness

In her lectures, Helen Mayberg uses a quote from the author William Styron to describe what treatment-resistant depression feels like. "In depression," Styron wrote in his 1990 memoir *Darkness Visible*, "faith in deliverance, in ultimate restoration, is absent. The pain is unrelenting, and what makes the condition intolerable is the foreknowledge that no remedy will come—not in a day, an hour, a month, or a minute." Mayberg reminds her audiences around the world that Styron responded to standard antidepressant therapy, that his own restoration from depression was relatively routine. But imagine if this weren't the case. Imagine that you've tried over a dozen different forms of antidepressants, suffered years of side effects, and seen countless psychiatrists and therapists, and still this unrelenting, intolerable pain finds no balm. This, Mayberg says, is the epitome of hopelessness, the extreme of depression that can't be captured by lists of symptoms but is the excruciating feeling that there is no help out there no matter where you look.

Although replicated by other laboratories using different methods, Mayberg's brain-based view of depression is still a work in progress. Even though the technology has improved in leaps and bounds since she began her research in the 1980s, an MRI scan is still not a direct insight into how the brain works. It uses blood flow as a surrogate for activity rather than measuring the actual firing of neurons, the flow of a complex mix of brain chemicals, which would require a level of technological acuity that no scanner yet possesses. All that said, there is no doubt that her research has brought some much-needed clarity to the treatment of depression. As was the hope for neuroanatomists of the nineteenth century, Mayberg has shown that there are biological markers of depression in the brain. They're not faulty neurons or lesions that can be seen under the microscope. Rather, depression is a disease of brain circuitry, a dysfunction in how interconnected regions communicate with one another and how this dysfunction becomes the new norm.

An accurate diagnosis doesn't always require a brain scan, a luxury that few health services or insurance companies would provide for everyone showing signs of depression. In clinical practice, Mayberg sees her work as being implemented after a few basic decisions. If someone is placed into CBT and can be guided into remission, then perhaps that was the correct decision. If they are still depressed, doctors or psychiatrists shouldn't try and push more psychotherapy into their heads in the hope that more equals better. It might be an expensive waste of time. Instead, they should have the option to switch to an antidepressant. The same can be said for a patient who first seeks treatment with medication but after three months doesn't recover. Instead of the usual switch to another antidepressant, the patient might first try a course of CBT. Although depression isn't always a binary set of outcomes, this strategy will increase the number of patients who, within six months of first meeting with a medical professional, will receive the treatment most suited to their illness, their brain type. For those who still show no sign of significant improvement after both classes of treatment, additional medications or other biological treatments might be considered, such as ECT, the newer approaches of transcranial magnetic stimulation, and ketamine infusions, or an experimental treatment such as deep brain stimulation. These are aggressive treatments that shouldn't be handed out indiscriminately. But for people whose symptoms are both severe and relentless, they can be worth the risk.

• • •

In 2019, in a paper from Mayberg's DBS team at Emory University, Andrea Crowell and her colleagues published the long-term outcomes of deep brain stimulation for treatment-resistant depression. Like the first patient to receive DBS, twenty-eight people with intractable depression had an electrode implanted in their brain, just next to area 25, that was controlled by a pacemaker in the chest. In total, they documented more than eight years of continuous stimulation. While there were untoward (but treatable) incidents—suicide attempts, infections, one non-lethal brain hemorrhage, and a seizure after surgery—there was also a staggering level of sustained improvement and recovery in such a hard-to-treat population. In the first year, there was a 50 percent improvement in their stubborn symptoms. By the second year, 30 percent of the patients were in full remission from their previously intractable depression. Although there was no control group and therefore no way of determining whether some recoveries were spontaneous, there was clear evidence that the ongoing stimulation in area 25 had a sustained antidepressant effect. When the electrode was switched off, the depression returned. When one patient relapsed, they found that the pacemaker that controlled the electrode had broken. Even when the non-rechargeable batteries of these devices started to wane, the depressive symptoms started to return in tandem. Battery replacement or device repair restored the previous antidepressant effects. Importantly, there have been no relapses while sufficient stimulation has been provided, a remarkable milestone given that relapse is a common occurrence in depressed patients with multiple levels of treatment resistance.

Mayberg's first patient ("DBS 01"), the woman who described a feeling of spring in 2003, died six years after her first surgical operation. She was healthy for much of that time, walking around with the electrode in her brain and the pacemaker in her chest, but had other issues in her life beyond the depression. Her death is a reminder that deep brain stimulation isn't a cure. It is an experimental treatment guided by decades of research that can help break a relentless cycle of mental anguish. Oftentimes, it isn't enough. Full recovery requires a program of social rehabilitation as well as surgical precision. To paraphrase one of Mayberg's patients: deep brain stimulation isn't perfect, it just makes life possible. When trying to help critically ill people, Mayberg has to accept that the road isn't going to be a straight line to success. "It's a journey where the final destination is a more complete understanding of depression," she says. "Only then can we truly work towards a cure."

Mind on Fire

Brian Leonard, a retired pharmacologist with a neat white beard, spends much of his time reading and writing in an old barn converted into an office. With thick stone walls, a slate roof, and double-glazed windows, it is a cozy refuge from the bitingly cold winds and pouring rains that sweep in—often unannounced—from the Atlantic Ocean, just a short walk back to the bungalow he shares with his wife and their elderly dogs. In 1999, Leonard retired from his position of professor of pharmacology at the National University of Ireland, Galway, a small campus of buildings that surround a Gothic quadrangle that first opened its doors to students in 1849 and once contained wings for arts, medicine, and law. In his eighties and frustrated with the sluggishness that comes with aging, Leonard wishes that he could return to the laboratory and continue the research he had started forty years ago—now suddenly in vogue. In the 1980s, Leonard was a prominent voice within a fringe theory in psychiatry: depression was a product of the immune system. To him, it was akin to rheumatoid arthritis, a disease stemming from low-grade, chronic inflammation.

Our immune system evolved to protect us from harm, but it can also lead to some of the most painful and destructive diseases. They are embodiments of a paradox, afflictions that haunt millions of people every year. In 2004, *Time* magazine called inflammation the "Silent Killer" and the cause of many of the twenty-first century's most problematic diseases, such as diabetes, cardiovascular disease, inflammatory bowel disease, and cancer. In recent years, research into the immune system has shifted from physical diseases of the body to the ailments of the mind. Depression provides some of the most compelling evidence that mental illness is intimately tied to our immune system. Patients who don't respond to conventional antidepressants commonly show high levels of inflammation in their blood, suggesting that so-called treatment resistance might be tackled with anti-inflammatories. In 2013, a randomized controlled trial found that the drug infliximab, an anti-inflammatory used to treat rheumatoid arthritis, had a similar level of efficacy as antidepressants in its reduction of depressive symptoms. Most persuasive, perhaps, is a longitudinal study that tracked the health of over three thousand individuals, from children to adults, and found that elevated levels of inflammation at nine years old predicted the onset of depressive symptoms twelve years later.

In 2018, Edward Bullmore, a professor of psychiatry at the University of Cambridge, wrote *The Inflamed Mind: A Radical New Approach to Depression*, a popular science book that pushed the link between inflammation and mental illness into the mainstream. The topic has become so normalized, so commonplace, that it is easy to forget where it came from and how it was initially seen as a fringe science. "Now everybody believes it, and always did believe it," says Michael Berk, a professor at Deakin University in Geelong, southeast Australia. "But history wasn't always so."

From the pharmacology department in western Ireland, a single-story building on the edge of campus that he still lovingly refers to as "the hut," Brian Leonard supervised forty PhD students and mentored many more scientists who are now leading this field (Berk included). Working on the capricious Irish coastline, it often felt like he was isolated and exposed. Slowly, however, the four scientific institutions of Ireland—Queen's University Belfast, Trinity College Dublin, University College Dublin, and the National University of Ireland, Galway—started to meet twice a year, sharing their ideas and allowing their students to present their latest work. Leonard's students seemed to speak in a different language from the rest, however. While other labs were working on brain chemistry and neurons, the scientists in

Leonard's "hut" were increasingly focused on markers of the immune system in the blood. They spoke of macrophages and lymphocytes, the Pac-Man-esque cells of the immune system that gobble up unwanted intruders such as bacteria. They discussed cytokines, a mysterious family of molecules that seemed to control a person's immune response like temperature gauges control a central heating system. They used terms like *pro-inflammatory* and *anti-inflammatory*. Secretly, they thought that serotonin might be a secondary consequence of a more central cause of depression, a subplot to an epic story that reached out of the synapse and encompassed the entire body.

Leonard and his students weren't the only ones to hold such radical views. "The lack of scientific insight into mental disorders can be conveniently attributed to the enormous complexity and inaccessibility of the human brain," one author working in San Jose, California, wrote in 1991. "On the other hand, over 100 years of unsuccessful research on the [potential causes] of mental illness could be telling us that there is something wrong with our research."

Until the 1980s, the study of the immune system, known as "immunology," had largely taken place in test tubes and on sterile laboratory tabletops. Macrophages, lymphocytes, and neutrophils—types of white blood cells—were understood in closed and tightly controlled settings. But a human is nothing like a test tube. Our bodies are huge bioreactors of molecules, cells, and organ systems that are in constant communication and flux. "It's time to put the immune system back in the body where it belongs," said Karen Bullock, an immunologist from the State University of New York at Stony Brook, in 1985. The role of the immune system doesn't stop at infectious microbes, but reaches into almost every aspect of our lives, from our guts to our brains. "We're at a stage where it is difficult to say definitively what is happening," said Nicholas Hall from George Washington University in Washington, D.C., referring to the broad topic of the immune system's connection to the brain. "We're putting together two kinds of black boxes and trying to make sense of what happens."

A few glimmers of light came from studying the blood of depressed patients. In the late 1980s, several labs based in the U.S. and the Netherlands found that the patients' immune systems showed the molecular markers of inflammation. Those molecules that cool the immune response, known as anti-inflammatories, were lacking. It was as if the temperature gauge of the immune system were kept slightly above normal. In his ramshackle pharmacology laboratory in Galway, one of Leonard's most productive and tireless students, Cai Song, had found similar patterns. After defrosting blood samples of depressed patients from a local psychiatric unit in Galway, she found that pro-inflammatory signals were consistently increased. Anti-inflammatory signals were decreased. A core group of inflammatory markers—C-reactive protein (CRP), interleukin 6 (IL-6), tumor necrosis factor (TNF-α)—proved to be significant; using different patients, blood samples, and equipment, the same findings emerged from a number of laboratories in the 1990s. In particular, a research group led by Michael Maes in Maastricht, the Netherlands, was pivotal in illuminating the consistent patterns of inflammation in depressed patients.

There was an imbalance. Rather than suggesting that depressed patients were deficient in one or two brain chemicals, a whole suite of immune signals pointed in the same direction: low-grade, chronic inflammation.

Inflammation is a bodily reaction often felt as heat and seen as swelling. It isn't always a bad thing. In fact, it is vital to our survival. When a threat to the body is detected—whether it is a bacteria, a virus, or an injury—the immune system responds by sending a swarm of specific white blood cells to the necessary site. These cells then produce a cocktail of proteins that sterilize the area. The bacteria and viruses are, hopefully, vanquished. Dead or damaged cells are digested and cleared away. The body can then start the process of recovery and repair. This is acute inflammation. It is as explosive as it is necessary. "That's not what we're talking about," Leonard says. "We're talking about chronic, low-grade inflammation. . . . It is absolutely different." In such cases, there is no immediate threat to be killed and no injury to sterilize. With nothing else to do, the white blood cells start to attack the very cells they were supposed to protect—the body's, yours. Diabetes, rheumatoid arthritis, and Crohn's disease are all rooted in chronic low-grade inflammation. Even a clogged artery, once believed to be caused by a buildup of sticky cholesterol, is more commonly the result of a localized swelling caused by a slight elevation of inflammation within a blood vessel.

As with Nathan Kline's hopes for an emotional pendulum in depression, this led to an obvious possibility in terms of treatment. Could anti-inflammatories be used to reduce the inflammation associated with depression and, in doing so, treat the depressive symptoms, as well? Although this remained untested, there was evidence from the contrary: pro-inflammatory drugs seemed to *induce* a state of depression.

Alpha interferon is a drug used to fight particularly stubborn forms of cancer. It works by increasing the activity of the immune system, turning its temperature up in order to help the body burn out those cells that are dividing out of control. In the 1980s, however, two research groups recorded that symptoms of anxiety and depression often emerged as side effects of this drug. "Patients ignored their eating and other regular daily activities. [They] lost their appetite, and two refused to eat and drink anything and therefore needed intravenous fluid infusions," researchers from the University of Helsinki wrote in 1988. "Patients had slowed thinking, and they ceased to speak." Their memory was affected, and they felt agitated and struggled to sleep. Although there wasn't a one-to-one relationship between pro-inflammation and depressive symptoms (one patient developed a phobia of pigeons), a large percentage of patients exhibited seven out of the nine symptoms listed in the *DSM-III* criteria for a major depressive episode. "A patient only needs to exhibit five of the symptoms to be diagnosed with major depression," one author noted.

When Andrew Miller saw his first patient on alpha interferon in 1997, he was stunned. He had been called to the oncology department of Emory University in Atlanta, the institution where he worked as a psychiatrist, and found a woman sitting down, listless, who ticked all the boxes for depression. Although she had come to terms with her cancer long ago, she suddenly felt like she might not be able to continue, that she had lost the love and affection she once felt toward her family. Every day was a battle. She couldn't sleep. Worst of all, she had no idea why she felt this way. "This was a patient that was really describing a syndrome that, for me, was a carbon copy of what I see in my clinic in patients presenting with depression," Miller says. The only difference was that this depression didn't follow some change in life like losing a job, divorce, or losing a family member. After a lengthy discussion, Miller was certain that it wasn't related to the stress of cancer and its window into mortality. "There was no psychological context," Miller says. "It was occurring in a vacuum. We saw this happen over and over again. It's a very eerie thing for a psychiatrist to see a patient with a syndrome like depression without a psychological context." This was a purely biological disorder. Once the alpha interferon drug was discontinued, the depressive symptoms disappeared within two weeks.

What did this mean for our understanding of depression? Was this a previously unknown form of the illness? Was there a specific "inflammation syndrome," associated more with fatigue, sleeplessness, and low mood? Some proposed that there was indeed an overlap with the so-called sickness behavior associated with other infections such as the common cold: social withdrawal, fatigue, a lack of motivation and drive. That was one possibility. The other was much grander in scope. Was inflammation the underpinning of depression, period? Rather than a brain disease, hatred turned inward, or an imbalance of serotonin, was depression actually an immune disorder?

After Cai Song submitted her chunky thesis, finished her PhD, and left the Gothic scenes of Galway, Brian Leonard felt as if there was a hole in his pharmacology department. Song's tireless work ethic—Leonard sometimes had to kick her out of the laboratory in the evenings—made her departure feel even more obvious. In 1995, John Cryan, a cherubic young Irishman had just finished his degree in biochemistry at the National University of Ireland and, like Song, was more than happy to test novel approaches to treating depression. His interest wasn't as controversial as the immune system, however. It involved serotonin, the latest SSRI drugs, and research on the so-called lag phase that had still failed to find a solution or an explanation. The few weeks before antidepressants fully take effect is more than just an academic puzzle. It is a dangerous window for depressed patients to take their own lives. "The annual rate of suicide in depressed patients is 3 to 4 times that of other psychiatric diagnostic groups and 20 to 30 times higher than that of the general population," Leonard wrote in 1994. "This delay in onset [of antidepressants] exposes the patient to an increased period of suicide risk, may extend hospitalization and certainly does little to alleviate the emotional distress of the condition."

Working in the Experimental Medicine Building that adjoined Leonard's pharmacology hut, Cryan hoped to fill in this gap. Was there some way to accelerate the antidepressants already available? Just as a catalyst speeds up a chemical reaction, could another substance boost the effects of these drugs? He was part of a global research effort in the late 1990s. As the number of new SSRIs being produced started to plateau, there was a desperate need for something new, even if it was just a chemical tweak of their effects, reducing their lag time from a few weeks to a few days. There were some persuasive findings. Beta-blockers and anti-cortisol (a stress hormone) drugs both showed early promise. But this line of research led to no breakthroughs. When taken outside of the lab and into human trials, they failed. The lag period remained, a gaping hole in the serotonin theory of depression and a dangerous reality for patients around the world.

At the suggestion of Leonard, Cryan finished his PhD and accepted a research position at a pharmaceutical firm. "I sold my soul for a while," he says. Working for Novartis in Basel, Switzerland, he watched as the psychiatry units were drained of money and closed down. By this time, in the early noughties, there were plenty of antidepressants, all with similar effectiveness in clinical trials, and very little interest in producing any more. It costs around $985 million to take a drug from tests on mice into clinical trials on humans. Most of them fail. For the first time since their invention, drugs like Prozac were considered financially unviable. The only success came from Connie Sánchez, a pharmacologist who developed escitalopram (Lexapro) while working at the pharmaceutical firm Lundbeck in Copenhagen.

Feeling like his research in the industry was a dead end, Cryan returned to Ireland in 2005 and accepted a position at University College Cork, a university at the heart of a multicultural city in the southeast of the island. It was here that his career took an unexpected turn. From studying drugs and how they might work in the brain, he began focusing his attention on the other end of the body: the gut. Cryan became interested in how the millions of bacteria inside the coiling mass of intestines that connect the stomach to the colon might influence mood. Still a field in its infancy, scientists saw the microbiome—that ecosystem of bacteria and other microbes that live on and inside our bodies—as a regulator of health and disease. An institution that was strong in microbiology, University College Cork had set up a microbiome unit that was attracting students and scientists from around the world. They could raise their own germ-free mice in order to study how the microbiome, and even specific bacteria, influence the growth, development, and behavior of mammals. Mainly studying rodents in the laboratory, Cryan paired up with Ted Dinan, a psychiatrist and pharmacologist who met face-to-face with patients, prescribed drugs (including MAO inhibitors), and could link the research from the laboratory in Cork to the clinic. It was an unusual situation: two pharmacologists with no experience in microbiology studying the little-understood microbial ecosystem within the human body. Both fluent in neuroscience and drug therapy, Cryan and Dinan would soon be talking about probiotics, bacteria, and fecal transplants.

Leonard, still publishing his own theories on inflammation and depression, thought they were both crazy. Only later did he realize

that their work might be the only safe and sustainable route to reducing low-grade inflammation in the body. A healthy microbiome is a powerful way to reduce inflammation.

After Leonard retired from the National University of Ireland, Galway, in 1999, he left Ireland to work with a leading figure in the inflammation theory of depression, Michael Maes, who had his own laboratory in Maastricht. Soon after his arrival, Leonard met Aye-Mu Myint, a medical doctor and surgeon from Myanmar who, in her mid-forties, moved to Maastricht to start a PhD in neuroscience. When they met, Leonard and Myint found that they had been thinking along the same lines for years. Specifically, they had been obsessing over the same molecular pathway inside the human body, and how it might link depression to the immune system.

It starts with tryptophan, an essential amino acid found in most protein-rich foods, from chocolate to poultry, sesame seeds to tofu. Tryptophan has long been known to be an essential amino acid because it is the precursor to serotonin, the monoamine that, even if it isn't the cause of depression, is pivotal in the functioning human brain and, more significant, the gut. But this is only a tiny glimpse into tryptophan's potential life cycle. In recent decades, another chemical pathway has been revealed and is taking the limelight at international conferences and in the psychiatric literature. It has Leonard hooked.

It is called the kynurenine (pronounced "ky-nur-a-neen") pathway, a family of chemicals that are produced from tryptophan and have wide-ranging effects on the body. Most commonly, tryptophan is broken down via the kynurenine pathway into a form of fuel used for energy production in our body's cells, literally allowing our muscles to move and our brains to flash with thought. This process takes place through enzymes in the liver, in the blood, and in muscle tissue. It accounts for 95 percent of tryptophan usage in our body. But there is a Jekyll and Hyde moment in this pathway, a switch that, crucially for this story, takes place when inflammation is elevated in the body in response to threat, disease, or pathogens. When this happens, fuel

production is stalled and kynurenine is broken down into quinolinic acid, a molecule that flows through the blood and reaches the neurons of the brain. There, it lives up to its acidic name. It breaks down neurons in the brain like hydrochloric acid in the gut kills bacteria. It is, in the parlance of immunology, neurotoxic.

In 2003, Myint and Y. K. Kim from Korea University's College of Medicine published a "neurodegeneration hypothesis of depression." With the kynurenine pathway at its core, they argued that depression was rooted in the death of neurons in the brain; a disease akin—and potentially a precursor—to dementia. "It's very complicated and I wouldn't say that we have the answers," Leonard admits as he looms over a figure of the kynurenine pathway in one of his textbooks. "But it's trying to look in a . . . well, what *I* consider to be a more creative way."

First studied in detail by the Russian scientist Slava Lapin in the 1970s, kynurenine is currently a hot topic in the field of neuroimmunology. "Everyone's talking about the kynurenine pathway and really very seldom in the U.S. or Europe [do scientists] give him credit," says Leonard. "He was the one who pinpointed this." Before the Berlin Wall was torn down and the Soviet Union disintegrated, Leonard managed to arrange a travel visa for Lapin, allowing him to present his work at a British Association for Psychopharmacology conference in Ireland. Leonard enjoyed his company like an old friend. "He was one of the old-style nineteenth-century Russian intellectuals," he says. He spoke French fluently, played piano to a grade suited to grand concert halls, and rubbed shoulders with Russian intelligentsia such as Aleksandr Solzhenitsyn, the novelist who won a Nobel Prize for his writings on the gulags of the Soviet Union. But what Leonard most admired about Lapin was his approach to science: a holistic view that makes interconnections between once disparate theories.

Today, Leonard is still mobile, both on foot and in his small car, and has the same booming voice that once "enlivened conferences too numerous to count." But he is slower, more forgetful, and becoming increasingly disappointed with where science is heading. People become so specialized, he says, so parochial, that they forget that the

brain is a part of the body. "The body doesn't work in bits," he says. He reminisces about the days when he studied pharmacology at the University of Birmingham, the chemistry of the body, and saw it as a huge puzzle that couldn't be understood without all the different parts. Whether the answer lies with the immune system or the microbiome, kynurenine or TNF, he is aghast at how psychiatry became so fixated on serotonin.

"For Life"

Although the term *microbiome* was made famous by Nobel laureate Joshua Lederberg in 2001, its history reaches back to the late nineteenth century, a time when Sigmund Freud and Emil Kraepelin were paving the future of psychiatry. While working in Paris, the microbiologist and Nobel laureate Élie Metchnikoff put forward the idea that the microbes in our gut needn't be avoided (as the "germ theory" of disease dictated). They could be beneficial and promoted by certain foodstuffs. In particular, he wrote, fermented foods that contain lactic acid–producing bacteria, such as *Lactobacillus bulgaricus*, could have far-reaching benefits for the mind and body. People who regularly drank fermented milk in Bulgaria lived longer and healthier lives, Metchnikoff noticed. He wanted the whole world to know their secret. "[T]here is hope that we shall in time be able to transform the entire intestinal flora from a harmful to an innocuous one," Metchnikoff wrote in an article titled "Why Not Live Forever?" for *Cosmopolitan* in 1912. "The beneficent effect of this transformation must be enormous."

Metchnikoff was an internationally recognized scientist, a cantankerous but fatherly figure of immunology. He discovered the most fundamental part of the immune response: macrophages, those cells that flow through our blood and gobble up microbial invaders. Growing up in rural Russia, Metchnikoff had a precocious talent for science and medicine. He finished a four-year degree in just two. When he became a lecturer at the age of twenty-two, many of his students were older than he was. In his later years, as his hair grew shaggier and his beard was so unkempt that it was once likened to "fields of wheat after a thunderstorm," Metchnikoff's brilliance was used to boost the sales of unproven probiotics. "Metchnikoff's great discoveries are now procurable in tablet form," a full-page newspaper advert from the Berlin Labs in New York stated. Intesti-Fermin (the brand name of probiotic pills), they claimed, "promotes physical and mental health and provides a truly scientific aid to high efficiency in everyday life." What they didn't mention was that there was no evidence that these bacteria actually enter and colonize our guts. Passing through the strong digestive acid of our stomachs is just the first step. They have to find conditions within our intestines that are suitable to their needs. Was the pH okay? Were there nutrients tailored to the microbe's needs? And, importantly, were the other species already living there going to outcompete any newcomers, like reef fish quarreling over a piece of coral? For bacteria in our intestines, the zones of habitability are clearly defined and each is a battleground of millions.

As it turned out, *L. bulgaricus* didn't survive and multiply in our digestive system. But one of the benefits of bacteria is that there are many more options available. In our guts, for instance, there are an estimated one thousand different species. In the 1920s, manufacturers replaced *L. bulgaricus* with *L. acidophilus*, a closely related species of bacteria that naturally inhabited our guts and was heralded as the latest probiotic fad for a healthy mind. "The results, as thousands of physicians and users testify, are nothing short of amazing," an advert in the *New York Times* stated. "Not only a banishing of mental and physical depression, but a flooding of new vitality throughout the system."

Such sensational adverts reach into the twenty-first century. "There's 'Metchnikoff's life,' a fermented milk drink made by South Korea's Hankuk Yakult that proudly displays Metchnikoff's patriarchal portrait on the cup," writes Luba Vikhanski, a science journalist and author, in her book *Immunity: How Elie Metchnikoff Changed the Course of Modern Medicine*. "A recent Russian-language commercial for Danone yogurt featured a radiant Metchnikoff, played by a bearded actor. And since 2007 the Brussels-based International Dairy Federation, or IDF, has been awarding an IDF Elie Metchnikoff Prize to promote research 'in the fields of microbiology, biotechnology, nutrition, and health with regard to fermented milks.'"

In 2013, over a century after Metchnikoff's article in *Cosmopolitan*, Kirsten Tillisch and her colleagues at the University of California, Los Angeles, provided the first evidence that bacteria can influence our brains. *Lactobacillus rhamnosus* is a species of lactic acid–producing bacteria commonly found in fermented foods and drink. After eating two small pots of fermented yogurt a day for four weeks, the brains of a group of healthy women showed a decreased activity in those regions associated with interoception (the inner sense of the self) and emotional reactivity. "These changes were not observed in a non-fermented milk product of identical taste," Tillisch and her colleagues wrote in the journal *Gastroenterology*. "[T]he findings appear to be related to the ingested bacteria strains and their effect on the host." By dampening the emotional response to the environment, could these probiotics be used as a treatment in mental disorders?

The term *probiotic* means "for life." They are health-promoting microbes. They are traditionally defined as "living micro-organisms that contribute to intestinal microbial balance and have the potential to improve the health of their human host." The same year that Tillisch published her landmark paper, Ted Dinan coined the term *psychobiotics*, a necessary break from the past of fad probiotics—such as Intesti-Fermin from Élie Metchnikoff's days—and toward a future of bacteria specifically used to influence our mental health. He defines a psychobiotic as "a live organism that, when ingested in adequate amounts, produces a health benefit in patients suffering from psychiatric illness."

To study a new treatment for psychiatric illness, to compare it to old pharmacological options, Dinan turned to Cryan for help. He had honed the so-called forced swim test in mice into a fine art. First developed in the late 1970s, the forced swim test involved giving mice an experimental drug, placing them in a jug of tepid water, and timing how long they swam. Longer times were seen as a good sign of increased drive and motivation in the face of a stressful environment. Although there is no such thing as a depressed mouse (guilt and remorse are particularly difficult to replicate in a rodent), this was often the first test that modern antidepressants had to pass. Well before they were put into prescription, many SSRIs were first given to mice (or rats) and found to increase the amount of time they would swim before giving up.

One of the first bacterial strains to undergo the forced swim test—inside the body of a rat—was *Bifidobacterium infantis*. In babies, *B. infantis* composes 90 percent of the microbes in the large intestine. As adults, that figure settles down to 3 to 5 percent, still a sizable chunk of the microbial ecosystem. When Lieve Desbonnet, a PhD student supervised by Dinan, gave it to rats as a probiotic solution she found that there was no change in how long they swam. One of the first probiotics tested for its effects on the mind had just sunk along with the rodent that it was given to.

Lactobacillus rhamnosus, the species of bacteria that Kirsten Tillisch studied in yogurt, passed the forced swim test. Mice swam for significantly longer after being fed this probiotic for two weeks, and it also showed anti-inflammatory properties. "It looked like it was antidepressant, anxiolytic, and every flippin' thing under the sun," says Dinan. These positive findings in animal models gave Dinan and Cryan the confidence, and the funding, to start human trials. Rather than forcing the patients to swim, they could test them on standardized health questionnaires for depression and anxiety. They would pore through their blood samples for markers of inflammation, their brain chemicals for changes in serotonin and glutamate (the most common neurotransmitter in the human brain and the molecular target of ketamine). It was an important step away from rodent models and into the first stages of using bacteria as potential anti-inflammatories and antidepressants.

"It was the most unbelievable result," Dinan says. "It did nothing. I mean, you've never seen such a tight relationship between probiotic and placebo in your life. We looked at immunology, endocrinology, behavior, loads of things. It did nothing in humans." As Cryan laconically summed up the results to a BBC reporter, "[Y]ou couldn't fit more negative data into that paper." Published in late 2016, they called their study "Lost in Translation." After years of study and millions of pounds of research funding, this was a stark reminder that what works in mice doesn't guarantee clinical success in humans.

The process of peer-reviewed science is founded on, and fueled by, positive results. The big journals such as *Science* and *Nature* want to publish new exciting discoveries. Scientists want a track record in big journals. Negative results are pushed to one side and left to collect dust on a bookshelf or slowly erode inside a computer's hard drive. "The reality is that we've had loads of negative studies over the years," Dinan says. But he thought that this particular non-finding was too important to ignore. "This was the most positive [project] we've had in rodents, and it's the least positive we've ever done in humans."

Large studies into the microbiome and mental health are few and far between. As Cryan joked at Mind, Mood & Microbes, an international conference held in Amsterdam in January 2019, there are more reviews of the scientific literature than there are original studies. On the second day of the conference, Kurosh Djafarian, a mild-mannered man with jet-black hair and neat stubble, presented the first randomized controlled trial into probiotics and depression. A professor of clinical nutrition at Tehran University of Medical Sciences in Iran, Djafarian and his colleagues found that *Lactobacillus helveticus* and *Bifidobacterium longum* significantly reduced depressive symptoms over eight weeks of treatment. With twenty-eight people in the probiotic group and twenty-seven controls (people were given a sachet of powder that was the same flavor and color as the probiotic mix), it wasn't a huge trial, but it was still one of the most persuasive at the time. In addition to reviews that show probiotics to have significant benefits for people with mild to moderate levels of depression, this work was a highlight of the conference for many people, a sign that "good bacteria" could be used in the treatment of depression.

Listening to Djafarian in the audience was Scot Bay, a doctor from the United States wearing a brown woolen suit and tie. He had been excited to hear this talk since he first arrived in Amsterdam a few days prior. In his suburban private practice outside of Atlanta, he often meets with depressed people who haven't responded to multiple rounds of antidepressants and therapy. Interestingly, the majority also had comorbid digestive complaints such as irritable bowel syndrome. In 2015, he had listened to a podcast and heard Laura Steenbergen, an assistant professor at Leiden University, discuss her research into probiotics and their ability to reduce the body's stress response. She had given her patients Ecologic Barrier, a probiotic mixture that strengthens the internal wall of the intestines and helps to dampen any inflammation or infection. Bay ordered a box for eight of his treatment-resistant patients. "They got better," he says, excitement and surprise making the words pop from his mouth. "They all got better."

At the conference, Cryan reminded his audience that the field of the microbiome can easily slide from scientific research into pseudoscience, as it did in Metchnikoff's days in the twentieth century. Acknowledging the promise of a positive randomized controlled trial, he added a necessary coolant to proceedings. "Are we there yet?" he asked, referring to the microbiome's maturity as a clinical science. "No, we're only a little up the road."

One of the most persuasive studies in this field was published shortly after Mind, Mood & Microbes had finished and the scientists had returned to their far-flung institutions. Studying kindly donated stool samples from over a thousand people living in Belgium, Mireia Valles-Colomer, Sara Vieira-Silva, and their colleagues from KU Leuven, the Catholic university in Belgium, found that people with depression had their own microbial "fingerprint." Although it didn't apply to all depressed people in their study, over a quarter of their volunteers were deficient in diversity, as if their gut ecosystem had suffered from a serious catastrophe that wiped out many of the species that usually lived there. In particular, the bacterial groups *Dialister* and *Coprococcus* were depleted compared to people without depression. This same pattern, interestingly, had been observed in Crohn's disease, a disorder of low-grade chronic inflammation in the intestines.

Although this study was based on general practitioners' reports of depression (rather than psychiatrist-derived diagnoses), Valles-Colomer and her colleagues' findings were bolstered by large sample sizes. In total, they compared 151 patients against 933 controls (people without depression), a huge number for a field characterized by studies of a dozen or so patients. Did this mean that the lack of diversity caused the depression? Or did the depression result in the deficient microbiome as a consequence? "Most people think that it's kind of a loop, that you tend to have a trigger of inflammation, whether it's having a bad diet for a long time or taking some medications or having very bad sleeping patterns can all contribute to low-grade inflammation," says Vieira-Silva. "This will then benefit 'bad' bacteria in your gut, and then bad bacteria benefits by triggering even more inflammation."

The obvious corollary to this work is finding out whether this loop can be broken. How can low-grade inflammation, and the microbiome "fingerprint" that it thrives upon, be persuaded into a more balanced state? "Our personal experience is that we tend to see more effects from prebiotics than probiotics," says Valles-Colomer, adding that fiber from fruit and vegetables is the most essential prebiotic anyone can add to their diet. "But if you have a very disturbed microbiota, then you might have lost key components of the microbiota, so then I would do a combined pre- and probiotic." Together, that would introduce bacterial strains that might have been lost while nurturing the low numbers of beneficial bacteria that are already present.

One of the most effective means of increasing the diversity of our microbial residents is through what we eat. What sustains us sustains them, too, like feeding trillions of microscopic nest mates. Changing a few meals won't budge the already stable populations. It requires long-term dietary changes. But the benefits can be dramatic.

In 2017, Felice Jacka, a professor at the Food and Mood Centre at Deakin University in southeast Australia, and her colleagues found that, over a three-month period, depressed patients who ate more vegetables, fruits, whole grains, legumes, fish, lean red meats, olive oil, and nuts, while cutting out "extras" such as sweets, refined cereals, fried food, fast food, processed meats, and sugary drinks saw a significant drop in their depressive symptoms compared to patients who had regular meetings with their counselors but had no change in diet. In short, a Mediterranean-style diet can act as an antidepressant. Although this undoubtedly altered the patients' microbiomes, it is still unclear whether this was the reason for the effect. Though inflammation may have been reduced, it was not assessed in the study. But, as of 2019, Jacka's trial has been replicated by independent research groups using different methods and patient groups, but, importantly, finding the same result. Whatever the mechanism, a varied diet, rich in fruit, vegetables, and a little bit of meat or fish, is a recipe for both physical and mental health.

After reading such studies, I changed my daily routine in 2018, around the time I was moving from citalopram to sertraline. After years of vegetarianism, I started eating fish. River cichlid in Zimbabwe. Cod on the blustery coast of Suffolk, east England. The occasional salty morsel of anchovies in a Caesar salad. Not only was each bite a small assault on my environmental morals, I've never really had an appetite for seafood. I didn't eat much of it as a child, even less as an adult. But, like reaching out for psychotherapy or antidepressants, there's no greater motivator than finding something that works. I realized that my changed diet was built on thousands of years of writings on depression, from Galen to Felice Jacka. What we eat can influence our mood.

Did this balanced diet reduce inflammation levels in my blood? Was it protecting my brain from some of the neurotoxic by-products of kynurenine? Was my microbiota changing into a less depressed footprint? I would never know. But eating a healthy diet is an approach that I can control and implement into my life almost immediately. I stopped eating family-size bags of chips myself. As alcohol is a known pro-inflammatory, I stopped drinking for two years and only rarely drink today. And, after realizing that omega-3 oils are actually produced by oceanic algae (which fish, in turn, eat), I returned to my vegetarian diet with vegan-friendly fish oils. Led by science, I was eating a Mediterranean-style diet without the fish.

Not everyone has access to cheap fresh groceries and omega-3 capsules made from algae. For me, a change in diet could happen almost immediately, a new treatment for depression after a walk to my local greengrocers. But for a growing number of people on shoestring budgets or in poverty, a change in diet can be an insurmountable task without external support. This isn't limited to low-income countries. It is a global problem. According to the Australian Health Survey of 2014 to 2015, for example, only 5.6 percent of the population have a healthy proportion of fruits and vegetables in their diet. In the UK, 19 percent of children live in households that frequently can't afford to buy the ingredients needed for a healthy diet. Cheap, processed, fatty foods that are full of salt and sugar are a common substitute, providing the sustenance that can be consumed in large quantities without satiation. High in fat and low in protein and fiber, this quick-fix Western diet is a growing health concern. For both the rich and poor, it has led to a frightening increase in obesity. Around the world, an estimated 13 percent of adults and 18 percent of children are obese. In the U.S., levels of adult obesity have risen from 10 percent in the 1960s to over 40 percent in 2018.

Defining obesity is an inexact science. It is based on a rough estimate of a person's volume (height squared) divided by their weight. And yet, this body mass index (BMI) is a reliable indicator of a person's health. An overweight person ranges from a BMI of 25 to 29. Obesity begins at a BMI of 30 and is known to increase the risk of some of the most prevalent diseases of the twenty-first century. Heart disease, cancer, hypertension, type 2 diabetes: every increase in kilogram per square meter pushes the body toward a state of disease. Someone with a BMI of 30, for example, is twenty-eight times more likely to develop diabetes compared to someone with a BMI of under

21. If the BMI increases to 35 or more—a state known as "morbidly obese"—the risk increases to ninety-three times.

These figures were listed in an oft-cited review in the journal *Nature*. Published in 2000, "Obesity as a Medical Problem" welcomed the new century with a warning. "Obesity should no longer be regarded as a cosmetic problem affecting certain individuals, but an epidemic that threatens global well being," wrote Peter Kopelman, a doctor then working at the Royal London School of Medicine. Although genetic susceptibility to obesity was a target for future studies (over fifty genes are thought to underlie this condition), the smoking gun behind this epidemic were rapid changes in diet and sedentary lifestyles. Wherever this Western lifestyle was introduced, obesity soon followed. Nowhere was this more evident than in the Pacific Islands, the twelve thousand or so pockets of paradise that freckle the waters of the western Pacific. After World War II, a shift from subsistence farming to urban living and the import of cheap, processed foods led to some of the highest rates of obesity in the world. "Traditional foods such as fish, taro, yams, and indigenous fruits and vegetables have been supplanted by imported rice, sugar, canned foods, soft drinks, and calorie-dense snack foods," wrote epidemiologists Nicola Hawley and Stephen McGarvey in 2015. "These imported foods are often of poor quality (fatty cuts of meat like turkey tails and lamb flaps which are often considered waste in their countries of origin, instant noodles, highly processed, high-sugar snack foods) but have come to represent prestige and cultural capital." With a genetic predisposition to weight gain shared by many Pacific Islanders, these sandy microcosms have come to represent a warning to the rest of the world. In American Samoa, over 70 percent of the population are obese. In Tonga, Nauru, and Samoa, these figures are still above 50 percent. Diabetes, stroke, and cardiovascular disease are on the rise, while life expectancy shrinks. People who would once live to their late seventies are now dying in their sixties.

In Peter Kopelman's review in 2000, physical diseases took center stage. There was no mention of mental health. But, in the two decades since it was published, depression has become intimate with obesity. It is now known as just another "obesity-related disease."

"With few exceptions," Michael Berk and his colleagues wrote in 2013, "studies have consistently shown a relationship between obesity and depression regardless of methodological variability." Being obese increases the risk of depression by over 50 percent, a rate that is mirrored by the risk of obesity for those who are depressed. It's another vicious cycle; like poverty, obesity is both a cause and a consequence of depression.

While the classical definitions of depression—melancholia, for example—are associated with a lack of appetite and weight loss, obesity is associated with a more "atypical" symptom profile. Overeating, anxiety, fatigue, and oversleeping: it is a resurgence of a diagnosis that first gained popularity in the 1950s. Once prescribed MAO inhibitors, atypical depression remains a particularly stubborn ailment to treat, especially when a patient is obese. Although there are myriad explanations for this (both social and physiological), one of the central areas of research in recent years is inflammation. Fatty tissue—especially the kind that surrounds the abdomen—releases inflammatory molecules into the blood. Not only has this been associated with depression, but such low-grade chronic inflammation is known to make standard antidepressants less effective.

Full disclosure: I'm slim, some might say skinny. I'm 178 centimeters tall (just over five foot ten) and weigh sixty-five kilograms (147 pounds). My BMI hovers around 21, in the "healthy" range that extends from 19 to 25. But I was an overweight child. I remember standing on the weighing scales at primary school and feeling a sense of shame. I hid my rolls of fat and always wore baggy sweaters. A T-shirt and a breezy day would reveal the contours of my body and make me feel like the whole world was watching, laughing. My parents tried to reassure me that it was just "puppy fat," an energy reserve to help me grow tall in my teens. But there were other explanations. In the late nineties and early noughties, desktop computers were becoming a common feature of the household. I sat and played video games—first with floppy disks and then CDs—and would watch television in the evenings. I played sports on the weekends, badminton, cricket, and football. But there was a lot of idleness and overeating. At school, I could eat pizza, hamburgers, chips, and, for dessert, a crunchy chunk of cornflakes baked with chocolate, syrup, and margarine that was aptly called tarmac. I would save up any loose change for the vending machine that sat in the musty-smelling corridor below the sports hall, a cheap dispenser of sugary drinks. Even in cooking lessons we were taught how to cook pizza instead of healthier alternatives.

From 2005 onward, Jamie Oliver, a baby-faced and exuberant chef who had moved from the sweaty confines of the restaurant to national TV in the late 1990s, helped change how a nation fed its schoolchildren. I was nearing the end of my school years when *Jamie's School Dinners* first aired. I saw the transformation take place in the cafeteria. Pasta replaced pizza. Hearty meals with vegetables sizzled where hamburgers were once stacked. The infamous Turkey Twizzlers disappeared altogether and only lived on through nostalgia, a ghost of a more processed past. A study in 2009 found that this healthier diet—although not accepted by everyone—had an immediate impact on school attendance and the number of children achieving the top grades. Jamie Oliver didn't mention depression in his "Feed Me Better" petition to the UK government in 2005. But some argue that his work had—and is still having—a significant impact on mental health. "I think he's very important," says Michael Berk, the former mentee of Brian Leonard and a colleague of Felice Jacka at Deakin University. "His work will play a role in preventing depression. Although the magnitude of the effect might be really small, it will have a big community impact because the entire population is involved."

Eating a diet full of fiber—particularly from fruit and vegetables—is just one way to reduce obesity, inflammation, and the vicious cycle that encompasses diabetes, heart disease, and depression. The other is exercise. Since the 1980s, dozens of clinical trials have found exercise to have an antidepressant effect. In a study published in 2005, Andrea Dunn, from the Cooper Institute in Colorado, and her colleagues found that three days of moderate exercise per week—as gauged by time spent on a treadmill or stationary bicycle—had the same efficacy in treating depression as first-line treatments such as SSRIs and cognitive behavioral therapy. Then, in 2013, a study from

Madhukar Trivedi, a professor of psychiatry at the University of Texas Southwestern Medical Center, and his colleagues found that the antidepressant effects of moderate exercise are most pronounced in people who have high levels of inflammatory markers in their blood, whether obese or not.

This presents an opportunity for precision medicine. Since current antidepressants are least effective in people who have high levels of inflammatory markers, exercise could be used both as a standalone treatment and a boost to subsequent drug therapy. If the depression is unmoved by physical activity, levels of inflammation in a person's body may have been reduced to a level at which antidepressants are more likely to work. There's even a choice between antidepressants, some being more suitable for people with inflammatory markers in their blood. A large clinical trial from 2014, for example, found that nortriptyline (a tricyclic antidepressant) was more effective than escitalopram (an SSRI) at treating depression in people with elevated levels of C-reactive protein, a marker of inflammation that can be assessed with a quick blood test. Three years later, a similar study found that another marker, interleukin-17 (IL-17), predicted a better response to a combination of an SSRI and bupropion—a dopamine-based drug that's also an anti-inflammatory—compared to an SSRI on its own. Whether it's exercise or a certain type of antidepressant, low-grade chronic inflammation can be used as a window into a more effective future.

It is easier to administer drugs than it is to prescribe exercise. Insurance companies and national health services often don't cover personal trainers or gym memberships in the same way that they subsidize drugs or therapists. Running shoes, yoga courses, and court fees can be costly. And finding the time for regular exercise is difficult even for people who are mentally healthy, never mind for those who might be suffering from a crippling lack of motivation, drive, and self-esteem. But, for me, the alternative is much worse. Depression is a huge motivator, one of the most powerful forces that gets me up in the morning and out for a run.

When I was withdrawing from sertraline in early 2020, I thought that the more I ran the better I would feel. As my dose tapered toward zero, I slowly increased the number of kilometers I was covering per week. After a few months, I was running the equivalent of a half-marathon once or twice a week. It was too much. I often felt fatigued and low in the days after exercise. And this was frustrating. *Why don't I feel over the moon after such a weekly achievement? Am I destined to be depressed no matter how hard I work against it?* Then I read a study that offered an explanation, and again it related to the new buzzword of psychiatry: kynurenine. In 2015, a group of Austrian researchers found that intense exercise—even in athletes—can activate the kynurenine pathway and shunt tryptophan away from serotonin, reducing the amount of this neurotransmitter available for brain function; that is, a reversal of an antidepressant effect. "One may conclude that sports performed occasionally with two or three days interval exerts a beneficial effect on general well-being as it is the case when it is performed as recreational activity," Barbara Prüller-Strasser, from the Medical University of Innsbruck, and her colleagues wrote, "whereas intense training will more and more achieve adverse effects on both mood and immune system."

Reading this, I tried running for shorter distances three times a week. Five or ten kilometers: it felt like a healthier routine and didn't leave me exhausted. Afterward, our dog Bernie is still panting, his tongue lolling out of his mouth as he waits patiently for breakfast. As I stretch my muscles and warm down, I contemplate what is happening inside my body. Is my brain being flooded with neurotransmitters and endorphins? Are its synapses making new connections, reversing the effects of stress and depression? Are the trillions of bacteria inside my gut sending a cacophony of healthy signals to my brain through the vagus nerve? Are levels of inflammation running on a cool norm? Even if it's a temporary high, I feel empowered by those treatments I can tweak and modify from one day to the next. The food I eat and the amount I exercise are antidepressants that I prescribe. They are as much a part of my treatment as drugs and psychotherapy, and they come with fewer side effects.

The Beginning

There is a chicken-or-egg question in the study of inflammation and depression. Which came first? Did the inflammation trigger the depression? Or was it the depression that caused the body to become inflamed? "We know that psychological stress activates our immune response," says Golam Khandaker, an epidemiologist at the University of Cambridge. "People who are stressed have higher levels of immune activity. So it's possible that the increased inflammatory proteins in depressed patients is just a manifestation of these people being psychologically stressed, because we know depression is a very stressful condition."

But there are multiple lines of research that show inflammation to be the cause—and not the consequence—of depression. First, longitudinal studies that have tracked the health of over one hundred thousand people in the UK, the Netherlands, Denmark, and the U.S. show that people with slightly elevated markers of inflammation in the blood are significantly more likely to develop depression later in life. Interestingly, people who have a genetic predisposition for lower levels of inflammation in life have been found to be at a lower risk of becoming depressed. Further, not only do pro-inflammatory drugs such as alpha interferon often lead to depression, but patients given anti-inflammatories for immune-related diseases such as psoriasis, Crohn's disease, or rheumatoid arthritis show a marked decrease in their depressive symptoms. Importantly, these reductions didn't correlate with whether their immune-related disease improved. Even if their joints were stiff, their intestines were inflamed, or their skin was covered in itchy rashes, their mental health showed remarkable transformations.

These studies have now paved the way for more direct clinical trials. "What's exciting about this field now is that on the back of a lot of research based on healthy volunteers, patients, genes, population studies, we have now reached the stage of clinical trials in patients with depression," says Khandaker, who, along with his colleagues at the University of Cambridge, is conducting one such clinical trial. The Insight Study begins with one of the most important facts about this field: not everyone with depression is inflamed. The latest estimates suggest that one-quarter to a third of people with depression show increased levels of CRP and IL-6, two inflammatory markers that are easily tested in the blood and that, importantly, don't change with the time of day or fluctuate after meals. Once such patients have been identified—through two blood tests and the identification of inflammation-related symptoms such as fatigue, sleep disturbances, and concentration problems—Khandaker and his colleagues will then either provide an infusion of a placebo or an anti-inflammatory. Such a rigorous selection process is paramount when it comes to patient safety. As they dampen the immune response, anti-inflammatories can lead to infections, disrupt brain function, and should only be given to people who show the hallmark signs of inflammation. If given to depressed patients without any signs of inflammation, these drugs can be harmful rather than helpful.

Even with the bulk of evidence that this field has accrued over the last two decades, Golam Khandaker still takes the ruthlessly skeptical worldview that only scientists can muster. "I am not drawing any firm conclusions based on what we know. And I think it would be foolish to draw any firm conclusions regarding whether anti-inflammatory drugs have a role in mental health or not," he says. "This is just the beginning."

From his converted barn in Galway, Brian Leonard watches as the field he helped create takes its first steps into the real world. In his textbooks and reviews, Leonard frequently quotes Arthur Schopenhauer, the wild-haired German philosopher from the nineteenth century. "All truth passes through three stages," Schopenhauer wrote. "First it is ridiculed. Second it is violently opposed. Third it is accepted as being self-evident." Twenty years ago, Leonard believed that the field of neuroimmunology was somewhere between one and two, still being ridiculed and beginning to experience serious criticism. By 2013, he thought it was at stage three: self-evident. "Unlike false hopes raised by earlier theories of mental illness," Leonard and Angelos Halaris from the University of Chicago wrote, "we are confident that this relatively young subspecialty will have staying power."

Surfing in the Brain Scanner

On the "Day of the Dead," November 2, 2005, Draulio Barros de Araujo, a neuroscientist from the University of São Paulo in Brazil, experienced his first psychedelic trip on ayahuasca, a traditional tea made from two or more plants found in the Amazon Basin. One of his PhD students was a member of the local Santo Daime Church, a religious institution that has been taking ayahuasca as a part of their ceremonies since the 1930s, and thought that Araujo would be interested in the experience. After all, he had spent his career using brain imaging technologies and developing mathematical models to "see" how the mind works. How might his scientific expertise be reshaped by a drug that can produce vivid images of nature, a tea that can summon gods down from the heavens to provide counsel to mortals? Science called these visions hallucinations. But members of the Santo Daime Church call them *mirações*: "seeings." These images are as real as any wavelength of light that tickles a retina.

Araujo felt honored at the invite. Ayahuasca ceremonies weren't just a central pillar of his student's spiritual world, but a rich part of Brazilian culture that reached back centuries, into the very heart of the Amazon rainforest and its people. A neuroscientist and former physicist with feet firmly rooted in peer-reviewed journals and statistical significance, Araujo replaced his scientific cap with a wide-brimmed hat of an explorer.

The active ingredient of ayahuasca, the molecule that is responsible for these *mirações*, is called N,N-Dimethyltryptamine, or DMT. It is naturally found in the bush *Psychotria viridis* and, when digested and shuttled into the brain's synapses, interacts with a specific type of serotonin receptor to produce psychedelic experiences. (Another form of DMT called 5-MeO-DMT is extracted from the venom of the Sonoran Desert toad, *Bufo alvarius.*) The same basic mechanism underlies the effects of psilocybin, the active ingredient of magic mushrooms, and the laboratory-made LSD. But ayahuasca isn't just rooted in *P. viridis* and the DMT it contains. There is a second ingredient that is vital to its effects. Extracted from the climbing vine *Banisteriopsis caapi*, harmine is a naturally occurring MAO inhibitor, a chemical akin to iproniazid and Parnate. By blocking the enzymes within the synapse, harmine prevents DMT from being broken down and flushed away in urine. In other words, MAO inhibitors weren't "discovered" with the first antidepressants. They had been a part of indigenous traditions for hundreds of years.

Ayahuasca is a potent brew. Half a cup of this tea often leads to severe nausea, stomach cramps, and diarrhea before inducing a dreamlike state that can last for four or more hours. The images the drinker sees are often surreal, majestic, and awesome. But they can also be frightening and unbearable. In the Santo Daime Church, it is known as going to work, a form of heavy lifting. Shamanic rituals refer to an ayahuasca trip as "labor."

Araujo's first experience was a confusing constellation of geometric images, animals, and plant life. It didn't make much sense. It was only after the experience that he realized how significant his trip had been. Although it felt like he was dreaming, these images were just as real as his everyday conscious thoughts. "The difference with ayahuasca is that when I woke up from the bizarre dream I was having, that sometimes resembled nightmares, it was not just a dream," de Araujo says. "It was real. For someone who was interested in the human mind, that was just astonishing."

Living in one of the most psychedelic countries in the world, one in which ayahuasca has been legal in religious ceremonies since the late 1980s, Araujo found himself in the perfect place to learn more about this experience. He turned to his local experts, churchgoers who had been using ayahuasca every two weeks for years or, in some cases, decades. "This is the level of experience that a lot of people in the church have," he says. Their familiarity with the psychedelic experience, he thought, was the perfect base on which to design an experiment. Not only would they be able to cope with the experience while lying down in the unfamiliar setting of a brain scanner, they might also be able to perform basic psychological tasks while doing so. "It's similar to having someone do a task while surfing," Araujo says. "If you don't know how to surf, then it's going to be really hard to do a task on top of surfing."

Araujo, a smiley and slim man with short gray hair, found ten volunteers who had surfed the psychedelic waves of ayahuasca for decades and were willing to be a part of his study. One man couldn't keep his head still during the scan, however, leaving nine in total (four men and five women). Inside the donut-shaped scanner, their brains showed wondrous things. In particular, when told to imagine a photograph they had just been shown, the parts of the brain involved in memory and vision were alight in activity, sometimes resembling the patterns seen in REM sleep or lucid dreaming. But the person was wide-awake. From Araujo's point of view, it seemed as if they were actually seeing those images right in front of them, as if the trees, animals, or people they had been shown were actually in the brain scanner. "By boosting the intensity of recalled images to the same level of [a] natural image," Araujo and his colleagues wrote, "Ayahuasca lends a status of reality to inner experiences." The title of their paper was "Seeing with the Eyes Shut."

Outside of the scanner and back in the Santo Daime ceremonies, it was obvious that the visions had benefits that far outlasted the ayahuasca trip. People with alcohol dependency or depression felt that they had worked through their problems. Again, this had been known since time immemorial. Although the psychedelic experiences on LSD and, more recently, psilocybin in Europe and the U.S. often focus on transcendental visions of God or an insight into an ultimate truth of the cosmos, the centuries-old use of ayahuasca has a slightly different modality: making sick people better. The visions of divine figures have, historically, been seen as real spirits or demons that have to be vanquished from the body through shamanic ritual. These weren't seen as mystical experiences, that is. They were purely medicinal, like removing a splinter from under the skin or an inflamed appendix. Araujo hoped to tap into this history of healing and see whether it could purge depression from people who hadn't responded to rounds of antidepressants, psychotherapy, and ECT.

"Turn On, Tune In, and Drop Out"

Extracted from climbing vines, leafy bushes, prickly cacti, or pimple-headed mushrooms, chemicals that can twist our sense of reality have been used by indigenous cultures for thousands of years, from Siberia to the Sahara and down into Mesoamerica. The word *psychedelic*, meaning "mind manifesting," meanwhile, is a recent invention. It was coined in the 1950s by a doctor working in Canada, Humphry Osmond, as an alternative to *hallucinogens*, a term that seemed overly negative for drugs that were hailed as direct lines to God and catalysts of love and peace. Writing to Aldous Huxley, the author of *Brave New World* and *The Doors of Perception*, Osmond summarized his choice in a poem:

> To fathom Hell or soar angelic,
> Just take a pinch of psychedelic.

As with cocaine in the nineteenth century, LSD was the chemical encapsulation of hundreds of years of South American history. First synthesized in 1938, it was seen as much a part of psychiatric treatment as the emerging antidepressants, such as MAO inhibitors and imipramine. At a time when psychoanalysis was still at its height, LSD was seen as a catalyst to psychotherapy, a key that could unlock the cargo of the unconscious.

Betty Eisner, a therapist living in Los Angeles in the 1950s, was amazed at how this hallucinogenic drug could uncover, layer by layer, repressed memories and traumatic experiences as if it were a knife slicing through the onion of the unconscious. "LSD makes available, from the very first session, other levels of consciousness which might require months or years of conventional therapy to effect," she wrote. After her own experiments with the drug (the second ending with a deep depression that felt like the "universe had collapsed" on her), she estimated that six hours of an LSD trip was equivalent to four years of psychoanalysis. This drug was so powerful that it could make the most seasoned analyst sit back and try not to get in the way of its progress. "I think the function of the therapist is to optimize conditions for the LSD to work," Eisner wrote.

In a standard hospital room in Los Angeles, Eisner's therapy sessions were simple but otherworldly. She provided her patients with a large handheld mirror, advised them to bring a few family photographs from their childhood, and then gave them a small dose of LSD. During the eight hours of an LSD trip, patients could reframe old memories and see themselves from a different—and hopefully more positive—angle. All the while, the phonograph in the room spun through scores of classical music that seemed to harmonize the whole experience, guiding patients through the ups and downs of their past as if they were strung along a wave of sound. "Concertos seemed to express and enhance the relationship of the individual to the environment as expressed by the interaction of the soloist with the orchestra," Eisner wrote. "I've found a Mantovani record of classical selections is good to start—and then Chopin's first piano concerto is better than anything."

Throughout the 1950s and into the early '60s, LSD was employed as an aid to psychotherapy for a range of mental illnesses, from anxiety to alcoholism and impotence to homosexuality (which was listed in the psychoanalysis-driven *DSM-I* as a "sociopathic personality disturbance"). Its usefulness in depressions was first tested in 1952, four years before imipramine and iproniazid were first given to depressed patients in New York and Münsterlingen. Charles Savage, a military physician working in New York, gave LSD to fifteen patients with depression—both endogenous and reactive—and found that three people fully recovered and four improved. But a comparable group of patients, matched for age, sex, and type of depression, showed similar rates of recovery with psychotherapy on its own. LSD's effects, in other words, seemed to offer no more than standard treatment. Further, while brief moments of euphoria were commonly observed, severe anxiety frequently surfaced and "encouraged reticence rather than confidence," Savage wrote. Rather than opening up the minds of depressed patients, LSD could just as easily close them shut. "However," Savage concluded, "LSD affords therapeutically valuable insights into the unconscious processes by the medium of hallucinations it produces." To him, LSD was a tool for understanding, not for recovery.

Although Betty Eisner didn't specialize in depressed patients, she furnished psychedelic therapy with its future surroundings. She realized that the setting of these experiences was just as important as the drug that induced them. LSD didn't just create random visions of color or the cosmos, but seemed to blur the human sensorium with past memories and thoughts, painting them onto an ever-changing tapestry. With music, photographs, and the occasional piece of advice, Eisner became a conductor within her little hospital room, working to channel her patients' experiences toward a therapeutic crescendo. She was good at it. Sometimes it felt as if she had a natural gift for making people feel better. Although she could speak fluently in psychological and pharmacological terminology, she never forgot that the patient before her was a human in need of support. "The one thing I have noticed is that the subject who takes LSD should be the whole center of attention for as long as he or she needs it."

Almost every trip on LSD led to moments of suffering and the potential for pain. After all, uncovering traumatic experiences is never easy. What might emerge? A dormant monster of their past? A memory that they thought was best left alone? After taking an inventory of her patients' lives as a barometer for what might emerge during the LSD session, Eisner told them to confront whatever fears they faced. "[T]he main technique which was found effective for basic problems which presented themselves in symbolic terms—such as raging fires, the void, dragons, a vortex—was to instruct the individual to move toward whatever appeared," Eisner wrote. Although they didn't walk about the room and could even lie completely still, in their minds they were taking some of the most difficult steps of their life. "[As] the individual under LSD 'walked' toward the fire in order to be consumed, the flames which appeared to be of hellish intensity, suddenly changed in the moment of impact, of stepping into their midst, and were transformed, as though miraculously, into a situation capable of resolution."

During a trip to Europe in 1958, Eisner met with many likeminded LSD researchers. In England, Germany, Czechoslovakia, and Italy, researchers were trying to grapple with the secrets that this drug seemed to uncover. As psychedelic trips often involved journeys through the stars and planets, were they an insight into our innate connection with the cosmos? Were the images projections of the internal workings of the mind? Or, as Eisner believed, was LSD a bridge from human to human, and, ultimately, from human to God? Whatever theories they held, Eisner found that her methods were unique. At Powick Hospital in Worcester, England, for instance, she found that patients were often left without a therapist—a guide—for long hours of their psychedelic trip. There was no music. No photographs or mirrors. And although art equipment was provided, it seemed as if their technique lacked the necessary element of individuality that she was accustomed to back in Los Angeles.

In September of that year, Eisner attended the first meeting of the International Congress of Neuro-Pharmacology, an academic summit that brought the leading figures of biological psychiatry to Rome. The three-day event, held in a building complex built under the fascist dictator Benito Mussolini, was dominated by the recent introduction of antidepressants and antipsychotics and what they might tell us about the chemical workings of the mind. But Eisner found enough researchers on LSD for a lifetime of discussion. "It was not only hearing what each one of us who was working with LSD had done, it was hearing what effect it had and why, what might have been a different and better way to use the drug, what each of us thought was the optimal method of dealing with different kinds of patients and situations, and basically and continually, the consideration of psychedelics in all their ramifications," Eisner wrote.

On the final day of the conference, Eisner presented her work into LSD and psychotherapy to a small audience. Still, she felt honored to be there, to have the opportunity to speak about her work in front of an esteemed audience. "I felt like a true pioneer, reporting the results of my explorations to far and unknown lands," she wrote. That evening, while walking through the mist and blurred lights along the River Tiber, Eisner spoke with Albert Hofmann, the chemist who first synthesized LSD in 1938 and, two decades later, isolated psilocybin, the active molecule of magic mushrooms, and felt as if she were dreaming. "[It was] a fitting end to a magical trip," she wrote. "But one must always come back to Earth."

While her early years of LSD research were defined by excitement and the hope of a new form of psychotherapy, Eisner soon found herself alone in an increasingly hostile world. While she tried to distance herself from Timothy Leary—the Harvard psychologist whose evangelical views on psilocybin led to his dismissal and exacerbated a general feeling of antagonism toward such psychoactive drugs—she nevertheless felt as if resistance to her own work was bubbling up. Sometimes, it felt like it wasn't about the drug at all. Many of her peers criticized her for giving LSD to patients when she "only" had a PhD in psychology from UCLA, and therefore wasn't a trained psychiatrist, an MD. "It does get discouraging to run into the prejudice which judges more from the initials after one's name (or one's sex, because I'm afraid this had some bearing, too) then by what the individual is and can do," she wrote to Humphry Osmond in 1961. "I get tired of carrying the torch and fighting the battle." Her former supervisor at UCLA, Sidney Cohen, a bright and prematurely graying pharmacologist, split from Eisner for other reasons. He couldn't agree with her theories of LSD that merged science with spirituality. While images of ancient Greece, Egypt, or India were just hallucinations to Cohen, Eisner saw them as evidence for a collective consciousness of the human species. To her, these visions weren't illusions. They were memories.

Cohen was a levelheaded pharmacologist, a scientist of chemical pathways and drug dynamics, and he worried that LSD was dangerous. As excitement over LSD-led psychotherapy was reaching its acme, he reached out to his peers from around the country and asked whether they'd had any problems during their LSD sessions. In total, forty-four replied. From a collective sample of five thousand individuals, Cohen felt reassured that there were no reported deaths from overdose, and while enduring psychotic episodes were a known risk, they were rare and often treatable with antipsychotics. Suicide was even rarer. Cohen only found two confirmed cases of users killing themselves "directly due to LSD." In 1960, he concluded that it was a relatively safe drug.

By 1962, he had changed his tune. "Recently, we have encountered an increasing number of untoward events in connection with LSD-25 administration," he wrote with his colleague in California, Keith Ditman. As LSD became more popular in the counterculture of California, and more suicides were found to have occurred during or shortly after an LSD trip, Cohen wrote several rebuttals to his own initial publication and tried to warn fellow researchers and the general public of the potential dangers of this psychedelic drug. He wasn't alone in his caution; the scientific literature was replete with similar warnings. "That it is a dangerous, toxic substance and not an innocent aid like milk, wheaties and orange juice . . . has been pointed out on different occasions," an editorial in the *New England Journal of Medicine* stated. "Among the ominous effects of LSD are chronic hallucinosis, panic, severe paranoid reactions, suicide, various other psychotic development and reappearance of LSD symptoms a month to over a year after the original use, without reingestion." Such "flashbacks" seemed to follow no pattern and no one could explain when or why they might occur. "At present," the editorial concluded, "further work with these drugs certainly should be undertaken only in controlled settings by scientists capable of impartial, critical judgment."

That already seemed an impossibility. Between 1959 and 1961, the Hollywood actor Cary Grant—one of the most famous people in the world at the time—had over sixty sessions of LSD therapy for his depression and, through an article published in *Good Housekeeping* in September 1960, helped push its use into the mainstream. Meanwhile, up and down the coast of California, LSD parties were being held in psychologists' homes and on university campuses. Psychedelic bootleggers managed to create their own homemade stash of LSD and push the cultural swing from the black sweaters of the beatniks to the tie-dyed T-shirts of the hippies. Ken Kesey, the author of *One Flew Over the Cuckoo's Nest*, held so-called acid tests across California in which live music—provided by bands such as the Grateful Dead—and LSD were mixed into powerful, mind-altering cocktails. Passing this test was like climbing a mountain: often strenuous with sheer drops at one side, but the view from the top was said to be spectacular. The whole scene was encapsulated by Timothy Leary: "Turn on, tune in, and drop out."

At an LSD party in Hollywood, Sidney Cohen met someone who soaked sugar cubes in doses of LSD that were ten times stronger than the average dose. After a ten-year-old boy accidentally ate one of these sugar cubes and had psychotic reactions for months afterward, the Food and Drug Administration heeded the warnings from medical professionals such as Cohen and tried to take control of this drug. Sandoz, the pharmaceutical company that produced LSD, restricted supply of its product to psychiatrists funded by the National Institute of Mental Health, cutting its customers from two hundred to seventy in one go. Charles Savage, the psychiatrist who first gave LSD to depressed patients, argued that the misuse in California shouldn't blight everyone's prospects of studying the drug. "LSD therapy should not be seen from the narrow vantage point of Southern California where it has been vastly misused." With her PhD in psychology, Betty Eisner was one of the researchers whose supply of LSD was cut, bringing her therapy sessions to an abrupt end.

In 1966, two U.S. Senate meetings were held to discuss LSD. Timothy Leary, recently sacked from Harvard, used Sidney Cohen's

1960 paper to argue that LSD was "remarkably safe." Cohen, meanwhile, put forward his updated perspective: "We have seen something which in a way is most alarming, more alarming than death in a way, and that is the loss of all cultural values, the loss of feeling of right and wrong, of good and bad," he said. "These people lead a valueless life, without motivation, without any ambition . . . they are deculturated, lost to society, lost to themselves." LSD was said to cause mental illness—turning a mild-mannered individual into a psychotic killer, for instance—and legislation against its use was destined for a signature. While not everyone was against Leary and LSD—Senator Robert Kennedy, whose wife had undergone LSD therapy, said, "[P]erhaps to some extent we have lost sight of the fact that it can be very, very helpful in our society if used properly"—LSD was made an illegal substance in the U.S. in October 1968. By 1970, it was banned in the UK and throughout much of Europe. Only in the Netherlands was it used into the 1980s as an aid to help concentration camp survivors explore and come to terms with their traumatic past.

Building a New System

In his late sixties, David Nutt is a mustachioed, grandfatherly figure of psychedelic therapy who wears knitted sweaters and swears a lot. A neuroscientist and pharmacologist by training, he has built his career on understanding drugs that politicians prefer to ignore: MDMA, ketamine, cannabis, and, most recently, psilocybin. He was once an adviser to the UK government but, in 2009, was expelled from the committee after he claimed that taking ecstasy was safer than riding a horse, a pastime that kills ten people per year in the UK and leads to thousands of permanent head injuries. (In the *Journal of Psychopharmacology*, Nutt sarcastically called this "overlooked addiction" Equasy, or "Equine Addiction Syndrome," and wrote, "Making riding illegal would completely prevent all these harms and would be, in practice, very easy to do—it is hard to use a horse in a clandestine manner or in the privacy of one's own home!") In his office in West London, Nutt has a metallic plaque advertising LSD and a giant mushroom made from a yew tree, a type of wood historically selected for the manufacture of longbows to defend churches in Britain. Nutt beams with delight in the knowledge that every church in England had at least one yew tree growing inside its perimeter. Such topical digressions distract him from the nonsensical reality that consumes most of his energy.

"It's the worst censorship of research in the history of the world," he says, referring to the existing legislation on psychoactive substances such as psychedelics and MDMA. Across North America and Europe, psilocybin and other psychedelics are listed as Schedule

1 drugs (substances that have no accepted medicinal use and a high potential for abuse). Such a classification infuriates Nutt. "Even if the war on drugs stopped people using drugs, which it doesn't, it still would not justify denying these drugs for treatment," he says. "And it doesn't even do that! As far as we can ascertain, [current drug legislation] doesn't affect use at all; other than making use more dangerous. It's outrageous, it's absurd. The whole thing is just a fucking farce."

Despite its legal classification, Nutt and his colleagues at Imperial College London were given access to psilocybin for their first study in 2011. As with Draulio Barros de Araujo in Brazil, they initially gave the drug to healthy volunteers and scanned their brains during their psychedelic trips. It was then that they saw the antidepressant potential of this drug shining through their data sets. Part of the frontal cortex that was often found to be overactive in depressed people was shut down. Nutt calls this the Mayberg center after Helen Mayberg and her research into area 25. After being silenced or quieted, other parts of the brain began to communicate with one another, creating a rich harmony of brain signals that reintroduced regions that had previously been rigidly isolated. It was from this finding that, Nutt says, depression seemed like a worthy target for psychedelic therapy. "Depression is classically a disorder where people are overengaged in internal thought. They feel guilty, they think about themselves, about what they've done wrong, they can't disengage from that inner-thought process, which is usually negative," he says. "And psychedelics usually disrupt that."

The first trial, published in 2015, was a familiar setting, one that transported the work of Betty Eisner into the twenty-first century. The hospital room at Imperial College London was kitted out with fake candles, essential oils, flowers, and a twirling model of the solar system floating above a padded gurney. A specially selected playlist filled the room with a suitable auditory landscape. The psilocybin was made in a laboratory. The experience was largely artificial; that is, a construct made from predetermined settings, sensations, and quality-controlled doses. But the experience was nonetheless magical. "I cried a lot during the experience, but I also laughed a lot," one patient, a thirty-six-year-old man who had been abused as a child and gripped by depression ever since, says. "I learnt so much in such a short period of time. Certain things I've thought before, but they weren't thoughts anymore. They were truth." When his depression returned six months after the trial, this patient contacted Amanda Feilding, the creator of the Beckley Foundation that partly funded the work at Imperial College, and asked for another dose. But she couldn't help him. This was a research project that cost millions of dollars and took years of planning. Psilocybin wasn't a legal prescription. It was still Schedule 1, a drug judged to have no medicinal value. As someone who remembers the days when LSD was legal in the 1960s, Feilding was furious. "It seems to be absolutely insane, and also criminal, that there's no way, under present regulation, where one can offer a booster for someone who has had a successful treatment," she says. "We must open this treatment up to suffering patients."

Not everyone is as eager to push psychedelics into mainstream medicine. "We have to establish whether it's a treatment," says James Rucker, who, in 2016, moved from the group at Imperial College to start his own research project at King's College London. "There's this thing called the winner's curse in research," he explains. "A research team has an idea: they think psilocybin is going to be a treatment for depression. Part of the reason why they have that idea is because they have a preconception in their head about the fact that the treatment's going to work." The scientists carrying out the work expect positive results and this expectation influences how patients respond to the new treatment. In a way, the winner's curse is the first step toward a placebo effect. "This is what happens for a lot of treatments for depression," Rucker adds. "The first trials with Prozac were quite similar. People who hadn't responded to anything else miraculously got better on this new and novel drug that was thought to be the next miracle cure." Today, Prozac is known to have no greater effects than any other antidepressant, and, according to a large meta-analysis on twenty-three different antidepressants published in *The Lancet* in 2018, is significantly less effective than escitalopram, another SSRI. Rucker, a quiet perfectionist born in 1978, a time when the first personal computers were beginning to hit the market, exudes a youthful manner that belies his clinical experience. He first became interested in the brain as a child, comparing its trillions of neurons to the electrical wiring in his computer's hardware. Now in his mid-forties, he has been a psychiatrist for over fifteen years and has struggled with his own mental health for much longer. "Depression runs through my family like a knife," he says. After trying several SSRIs, he was eventually given the same drug his father was once prescribed: imipramine. "It actually worked brilliantly for me. I stayed on it for about ten years." He felt no real side effects other than slight constipation, but that was a small price to pay compared to the libido-sapping effects of some SSRIs.

Today, Rucker is part of an international collaboration that aims to assess whether psychedelics might add to the pharmacological approach to depression, or whether they are a colorful flavor of the winner's curse. Funded by COMPASS Pathways, a research initiative set up by George Goldsmith and Ekaterina Malievskaia in 2016 after their son didn't respond to conventional antidepressant therapy, this multicenter trial includes 216 patients from eleven cities dotted across Northern Europe and the eastern United States and Canada. "We came at this not from a position of favoring the legalization of psychedelics," Malievskaia told a reporter at *International Business Times* in 2017. "We came from a perspective of creating another option for patients who have exhausted all the others."

"They've done what no one else has done," Rucker says, referring to Goldsmith and Malievskaia. "There are very strict rules about how pure you have to manufacture a compound in order to use it as a medicine in humans. It's difficult. And it's particularly difficult for a drug that is Schedule 1. And they've done it." Although COMPASS Pathways provides the drug free of charge to psychiatrists like Rucker, it is still the most expensive component of the clinical trial. The strict legislation that surrounds the drug means that its transportation—from manufacturers in the UK to King's College London, for example—requires a series of high-security stop-off points. The capsules are delivered by a special courier under constant surveillance, have to be triple signed during any handover, and are kept in safes bolted into a concrete floor. "All of that has a cost implication," says Rucker, "and we have to pay for that." At New York University, scientists have to weigh their supply of psilocybin twice a day, just in case someone decided to steal some. Psilocybin is a chemical found in a smorgasbord of mushrooms around the world and has been used in spiritual ceremonies for millennia. In academic research, however, it is being treated as if it is a nuclear weapon.

Unlike many other researchers in this field of psychedelic medicine, Rucker is grateful that it is so strictly monitored. Schedule 1 doesn't reflect its potential therapeutic qualities, but it reduces the chance of history repeating itself, he says. In the 1960s, LSD filtered into the counterculture and its promiscuous use led to a widespread ban. Today, with strict laws prohibiting the use of psychedelics in the public, Rucker thinks that this will help keep the drug within psychiatric facilities, a restriction that keeps its use open to people who might benefit from the drug. "Of course, recreational use still happens," he says. "It happens anyway," regardless of what schedules or laws are placed on a substance that can be harvested in a local park or from a cowpat in an open meadow.

In late 2017, I had recently turned twenty-seven years old and was more desperate than ever. Still on the highest dose of citalopram, I felt nauseous and fed up, and I was on the lookout for alternatives. I had started researching depression treatment and I felt like the medical literature had more to offer than my doctor could provide. I quickly realized that one class of drugs that was getting a lot of media attention—in brain imaging studies and on the front cover of magazines like *New Scientist*—psychedelic therapy, was the next big thing in treatment. Early trials with psilocybin in people with cancer and depression suggested that this was *the* treatment for people who had failed to respond to other treatments. In a less rational, more desperate mood, I was willing to try anything to feel better. I decided to harvest some mushrooms of my own.

It wouldn't be my first experience with psychedelics, but it was the first time I was hoping for more than just some trippy images with friends. In what's called a washout period, I reduced the dose of citalopram gradually over a few weeks, hoping this would be the last time I would ever have to request a prescription from my doctor.

The mushrooms tasted earthy and, although there was a slightly bitter aftertaste, weren't wholly unappetizing. Their stems were white and had bruises of bright blue, the hallmark sign of psilocybin. Though they were listed as Schedule 1, illegal to buy, grow, or consume, I didn't care. If the studies were correct, this might actually be an antidepressant that worked. I prepared music that I enjoyed and thought would be soothing under the influence of the drug.

I kept a notebook of my experience, noting down anything that I felt and saw. With a hologram of stormtroopers from *Star Wars* on its front cover, I wrote down what might emerge during my trip: my childhood anxieties, the separation of my parents, the absence of self-esteem and confidence, suicide. Then the psilocybin started to take effect.

"I don't want to throw up," I wrote. "A warmth inside my body. A glow like a heating filament. Tiny glowing lights behind closed eyelids—purple and green. Illuminating and then fading."

At one point, I appreciated a cup of tea, not only for its taste but how it seemed to move. "The tea is breathing," I wrote. "As the steam comes out of the top, the tea level seems to move, tidally."

I then had a "moment" with a squirrel I saw out the window. "Beauty," I wrote. "Soft skips from one moment to the next. Cleaning of the face and tail, all otherworldly cuteness. The neighbor walks down his garden, his formal shoes clapping on the stone flags. The squirrel freezes. Two front legs braced like furniture. I feel it. Everyone gets anxious." Then: "I'm going to close the curtain, to take the sharpness out of the sun."

After further ramblings on the dustiness of our apartment, how my pen is a tool for ink, and the stressful, irritating, and then beautiful moments of Peter Green's Fleetwood Mac, I drew a line under my notes. They stopped for a few hours. At three o'clock in the afternoon, I summarized what had just happened. "A tremendous force was pushing down on me, reducing me to insignificance. My mouth was relaxed, open. Only air coming in, air coming out." But it wasn't me breathing, I was a mere apparatus for air, like a wind instrument, that someone else was playing through the rhythms of life. "I feel small, fragile, something to be preserved and taken care of. My shoulders are only so wide and I can feel my bones through a thin layer of skin."

As I wrote during this trip, I thought this was the end of my citalopram prescription. What if I could take smaller doses of magic mushrooms in order to adjust my perspective on life, to see beauty rather than blight?

After six or so hours, I moved into the kitchen, sat on a footstool, and ate some vegetable soup.

It was soon clear that this day had little effect on my well-being. There were moments of peace and periods of darkness, but there were few long-lasting benefits or nuggets of understanding that I could employ in my everyday life. I was missing a crucial part of psychedelic therapy. The drug is only useful in the right setting. I had no guidance from a trained therapist, and the music I chose was sometimes terrifying. Despite my intentions, it was a recreational experience and not medicinal.

Wanting to know more about psychedelic therapy, I reached out to Ros Watts, a silver-haired clinical psychologist at Imperial College London who guides people with depression through psilocybin trips. The guidance begins long before any pill of psilocybin is taken. Watts instills a few important maxims into her patients during their first checkups. "Trust, let go, be open," for example. "'In and Through,' which is a very good one," Watts says. "If you see a monster, don't run away from it, don't try and hide, look it in the eye and try to go through it."

Formerly a psychotherapist for the National Health Service in England, Watts became tired and deflated by the quality of the services provided and the long waiting lists to receive them. She thought that the standard five or six sessions of cognitive behavioral therapy weren't sufficient for many of her patients. "Therapy takes a really long time," Watts says. "But if you've got five sessions with someone and you've got a complex history of trauma, then it's really hard—it's just not enough." Plus, with patients far outnumbering the number of therapists, the waiting lists for therapy, for instance, was already eight months to a year. Who knows what could happen in that stretch of time? Inspired by one of her friends who found relief from her own depression after visiting an ayahuasca retreat, Watts moved to Imperial College London in 2015 and became the latest member of a psychedelic revolution. "Either you stay within a system that isn't working," Watts says, "or build a new one that does work."

Seeing with Eyes Shut

In 2018, Draulio Barros de Araujo, his PhD student Fernanda Palhano-Fontes, and their colleagues in Brazil published the first randomized placebo-controlled trial into psychedelic therapy for depression. Although legal to use in religious ceremonies, it is still difficult to obtain ethical licenses to test ayahuasca on critically ill people. It took Araujo years to obtain approval for a trial encompassing just twenty-nine patients.

But the results were worth the wait. Just a week after their ayahuasca experience, around 57 percent of the participants showed a significant drop—more than a 50 percent cut in their scores on standardized health questionnaires—in their depressive symptoms, a rapidity that far exceeds the three to four weeks of standard antidepressants. Although it is impossible to create a placebo of a drug that leads to visions and distortions of space, Araujo used an "active placebo" that led to the familiar gastrointestinal distress of ayahuasca. A mixture of water, yeast, citric acid, zinc sulphate, and caramel food coloring, this drink had no psychedelic effects whatsoever, but could be mistaken for the real thing by people who have never tried ayahuasca before, which was true for all of the patients in this trial. Using this clever trick, Araujo found that only 20 percent of patients given this active placebo showed significant improvement, less than half the rate of those who were given ayahuasca. Importantly, the only people who became substantially more depressed during this time were four people who were given the placebo.

Based at the Brain Research Institute of Natal—a three-story, whitewashed building set on a hillside in one of the least developed regions of northeast Brazil—it was certainly a change in scenery from the metropolis of São Paulo, where Araujo had worked until 2009. His patients, meanwhile, had never seen anything like the state-of-the-art institute and the Onofre Lopes University Hospital nearby. "A lot of our patients come from very difficult realities," Araujo says. "We are talking about one of the poorest parts of Brazil. [And] Brazil is already a poor country." One of the most common social problems was losing sons, brothers, or fathers to drug trafficking. Either they were imprisoned, disappeared, or dead. Although they were only meant to stay from Tuesday to Friday, several patients asked if they could stay for a few more nights or over the weekend.

Araujo will never forget the sixty-three-year-old man with intractable depression and a stomach as rigid as wood. He told de Araujo that he had been a fisherman all his life but, over the last ten years, he had rarely left his home. He hadn't seen the ocean in years. After the ayahuasca trial, he had a dream that brought these two aspects of his life into perspective. What if, after spending all of his working life at sea, the ocean had become the only place he felt comfortable, content, at peace? Could this explain his depression? Could the sights and smells of the ocean be curative? "You know what, Doctor?" the patient told Araujo. "Your tea of the Indians [ayahuasca] just relieved me from something that I had for years. I woke up thinking that I want to go to the ocean." After leaving Araujo's lab, he went straight to the shore and peered over the endless blue toward the horizon.

The old fisherman had never taken ayahuasca before and didn't know where the experience would lead him. Unbeknownst to him, he still hadn't taken ayahuasca, even after the trial had finished. He was given a placebo. "For me, it was the most beautiful moment I've had in the trial," says Araujo, chuckling at the thought of it coming from a fake ayahuasca drink.

"Psychiatrists treat the placebo effect as something that's bad," Araujo says. "After the experience we had in our trials, the placebo effect is one of the most beautiful things I've ever seen." Some patients had been depressed for decades, had lists of medications that were two pages long, and had been given several sessions of ECT. Easily making the cut for a diagnosis of treatment-resistant depression, some of them, like the retired fisherman, responded beautifully to a drink that was little more than lemon juice turned brown.

A week after his fake ayahuasca drink, the retired fisherman returned to Onofre Lopes University Hospital in Natal. At a meeting with a psychiatrist, he lifted up his shirt and revealed his belly. It was no longer stiff, as it had been when he was depressed. Though he had never met this psychiatrist before, he was delighted to show her, rolling his belly in and out like the undulating waves of the open ocean.

Whether they are swallowed as a pill, eaten as mushrooms, or drunk as a tea, psychedelics force us to rethink the placebo effect. "The placebo effect is all about the expectations we bring to a treatment, and that is all cognitively mediated—in our brains," says James Rucker, the soft-spoken psychiatrist from King's College London. "And it's precisely that mechanism that psychedelics seem to effect, they dissolve our opinion structure and expectation-driven models of reality." Whether it is ayahuasca, psilocybin, mescaline, DMT, or LSD, psychedelics are catalysts of expectation. The thoughts that someone brings into a psychedelic session can be manufactured into a sense of reality, with an added flash of color. If someone has heard that they might see a glimpse of the cosmos, they might end up flying at hyperspeed toward the stars, floating among the Milky Way, or getting sucked into a black hole. If someone thinks that these drugs can bring them into greater harmony with the trauma of their past, they might meet a former abuser or see a vision of themselves happy and at peace. As Jules Evans, the policy director at the Centre for the History of the Emotions at Queen Mary, University of London, wrote in *Aeon* magazine, "Ayahuasca reflected my beliefs back to me, in glorious technicolour, and made them feel transcendentally true." If someone *wants* to recover from depression, in other words, maybe these drugs can make that into a reality.

• • •

If the early trials into psychedelics are repeated and confirmed, what can explain their antidepressant effect? Although the details are somewhat nebulous, there are two leading explanations. The first, frequently promoted by European and American research groups working on psilocybin, is that they decrease the amount of "mind wandering." When not engaged in a task such as conversation or manual labor, the brain shows a typical pattern of activity called the default mode network, or DMN, a neuronal highway that connects disparate parts of the brain involved in intention, memory, and vision. When the body is not doing anything, the mind is actively thinking about the future, musing over ourselves and how we feel, and dwelling on the past. As Palhano-Fontes and her colleagues wrote in 2015, it is "a private, continuous and often unnoticed phenomenon." But, over the last decade or so, brain imaging techniques have allowed scientists to take note of this phenomenon. A famous study from 2012, for example, found that experienced meditators who were trained in either concentration, loving-kindness, or choiceless awareness practices could shut down the DMN by reducing the amount of mind wandering that takes place when focusing on their breath and allowing thoughts to flow through their minds without friction.

Similar effects have been found for all the psychedelics so far tested. The activity of the default mode network is reduced. In 2015, for example, Araujo and his colleagues found that eight brain regions that compose the DMN showed significant drops in activity after ayahuasca intake. But he is hesitant to jump to conclusions about what this might mean in terms of how psychedelics work. "I have a lot of issues with the hype around the default mode network," he says. Many studies don't actually show that the network is downregulated, even the study into meditation from 2012. Instead, they focus on one or two parts of the brain that are associated with the network. But a network, by definition, is an interconnected series of parts that can't be seen in isolation. "If you look at [the psilocybin] data, if you look at the meditation data, you will see that whatever is being called the 'default mode network' is actually one single area, which is the posterior cingulate cortex," Araujo says.

Then there's the simplest explanation, the Occam's razor of psychedelic therapy. Since ayahuasca trips have long been referred to as "labor," perhaps any reduction in the default mode network is not because people are reducing their mind wandering but because they are hard at work. The heavy lifting might naturally switch off the DMN, as any form of activity does in normal consciousness. People are so focused on their trip, so dedicated to the labor that it entails, that their brains aren't in default mode.

After spending two years on sabbatical in Santa Barbara, California, Araujo has sifted through his data sets from Natal and has found a more illuminating theory that doesn't involve DMN. Suitably, it relates to the original descriptor of the hallucinations that this tea creates: *miraçôes*, "seeings." While under the influence of ayahuasca, the visual cortex of Araujo's original volunteers, his psychedelic "surfers," showed the same patterns of activity as when someone is awake with their eyes open. Even when their eyes were closed, these same patterns, known as alpha waves, remained stable. (The same result was found by researchers in London using DMT.) Whatever visions ayahuasca generates, that is, the brain interprets them as reality, not as dreamlike hallucinations. "The interpretation is that when you are under the influence of psychedelics, you lose the mechanism that [occurs] when you have your eyes closed," Araujo says. "In other words, you gain the ability to 'see' while you have your eyes closed. And what is it that you see? In our hypothesis, what you see is your thoughts."

We usually experience our thoughts. We don't see them. "So what we think is that what these [psychedelic] substances do is increase your awareness to this process of spontaneous thinking," Araujo adds. And as humans we believe what we see. This might even explain why these substances can be so effective in the treatment of depression.

If the old fisherman's insights from the active placebo were beautiful, the patients that actually received ayahuasca brought a harrowing insight into this "seeing with the eyes shut" theory. One patient was walking over a meadow of wildflowers and luscious green grass on a sunny day. It was a peaceful paradise. Then she saw her dead sister standing in front of her. As real as any conversation she has ever had, her sister told her that she is fine, that she should look after herself rather than worrying over those who have passed away. It was as if her sister had been resurrected to give her this message. When she returned from her trip, this didn't feel like a dream at all. It felt like ayahuasca had summoned her sister from the heavens to show her that she was okay. "That's gotta be powerful," Araujo says. A second patient was a woman who had tried to kill herself. During her ayahuasca trip, she was lying in a coffin as her mother cried inconsolably at its side. But, in this vision, she is conscious and is trying to tell her mom that she is okay. She can't. When she came back to her everyday life, this image provided an alternative perspective on her potential absence, and made her realize just how missed she would be.

Whether it is down to the default mode network, the catalysis of the placebo effect, or the photographic realism of dreams, the exact mechanism of how psychedelics work remains a mystery. But this is true for a range of standard prescriptions from aspirin to antidepressants. What brings the nascent field of psychedelics a little closer to a clinical reality is that there is already a psychedelic drug that has been approved by the European Commission and the U.S. Food and Drug Administration.

In the late 1990s, as John Cryan was searching for novel ways to reduce the lag time in serotonin-based antidepressants, a very different drug emerged onto the scene. It took effect within hours rather than weeks. It didn't work through serotonin, instead blocking the receptor of glutamate, the most common neurotransmitter in the human brain. And it had been on the World Health Organization's Essential Medicines List since 1985, a drug that was well stocked in medical cabinets around the world. This latest breakthrough was the anesthetic known as ketamine. After finding promising results in animal studies, researchers from a small health center in New Haven, Connecticut, found that, in just four hours, seven people with depressions that hadn't responded to typical antidepressants started to improve on this drug. With a single injection of ketamine solution, their improvements lasted for up to three weeks. "These findings represent one of the most significant advances in the field of depression over the past 50 years," two psychiatrists later wrote, "a novel, rapid acting, efficacious antidepressant agent with a mechanism that is completely different from currently available medications."

Although this clinical trial—the first with ketamine and depressed patients—was published in 2000, the drug wasn't unknown to the field of psychedelic therapy. In 1973, Betty Eisner read that ketamine held some of the key features that she found so useful in LSD. "[Ketamine] led to a loss of time sense and detachment from the environment," two doctors working in Shiraz, Iran, wrote that year. "It activated the unconscious and repressed memories, while it temporarily transported the patient back into childhood with frightening reality, reviving traumatic events with intense emotional reaction. Some had recall of events leading to their illness." One patient reported that, after a dose of ketamine, a "heavy burden of sin is now gone." Another felt "carefree with no worries."

Ketamine is a controversial drug. Nicknamed "Special K" or simply "ket," it has been a popular drug of choice at rave parties since the 1990s. Prolonged daily use can lead to irreversible damage to the bladder and kidney damage, as well as intense stomach cramps colloquially known as "K-cramps." And although it undoubtedly has some remarkable effects on depressed patients in the short term, its suitability for long-term treatment is still a matter of concern. "Currently, there is limited evidence to recommend ketamine as a viable treatment option for treatment resistant depression," a report from the Royal College of Psychiatrists stated in February 2017. "Short term efficacy has been demonstrated after a single treatment, but benefits are not lasting for most patients, and mood can rapidly decline after initial improvement, potentially increasing suicide risk." In the U.S., the FDA approved esketamine (a drug that is the molecular mirror image of ketamine and has very similar effects) for treatment-resistant depression in March 2019. "There has been a long-standing need for additional effective treatments for treatment-resistant depression, a serious and life-threatening condition," said

Tiffany Farchione, acting director of the Division of Psychiatry in the FDA's Center for Drug Evaluation and Research, in a press release. "Because of safety concerns, the drug will only be available through a restricted distribution system and it must be administered in a certified medical office where the health care provider can monitor the patient." For people who have received at least two antidepressants at an adequate dose and duration and didn't respond to either, esketamine provides a sense of hope, a shift from serotonin. For some patients, it can be a long-awaited lifeline.

Manufactured in 1962, used to treat nerve pain in American soldiers during the Vietnam War, and given to farm animals as a sedative before surgical operations, ketamine also has none of the historical and cultural foundations that other psychedelics have accrued over centuries. Nonetheless, Araujo sees an opportunity in its example. By using a fast-acting psychedelic with a short duration of action—such as DMT, one of the core ingredients of ayahuasca—could intensive therapy sessions be provided in a clinic setting on an outpatients basis? Instead of spending an entire night at a religious retreat, psychedelic therapy could be provided within one hour. "One thing I have to do as a scientist, and believing that these substances might be helpful for a lot of people, is to incentivize the opening of places that different types of people could have access to," Araujo says. "So we already have the [Santo Daime] church. Now I think we need therapeutic centers for people that might find other [nonreligious] settings more helpful."

As with moral treatment in the eighteenth century, the secrets and traditions of religious practices are still offering new avenues for therapy. Although translated into the jargon of science and brain imaging, this rich history shouldn't be ignored as psychedelic therapy grows in popularity. Araujo never loses sight of the cultural roots of his research. "It bothers me that part of my science will be interpreted as a 'new knowledge,'" he says, "when I know from the bottom of my heart that we didn't do anything. We're just playing. We're hardly scratching the surface."

As scientists around the world delve deeper into the potential for psychedelic therapy in the treatment of depression, they push the two fields of psychiatry into closer union. Whether it's a cup of ayahuasca tea or a pill of psilocybin, each trip crashes psychological and biological treatments together in a flash of color. There's the use of a drug with antidepressant qualities, a welcome addition to the chemicals that first amazed Nathan Kline and Roland Kuhn in the late 1950s. As each trip unfolds, there's a potential recalibration of those brain circuits that Helen Mayberg first imaged in the 1980s. Crucially, there's the psychotherapy that combines the work of Sigmund Freud, Myrna Weissman, and Aaron Beck, guiding patients through repressed thoughts, problematic relationships, and traumatic memories that surface during the session. The legacy of Freud is also seen in the comfortable and confidential room, the couch-like hospital bed. On top of this basic setting, the specifically chosen music, fragrances, and ornaments are a reminder of the little-known work of Betty Eisner. And then there are the scientific methods—brain scans, blood tests, health questionnaires—that call to mind the laboratories of Emil Kraepelin's clinic in Munich. Within a few hours, a century of psychiatry unfolds.

EPILOGUE

New Life

In March 2020, a few days after I had weaned myself off sertraline, I discovered that a new phase of my life had already begun. Lucy was pregnant. The two lines on the pregnancy tests were clear. In just under nine months' time, all going well, we would be expecting our first child. As the embryo grew, the pregnancy app we had downloaded suggested some strange developmental comparisons. One week it was the size a ladybug, the next a raspberry. At the twelve-week scan, the very real baby appeared instantaneously on the screen in front of us. I was silent, stunned at how obviously human this image was. Here was our future. My glasses were steamed up from wearing a face mask—fashioned from material with dog prints on it—that I hoped would help reduce the spread of COVID-19. My tears flowed down my cheeks and soaked into the fabric. At the twenty-week scan, with the peak of the pandemic just a few months behind us, I wasn't allowed into the hospital. I sat on a wall outside and waited for any news.

After half an hour or so, Lucy texted me to say that "everything is perfect," adding that she thought the baby looked like me. As a concession for COVID-19 restrictions, the hospital was handing out envelopes to those couples who wanted to know the sex of their baby. In a time of crippling uncertainty, we coveted this small insight into our future. We opened ours at a wildflower garden at the university campus nearby. We were having a girl.

Depression often coincides with new life. Magical, mesmerizing, and crazy to even contemplate, the growth of a single cell into a human baby can be a time of darkness for many soon-to-be mothers. (Depression runs through Lucy's family as well as mine, and she has a close network of relatives who can offer support and share their own lived experiences.) For an estimated one in every ten pregnant women, antenatal depression is a burden that can far outweigh that of their expanding bump. Once the baby is born, the risks are increased further. By some estimates, postnatal depression affects two out of every ten women. In some countries around the world, that figure can rise to one in two—half of all new mothers. "It's called postpartum depression, but actually a lot of cases started in pregnancy," says Louise Howard, professor of women's mental health at King's College London. "So the perinatal period is a time when interventions do need to be tailored to the women."

I have also worried how a child might affect my own mental health. In my most desperate moments, I told Lucy that I never wanted to have children, not only to prevent my depression being passed on to future generations but also to reduce my risk of relapse. I worried that the sleepless nights and the stress of caring for a child might exacerbate my anguish. In the worst-case scenario, I saw the stress building to such a degree that suicide left a permanent hole in the family, inevitably causing my child to suffer. Nature and nurture combine, the cycle is renewed. Depression takes root.

I regretted my comment to Lucy, both for how much pain it brought her as someone who wanted children, and also for how inaccurate it was. Some people are at higher risk than others, but depression isn't solely determined by the genome. While parenting is stressful, it also can be life-affirming. And my worries over how I might deal with the life transition of becoming a father were overly pessimistic. There is every chance that I'll be a great parent. After all, I have loved watching my baby nephew grow from a gurgling blob into a child who can smile, finish twenty-piece jigsaw puzzles in seconds, and talk. When it's my own child developing, surely the joy of such achievements will be even greater. Reality testing, overturning negative thought patterns: this is no longer cognitive therapy, but a part of my everyday life.

Despite our best efforts, the risk of depression can only be reduced; it can't be removed entirely. To remove depression from society would require the removal of all of its triggers—both social and biological, known and unknown. We can't kill depression. But, with treatment, we can stop it from killing us.

In April 2020, a month after my last dose of sertraline, my motivation started to wane. Thoughts of suicide became more appealing, seductive. Ideas of failure became all-encompassing. The pain inside my skull was often unbearable. Was this a temporary dip caused by the unforeseen stress of life in lockdown? Or, despite my best efforts, was I still not ready to live life without the support of antidepressants? Each day I wondered whether sertraline was key to my contentment in life, my ability to function. Was I only interested in life when I was on drugs? I tried to rationalize this latter question: being on antidepressants isn't a sign of weakness or failure, I told myself. I am not a machine with broken parts. I am someone who has a family history of mental disorder, has experienced life events that are known to trigger depression, and is currently living through a global health crisis and the second recession in my adult life.

In early June 2020, three months after my last dose, I returned to sertraline. My condition had been worsening for weeks, I was crippled by a lack of motivation, and my suicidal thoughts were becoming more threatening. I started to believe that our unborn child would be better off without me. I told Lucy all this, not to upset her but so we could discuss our options. We agreed that both psychotherapy and antidepressants were still a necessary part of my life. I applied for a course of "high-intensity" CBT and, as the waiting list was at least four months, swallowed my sertraline before bed.

I had forgotten how sick these drugs make me feel. The nausea, at least for the first couple of weeks, made the sides of my tongue tickle and the pit of my stomach swirl. I yawned a lot. Diarrhea was as unwelcome as it was frequent. A headache sat right behind my forehead and wouldn't budge. And yet, even those first few days were a relief. The initial dizziness and drowsiness of these drugs had a calming effect on my mind. Although I wasn't feeling good, it was still a refreshing way of feeling bad. Once the nausea and confusion subsided, I slowly noticed myself changing, just as I had when I first took these drugs in early 2018. I felt calmer, less anxious. I enjoyed sleep. I could read and write—two important pastimes for an author—and not feel like I was a failure. I came to the conclusion that antidepressants are effective treatments for my depression and feel overwhelmed with relief that they exist.

I wonder what treatments will be available for our child if she ever needs support. Depression is more common in women than in men, and we know that mental disorders run in families. I hope that unlike when I was growing up in the 1990s, talk therapy will be seen as a natural part of a health care system, like going to the dentist or seeing your doctor for vaccinations. Perhaps it won't be seen as something negative, as the word *treatment* suggests, but as a restorative activity that can tease apart the stresses that everyone faces. Whether it is in schools or at doctors' practices, might we call it health promotion rather than psychological therapy? Will treatments be tailored to the symptoms that are most intrusive on a person's mental health? Will there be blood tests for levels of inflammation? Will specific diet and exercise regimens be prescribed for their antidepressant qualities? Will brain scans—or the symptom profiles with which they are associated—be used to predict who might respond to biological treatments or talking therapies from day one?

Will we even use the word *depression*? Over the last decade, there has been a trend in psychiatry to do away with diagnoses all together. There is an emerging argument that all mental illnesses are interconnected, like the branches of a tree growing from the same trunk. Whether it's bipolar and schizophrenia, anxiety and depression, or autism and ADHD, there is a high level of overlap between one mental disorder and another. Symptoms are shared. Patterns in brain activity overlap. And the genetic risk factors for each disorder—while diverse and small in effect—can be the same whether someone is diagnosed with major depression or generalized anxiety disorder. These common threads extend into treatments, as well. People with anxiety disorders respond to antidepressants. People with severe depression respond to antipsychotics. Cognitive behavioral therapy can help people manage pretty much any mental concern, from insomnia to psychosis.

One common thread that might connect mental disorders such as depression and anxiety is called neuroticism, a gauge of a person's personality and reactivity to stress. Not only does this personality trait have a high genetic component, there is evidence that it is also a precursor to depression in adolescents. Childhood influences, personality traits, and an absence of strict diagnoses: psychiatry might be moving away from the legacy of Emil Kraepelin and toward that of Sigmund Freud. Although this idea of an interconnected tree of psychiatric possibility is in its infancy, it might be another turning point in the history of mental health care.

Whatever paradigm shifts lie in the future, there are likely to be revolutions in unlikely places. As history teaches us, treatments can emerge from sanatoriums in New York, slaughterhouses in Rome, and grandmothers in Zimbabwe. The treatment of depression is a story that connects us all, no matter our sex, age, or where we live.

Six months after his trip to the United States, Dixon Chibanda walked into a glitzy hotel in Central London wearing sneakers, smart trousers, a shirt, and his favorite leather jacket. Along with eleven other leaders in the field of global mental health, Chibanda was presenting his work to the general secretary of the United Nations, António Guterres. Vikram Patel, an elder of the field of global mental health, was presenting his case for scaling up mental health care systems that are culturally relevant. In Zimbabwe, there are grandmothers. In India, there are social workers. And in the United States, there are peers like Helen Skipper. Waiting in the gold-trimmed hotel lobby in Central London, besuited staff buzzing around the floor as they carried the bags and coats of arriving guests, both Chibanda and Patel thought that this might be a turning point in their field. Unlike a meeting with a member of the World Health Organization, Patel says, the UN doesn't just advocate around health issues, but also political and social crises. If such an intergovernmental organization is on board and funding global mental health projects, he adds, they will have gained a much-needed champion. Without this, global mental health may remain a series of isolated, small projects that can't be scaled up.

Chibanda was more optimistic than ever before. "The whole global mental health community is really pushing in the same direction, together. It's taken years [and] a lot of work from quite a number of different people," he says. "I would say that now we have a critical mass to really bring about change, and at a global level. I really feel that we're onto something big. The last five years or so, for me, have been quite the turning point. Observing my own work [on the Friendship Bench], and others, that's when I started to realize that this was going viral, so to speak."

Although it might take years for the UN and other organizations to make a decision on how to support projects like Chibanda's Friendship Bench and Patel's Healthy Activity Program in Goa, India, the current state in global mental health continues to be pretty desperate. In many low-income countries, nine out of ten people with mental illness still don't have access to any form of psychiatric care, whether it be in the community or otherwise. "It's really pretty grim," Patel says. "The situation right now is probably grimmer than any other area of health—not probably, it is grimmer."

Without significant investment, this gloomy situation is likely to become much worse following the COVID-19 outbreak. While loneliness, distress, and grief are normal reactions to a global health crisis, they are also the fertile soil in which long-term mental disorders can take root and grow. Mental health services—whether it's phone-based therapy, community health workers, or improving access to medication—require international attention before these roots grow too deep, before mental disorders become the next invisible disease to bring the world to a standstill.

Long-term epidemiological studies will soon reveal the people at highest risk of depression in the wake of this crisis, and what can be done to protect them. As Myrna Weissman's trans-generational cohort entered its fortieth year during the pandemic, she saw an opportunity to understand how COVID-19 impacts mental health. It will take years before any robust conclusions can be drawn. "The final story will never really be the final story, except for the overall message: COVID-19 is going to have a big impact on people's mental health," Weissman says while working from her home in New York. "I mean, how couldn't it? Losing your job, not being able to take care of your children, uncertainty, social isolation . . . I mean, that's like every known risk factor for depression and anxiety disorder."

And yet, there is hope in every crisis. Since the start of the pandemic, science has come to the fore in the search for a vaccine. Epidemiologists and science writers have skipped into the spotlight to offer rational, evidence-based guidance for an unpredictable future. Regular exercise became a mainstay of life in lockdown. And, whether they occur in person, over the phone, or through a screen, discussions on mental health have become everyday, normalized. Psychiatrists and psychologists from around the world are hoping that COVID-19 provides the wake-up call that governments need, a death knell for inaction. "Despite the evidence, known for decades, that depression is a leading and growing cause of avoidable suffering in the world, it has attracted relatively little attention from policy makers," The Lancet Commission for Depression 2021 states. "It has generated a 'perfect storm' that requires responses at multiple levels. It underscores the need to make the prevention, recognition, and treatment of depression an immediate global priority. This is a historic opportunity to transform not just the ailing health care systems of the world, but society itself."

If there is a future in global mental health then it is embodied by Dixon Chibanda and the grandmothers who helped make his project a success. Since the clinical trial in 2016, he has established benches on the island of Zanzibar off the eastern coast of Tanzania, in Malawi, and in the Caribbean. He now plans to train community health workers and set up benches in the UK and across the U.S. He has graced international conferences and rubbed shoulders with world leaders. At the World Economic Forum he sat next to Jacinda Ardern, the prime minister of New Zealand. At the inaugural Global Ministerial Mental Health Summit in London in 2018, Prince William and the Duchess of Cambridge sat on a Friendship Bench that was assembled for the occasion. At Davos, Chibanda took a seat along with Elisha London, CEO of United for Global Mental Health. Casually dressed in jeans, a padded gilet, and a jangle of copper wristbands, he smiled as he delicately held up a banner that read "Everyone, everywhere should have someone to turn to in support of their mental health."

Notes

Introduction

2 *the social triggers for depression*: E. S. Paykel, et al., "Life Events and Depression. A Controlled Study," *Archives of General Psychiatry* 21 (1969): 753–60. And G. W. Brown and T. Harris, *Social Origins of Depresssion* (Tavistock Publications, 1978).

3 *recalibrate the immune system*: D. S. Black and G. M. Slavich. *Annals of the New York Academy of Sciences* 1373 (2016): 13–24.

3 *running . . . as effective . . . as first-line treatments*: A. L. Dunn, et al., "Exercise Treatment for Depression. Efficacy and Dose Response," *American Journal of Preventative Medicine* 28 (1) (2005): 1–8.

3 *share nutrients with their neighbors*: T. Klein, R. T. W. Siegwolf, and C. Körner, "Belowground carbon trade among tall trees in a temperate forest," *Science* 352 (6283) (2016): 342–44.

7 *risk of depression in children threefold*: M. M., Weissman, et al., "A 30-Year Study of 3 Generations at High Risk and Low Risk for Depression," *JAMA Psychiatry* 73(9) (2016): 970–77.

Part One. Cutting Steps into the Mountain

The Anatomists

17 *women in elegant English dresses . . . smelling of patchouli*: L. Gandolfi, "Freud in Trieste: Journey to an Ambiguous City," *Psychoanalysis and History* 12, no. 2 (July 2010): 129–51, p. 135.

17 *"five seconds from the seashore"*: Ibid.

17 *"lobed organs"*: M. Weiser, "From the Eel to the Ego: Psychoanalysis and the Remnants of Freud's Early Scientific Practice," *Journal of the History of the Behavioral Sciences* 49, no. 3 (May 2013): 1–22.

17 *"Recently a Trieste zoologist claimed to have discovered the testicles"*: Ibid.
17 *four hundred dead eels*: L. C. Triarhou and M. del Cerro, "Freud's Contribution to Neuroanatomy," *JAMA Neurology* 42 (1985): 282–87.
17 *"stained from the white and red blood of the sea creatures"*: L. Schröter, *The Little Book of the Sea* (London: Granta Books, 2009), 165.
18 *his workbench with a view over the seashore*: Gandolfi, "Freud in Trieste: Journey to an Ambitious City," p. 138. "The sea may be seen at all times from my window [. . .] in front of which is my worktable."
18 *Having excelled at school*: E. Jones, *The Life and Work of Sigmund Freud* (New York: Penguin, 1967), 72. "30 March 1881—grade 'excellent.'"
18 *top of his class*: Ibid., 48.
18 *seven languages*: Ibid., 49. Freud spoke Latin, Greek, French, English, Italian, Spanish, and Hebrew.
18 *read Shakespeare from the age of eight*: Ibid.
19 *"kymographs, a myograph, compasses, ophthalmometers, scales"*: E. Lesky, *The Vienna Medical School of the Nineteenth Century* (Baltimore: Johns Hopkins University Press, 1976), 237.
19 *sewage flowed uncovered*: Ibid., 248. "No other great European city was as much in need of thorough sanitary organization as Vienna."
19 *"In the streets, in the squares, in the public buildings"*: Ibid., 248, quoting Jaromir von Mundy.
19 *anti-Semitism cast a dark shadow*: B. Koller, Carl Koller. In S. Freud, *Cocaine Papers* (New York: Stonehill, 1974). "An evil-smelling vine that twined about the whole social structure of Vienna, choking so many green hopes to death."
20 *Freud had zero tolerance*: Jones, *Life and Work of Sigmund Freud*. "[He] felt Jewish to the core." "Sensitive to the slightest hint of anti-Semitism."
20 *dreamed of England*: Ibid., 50. "Land of his dreams."
20 *hat flung into the gutter*: P. Gay, *Freud: A Life for Our Time* (London: Papermac, 1988), 11.
20 *Freud's primary tool of observation*: Jones, *Life and Work of Sigmund Freud*, 70. "The microscope was his one and only tool."
20 *Crafted from brass and steel*: "Sigmund Freud's Microscope—On the 150th Birthday Anniversary of the Histologist, Microscopy-UK," http://www.microscopy-uk.org.uk/mag/indexmag.html?http://www.microscopy-uk.org.uk/mag/artoct06/mc-freud.html.
20 *built by Edmund Hartnack*: L. C. Triarhou, "Exploring the Mind with a Microscope," *Hellenic Journal of Psychology* 6, no. 1 (2009): 1–13.
20 *life blown up to three hundred or five hundred times*: "Sigmund Freud's Microscope," Microscopy-UK.org.
20 *rifle factory and, before that, a stable*: Lesky, *Vienna Medical School of the Nineteenth Century*, 231.
20 *"[N]othing else than spinal ganglion cells"*: Jones, *Life and Work of Sigmund Freud*, 66.
21 *"three-dimensional, alive, alien eyeballs bobbing in space"*: B. Carey, "Analyse These," *New York Times*, April 25, 2006.
21 *"[B]y this method . . . the fibres are made to show in pink"*: S. Freud, "A New Histological Method for the Study of Nerve-Tracts in the Brain and Spinal Cord," *Brain* 7 (1884): 86–88.

21 *"I see your methods alone will make you famous yet"*: E. Freud, ed., *Letters of Sigmund Freud* (New York: Dover, 1992), 73 (Letter 27).

21 *silver-based stain that was far simpler*: M. Solms, "An Introduction to the Neuroscientific Works of Sigmund Freud," in G. van de Vijver and F. Geerardyn, eds., *The Pre-Psychoanalytic Writings of Sigmund Freud* (New York: Routledge, 2002), 22. "It seems that it was a difficult and fragile method—temperamental, even. In Freud's hands it yielded a wealth of observations and discoveries, but few of his colleagues had the sensitivity and patience to achieve comparable results."

21 *he came excruciatingly close to the theory that nerves weren't interconnected*: Triarhou and del Cerro, "Freud's Contribution to Neuroanatomy," 282–87. "On one image where two neurons seem to be continuous: a question mark. Did Freud doubt the continuity of nerve nets . . . Favoring the individuality of the neuron."

22 *Cajal cited Freud's work*: Triarhou, "Exploring the Mind with a Microscope," 1–13.

22 *go hungry from one day to the next*: R. K. Teusch, "More Courtship Letters of Freud and Martha Bernays," *Journal of the American Psychoanalytic Association* 65, no. 1 (2017): 111–25. "I need again to give more money to my family who suffer real hunger on many days . . . I am so worried about them all."

22 *"It increases my self-respect to see how much I am worth to anyone"*: Jones, *Life and Work of Sigmund Freud*, 156.

22 *stains on young Freud's suit remained unchallenged*: E. Freud, ed., *Letters of Sigmund Freud*, 122 (Letter 50). "My best suit is so worn I hardly dare to travel in it. . . . It is really the stains that mostly spoil it."

22 *diet of corned beef, cheese, and bread*: Jones, *Life and Work of Sigmund Freud*, 154.

23 *Freud used scraps of paper, pages torn from his lab notebook, and old envelopes*: E. Freud, ed., *Letters of Sigmund Freud*, 126 (Letter 54).

23 *"Marty, and in return I will love you very much"*: Ibid., 94 (Letter 36).

23 *"I have been so caught up in myself"*: Ibid., 89 (Letter 33).

23 *"I think I can feel happy nowadays only in your presence"*: Ibid., 123 (Letter 51).

23 *centered around a single delusion*: A. J. Lewis, "Melancholia: A Historical Review," *Journal of Mental Science* 80 (1934): 1–42. "The essence of melancholy lies in an obstinate fixation of attention on one object and in mistaken judgment about it."

24 *world was about to fall on top of them*: S. W. Jackson, "Melancholia and the Waning of the Humoral Theory," *Journal of the History of Medicine* 33, no. 3 (July 1978): 367–76.

24 *gifted in the arts and sciences for their ruminative pastimes*: J. Radden, *The Nature of Melancholy: From Aristotle to Kristeva* (New York: Oxford University Press, 2000), 87. "The Aristotelian link between brilliance and melancholy, a link that . . . was to become a resounding theme throughout the Renaissance and later writing on melancholy."

24 *"The word melancholia . . . consecrated in popular language"*: G. E. Berrios, *The History of Mental Symptoms: Descriptive Psychopathology Since the Nineteenth Century* (Cambridge, UK: Cambridge University Press, 1996).

24 tristimania: Ibid.

24 *"nerves" was sweeping through the Western world*: E. Shorter, *A History of Psychiatry: From the Era of the Asylum to the Age of Prozac* (Hoboken, NJ: John Wiley & Sons, 1997).

25 *For women, meanwhile, caring for sick relatives*: E. L. Bassuk, "The Rest Cure: Repetition or Resolution of Women's Conflicts?" *Poetics Today* 6 (1985): 245–57; S. Poirier, "The Weir Mitchell Rest Cure," *Women's Studies* 10 (1983): 15–40.

25 *it made a visit to mental asylums . . . unlikely*: Shorter, *History of Psychiatry*, 113. "Patients found the notion of suffering from a physical disorder of the nerves far more reassuring than learning that their problem was insanity. . . . Unlike psychiatric illness, nervous illness was thought for the most part to be non-inheritable and thus non-stigmatizing."

25 *"Private Institution for Brain and Nervous Disease"*: Ibid., 118.

25 *"I do not permit the patient to sit up or to sew or write or read"*: Poirier, "The Weir Mitchell Rest Cure," 20.

26 *their mail was carefully checked and censored*: Bassuk, "The Rest Cure," 245–57. "news from home carefully censored."

26 *lashings were used*: A. Stiles, "The Rest Cure, 1873–1925," Branch, October 2012, http://www.branchcollective.org/?ps_articles=anne-stiles-the-rest-cure-1873-1925.

26 *dozens of eggs*: D. Martin, "The Rest Cure Revisited," *American Journal of Psychiatry* 164, no. 5 (2007): 737–38.

26 *a mild electric current was applied*: T. H. Weisenberg, "The Weir Mitchell Rest Cure Forty Years Ago and Today," *Archives of Clinical Neuropsychology* 14, no. 3 (1925): 384–89.

26 *Regular massages helped encourage blood flow*: P. Vertinsky, "Feminist Charlotte Perkins Gilman's Pursuit of Health and Physical Fitness as a Strategy for Emancipation," *Journal of Sport History* 16 (1989): 5–26, p. 14.

26 *"The Yellow Wallpaper," a harrowing short story*: "The Yellow Wall-paper," *New England Magazine* (January 11, 1892): 647–57.

26 *"I have never spent such a wretched eight months in my life"*: Virginia Woolf, letter to Violet Dickinson, October 30, 1904, in V. Woolf, *The Complete Collection* (Century Book; Oregon Publishing, 2017) (Letter 186).

26 *"doctors had not the slightest idea of the nature"*: Stiles, "The Rest Cure, 1873–1925."

27 *"What she needed [was] a means of humanitarian action"*: Poirier, "The Weir Mitchell Rest Cure," 15–40.

27 *the "west cure"*: A. Stiles, "Go Rest Young Man," American Psychological Association, January 2012, https://www.apa.org/monitor/2012/01/go-rest.

27 *"[Von Gudden] thought that the only entrance"*: E. Kraepelin, *Memoirs* (Berlin: Springer-Verlag, 1987).

28 *"[I]n eight days almost an entire rabbit brain"*: A. Danek, W. Gudden, and H. Distel, "The Dream King's Psychiatrist Bernhard von Gudden (1824–1886): A Life Committed to Rationality," *History of Neurology* 46 (1989): 1349–53.

28 *Kraepelin was handed the brains of reptiles*: H. Hippius et al., *The University Department of Psychiatry in Munich: From Kraepelin and His Predecessors to Molecular Psychiatry* (New York: Springer, 2008).

28 *"With the means available . . . it was almost completely impossible"*: Kraepelin, *Memoirs*, 27.

29 *"In spite of the Spartan simplicity"*: Ibid.

29 *a yellowish complexion and a thick Vandyke beard*: E. Engstrom, "Emil Kraepelin:

Psychiatry and Public Affairs in Wilhelmine Germany," *History of Psychiatry* 2 (1991): 111–32.

29 *a dollop of raspberry syrup*: U. Muller, P. Fletcher, and H. Steinberg, "The Origin of Pharmacopsychology: Emil Kraepelin's Experiments in Leipzig, Dorpat, and Heidelberg (1882–1892)," *Psychopharmacology* 184 (2006): 131–38, p. 135.

29 *"higher doses caused an earlier onset"*: Ibid., 133.

29 *carbonated water as a control*: Ibid., 135.

30 *"And the danger of developing morphinism"*: Ibid.

30 *"We felt that we were trailblazers"*: Kraepelin, *Memoirs*.

31 *Wundt, didn't even mention his experiments in a comprehensive review*: Muller, Fletcher, and Steinberg, "The Origin of Pharmacopsychology," 131–38.

31 *during the Easter holidays of 1883*: Hippius et al., *University Department of Psychiatry in Munich*.

31 *hesitant to write anything at first*: Kraepelin, *Memoirs*.

31 *In its three hundred pages*: R. Blashfield, *The Classification of Psychopathology: Neo-Kraepelinian and Quantitative Approaches* (New York: Springer, 1984).

31 *embarrassing lack of effective treatments*: Hippius et al., *University Department of Psychiatry in Munich*. Chapter 7, "The Munich Hospital Managed by Emil Kraepelin." "Powerlessness of medical treatment."

31 *"confusing mass" of mental illness*: Ibid.

31 *Liters of alcohol were used*: Kraepelin, *Memoirs*, 11. "Beer played an exceedingly important part in the routine of the asylum."

32 *Kraepelin's* Compendium *would grow in size and popularity*: Shorter, *History of Psychiatry*, 103. "So startling were the views Kraepelin started announcing that the entire psychiatric world would follow each new edition with rapt interest."

Über Coca

33 *"A German has tested this stuff on soldiers"*: S. Freud, *Cocaine Papers* (New York: Stonehill, 1974). Chapter 11, "On the General Effect of Cocaine." March 1885. Aschenbrandt prompted Freud "to study the effects of coca on my own person and on others."

33 *"rendered them strong and capable of endurance"*: E. Freud, ed., *Letters of Sigmund Freud* (New York: Dover, 1992), 107 (Letter 43).

33 *"Perhaps others are working at it"*: E. Jones, *The Life and Work of Sigmund Freud* (New York: Penguin, 1967), 90.

34 *similar to caffeine in tea*: S. Freud, *Cocaine Papers*, 58.

34 *German pharmaceutical company Merck*: Jones, *Life and Work of Sigmund Freud*, 90.

34 *one part cocaine to ten parts water*: S. Freud, *Cocaine Papers*, 58. "0.05g in 1% water solution."

34 *floral and not at all unpleasant*: Ibid. "pleasant aromatic flavors."

34 *A sensation of heat spread through his skull*: Ibid., 59. "Intense feeling of heat in the head."

34 *"In my last severe depression I took coca"*: Ibid., 10.

34 *"a song of praise to this magical substance"*: Jones, *Life and Work of Sigmund Freud*, 95.

34 *well-stocked library in Vienna*: Ibid. Vogl's Library. "to assemble everything worth knowing about coca."

34 *rain was pouring outside Freud's bedroom window*: E. Freud, ed., *Letters of Sigmund Freud*, 119. July 1, 1884. "The new month has begun with rain." "Coca appeared today, but I haven't seen it yet."

34 *"It was inevitable that a plant"*: S. Freud, *Cocaine Papers*, 63.

34 *In 1860, a young graduate student*: A. Niemann, "Ueber eine neue organische Base in den Cocablättern," *Archiv der Pharmazie* 153, no. 2 (1860): 129–256.

35 *"almost all the maladies the flesh is heir to"*: S. Freud, *Cocaine Papers*, 128.

35 *"frequently witnessed the morose, silent, taciturn"*: J. Spillane, *Cocaine: From Medical Marvel to Modern Menace* (Baltimore: Johns Hopkins University Press, 2002).

35 *"Cocaine changed the scene as if by magic"*: S. Freud, *Cocaine Papers*, 145.

35 *beard trimmed*: Ibid. Chapter 8, "Papers on Coca and Other Prospects." January 1885.

35 *"Psychiatry is rich in drugs"*: Ibid.

36 *offering psychiatry the first antidepressant*: During a 1970 symposium on psychomimetic agents, Dr. Larry Stein said, "Cocaine is almost the perfect tricyclic antidepressant . . . but we heard that from Freud 80 years ago." Ibid.

36 *A popular and oft-used portrait*: *Sigmund Freud* by Max Halberstatt, 1932. Freud Museum, London.

36 *first article on psychoanalysis in 1896*: S. Freud, "L'hérédité et l'étiologie des névroses," *Revue neurologique* 4 (1896): 161–69.

36 *widely known for his work twenty or so years later*: G. Makari, *Revolution in Mind: The Creation of Psychoanalysis* (New York: Harper, 2008), 299. "The great flowering of psychoanalysis occurred between 1918 and 1938."

37 *Drinking cocaine to calm his insecurities*: Ibid., 71. "Fueled by cocaine, he hoped to make a psychology for neurologists."

37 *"with a new perception of perfection"*: P. Gay, *Freud: A Life for Our Time* (London: Papermac, 1988), 49.

37 *a disorder of women*: Makari, *Revolution in Mind*, 15.

37 *He wasn't the first doctor to propose this*: The physician and psychologist Paul Briquet proposed that men could suffer from hysteria in 1859.

38 *study cerebral paralysis and aphasia*: Gay, *Freud: A Life for Our Time*, 48.

38 *"wake" up and make this into a reality*: Makari, *Revolution in Mind*, 18.

38 *"I had the profoundest impression"*: S. Freud, *An Autobiographical Study: Inhibitions, Symptoms, and Anxiety, the Question of Lay Analysis* (London: Hogarth Press, 1925).

38 *"It induces an artificial form of alienation"*: Jones, *Life and Work of Sigmund Freud.*

38 *"The more psychiatry seeks, and finds, its scientific basis"*: E. Shorter, *A History of Psychiatry: From the Era of the Asylum to the Age of Prozac* (Hoboken, NJ: John Wiley & Sons, 1997), 77.

39 *"These unfortunate beings lead a miserable life"*: L. Lewin, *Phantastica: A Classic Survey on the Use and Abuse of Mind-Altering Plants* (Rochester, VT: Park Street Press, 1988), 72.

39 *"third scourge of humanity"*: S. Bernfeld, "Freud's Studies on Cocaine, 1884–1887," *Journal of the American Psychoanalytic Association* 35 (1954): 445, 603.

40 *saw white snakes and insects crawling over his skin*: S. Freud, *Cocaine Papers*, 158.

40 *"a thoroughly excellent person"*: Ibid.

40 *"Craving for and Fear of Cocaine"*: S. Freud, "Bemerkungen über Cocainsucht und Cocainfurcht, mit Beziehung auf einen Vortrag W. A. Hammonds," *Wiener klinische Wochenschrift*, no. 28 (1887): 929–32.

40 *He decried Albrecht Erlenmeyer's statement*: Bernfeld, "Freud's Studies on Cocaine, 1884–1887," 445. "The so-called third scourge of humanity, as Erlenmeyer most pathetically labels cocaine."

40 *"On the contrary . . . an aversion to the drug took place"*: Ibid., 604.

40 *"It made no impression on his contemporaries"*: Ibid.

41 *"the Haven of happiness"*: Jones, *Life and Work of Sigmund Freud.* Chapter 8, "Marriage."

41 *Rembrandt's* Anatomy Lesson *and Fuseli's* Nightmare: W. Johnstone, *The Austrian Mind: An Intellectual and Social History* (Berkeley: University of California Press, 1983), 240.

41 *"unknown central agent, which can be removed chemically"*: Bernfeld, "Freud's Studies on Cocaine, 1884–1887," 601.

"Psychiatry's Linnaeus"

42 *at the University of Dorpat . . . in Estonia in 1886*: H. Steinberg and M. C. Angermeyer, "Emil Kraepelin's Years at Dorpat as Professor of Psychiatry in 19th-Century Russia," *History of Psychiatry* 12 (2001): 297–327.

42 *over the River Embach, to his new home*: Ibid.

42 *"kind of exile"*: E. Kraepelin, *Memoirs* (Berlin: Springer-Verlag, 1987), 57.

42 *"trains came so rarely and were slow"*: Ibid., 47.

42 *Hans, his first son, died of sepsis*: See chapter 7, "The Munich Hospital Managed by Emil Kraepelin," in H. Hippius et al., *The University Department of Psychiatry in Munich: From Kraepelin and His Predecessors to Molecular Psychiatry* (New York: Springer, 2008).

42 *"little book"*: Kraepelin, *Memoirs.*

42 *3,000-page, four-volume tome by 1915*: The exact count was 3,026 pages. R. Blashfield, *The Classification of Psychopathology: Neo-Kraepelinian and Quantitative Approaches* (New York: Springer, 1984).

43 *"Dream King"*: A. Danek, W. Gudden, and H. Distel, "The Dream King's Psychiatrist Bernhard von Gudden (1824–1886): A Life Committed to Rationality," *History of Neurology* 46 (1989): 1349–53.

43 *grip the city of Dorpat in its vise*: Steinberg and Angermeyer, "Emil Kraepelin's Years at Dorpat as Professor of Psychiatry in 19th-Century Russia," 297–327.

43 *"My wife and I are fine now"*: Ibid.

43 *"I fled into the fresh air"*: Hippius et al., *University Department of Psychiatry in Munich*, chapter 7.

43 *ruby-tailed wasps, iridescent wild bees, and wall lizards*: Research trip to Heidelberg by author, October 2018.

44 *"Psychiatry's Linnaeus"*: E. J. Engstrom and K. Kendler, "Emil Kraepelin: Icon and Reality," *History of Psychiatry* 172, no. 12 (2015): 1190–96.

44 *never been so treasured by one man before*: M. M. Weber and E. J. Engstrom,

"Kraepelin's Counting Cards," *History of Psychiatry* 8 (1997): 375–85. "German psychiatrists deemed this highly sophisticated system impractical for their everyday work."

44 *"place in the country"*: E. Kahn, *The Emil Kraepelin Memorial Lecture*, New York Meeting, 1956.

44 *a villa on the tranquil shores of Lake Maggiore*: Hippius et al., *University Department of Psychiatry in Munich*, fig. 7.8.

44 *"[A]gain and again [he] worked through thousands"*: Weber and Engstrom, "Kraepelin's Counting Cards," 375–85.

45 *arose in young adulthood (before the age of twenty-five)*: E. Kraepelin, *Manic Depressive Insanity and Paranoia* (Edinburgh: E. & S. Livingstone, 1921).

45 *Bipolar . . . distinct mood disorder until 1957*: M. Mendelson, *Psychoanalytic Concepts of Depression* (Hicksville, NY: Spectrum Publications, 1974), 25. Leonard (1957).

45 *"are but manifestations of one disease process"*: Kahn, *The Emil Kraepelin Memorial Lecture.*

45 *"defective heredity is the most important, occurring in seventy to eighty percent of cases"*: Kraepelin, *Manic Depressive Insanity and Paranoia.*

45 *"[S]o-called psychic causes"*: M. Shepherd, "Two Faces of Emil Kraepelin," *British Journal of Psychiatry* 167 (1995): 174–83.

46 *"sources of emotional disturbance should be avoided"*: Kraepelin, *Manic Depressive Insanity and Paranoia.*

46 *"Modern psychiatry begins with Kraepelin"*: E. Slater and M. Roth, *Clinical Psychiatry* (London: Bailliere, Tindall, and Cassell, 1969), 10.

46 *"an icon who helped guide us"*: Engstrom and Kendler, "Emil Kraepelin: Icon and Reality," 1190–96.

46 *"cutting steps in the mountain"*: N. Hansson et al., "Psychiatry and the Nobel Prize: Emil Kraepelin's Nobelibility," *TRAMES* 20 (2016): 393–401, p. 398.

A Melancholic Humor

47 *"both in strength and quantity, and are well mixed"*: G. E. R. Lloyd, ed., *Hippocratic Writings*, translated by W. N. Mann and J. Chadwick (New York: Penguin, 1983), 262.

48 *a physician to gladiators*: C. Galen, "On the Affected Parts" (165 CE), in J. Radden, ed., *The Nature of Melancholy: From Aristotle to Kristeva* (New York: Oxford University Press, 2000), 61.

48 *died in 210 AD when he was in his seventies*: Ibid.

48 *"He was a great borrower"*: F. Alexander and S. Selesnick, *The History of Psychiatry* (New York: Harper & Row, 1966).

49 *the "gassy disease"*: Radden, ed., *Nature of Melancholy*, 66.

49 *"I can tell you . . . that atrabilious blood is generated by"*: Ibid.

50 *"If the outflowing blood"*: Ibid.

50 *In 849 AD, Abu Zayd al-Balkhi was born*: M. Badri, *Abu Zayd al-Balkhi's Sustenance of the Soul: The Cognitive Behavior Therapy of a Ninth Century Physician* (Herndon, VA: International Institute of Islamic Thought, 2014).

52 *two missionaries, Constantinus Africanus and Gerard of Cremona*: Radden, ed., *Nature of Melancholy*, 75.
52 *"came into being at the very beginning out of Adam's semen"*: Hildegard of Bingen, *Book of Holistic Healing* (1151–58 CE), in Radden, ed., *Nature of Melancholy*, 81.
52 *"Spirit is a most subtle vapor"*: R. Burton, *The Anatomy of Melancholy* (1924; New York: New York Review of Books, 2001), 146.
53 *"arise from a Melancholick humor"*: S. Jackson, "Melancholia and the Waning of the Humoral Theory," *Journal of the History of Medicine* 33, no. 3 (1978): 367–76.
53 *seventeenth century that the circulation of blood*: D. Ribatti, "William Harvey and the Discovery of the Circulation of the Blood," *Journal of Angiogenesis Research* 1, no. 3 (2009).
53 *microscopic cells hadn't yet been proposed*: T. Schwann, *Mikroskopische Untersuchungen über die Übereinstimmung in der Struktur und dem Wachsthum der Thiere und Pflanzen* (Berlin: Sander, 1839).

Instruments of Cure

54 *from ten and a half million to twenty-one million*: See specifically chapter 1, "The UK Population: Past, Present, and Future," in J. Jefferies, *Focus on People and Migration* (London: Palgrave Macmillan, 2005), 3.
54 *"melanic" mutant that was better camouflaged*: H. B. D. Kettlewell, "Selection Experiments on Industrial Melanism in the Lepidoptera," *Heredity* 9 (1955): 323–42.
54 *death of loved ones (either from disease or accidents at work)*: K. Jones, *Asylums and After: A Revised History of the Mental Health Services: From the Early 18th Century to the 1990s* (New York: Continuum International Publishing Group, 1993).
54 *Mania and melancholia . . . were the two diagnoses*: J. Haslam, *Observations on Madness and Melancholy, Including Practical Remarks on Those Diseases* (J. Callow, 1809), 19. "Insanity is now separated into mania and melancholia."
55 *"[T]hat has been the practice invariably for years"*: M. Glover, *The Retreat, York: An Early Quaker Experiment in the Treatment of Mental Illness* (York, UK: William Sessions, 1984).
55 *"were treated worse . . . than the beasts of the field"*: J. Connolly, *Treatment of the Insane Without Mechanical Restraints* (1856; London: Dawsons of Pall Mall, 1973).
55 *Hannah Mills, a local widow and Quaker, died in suspicious*: Jones, *Asylums and After.*
55 *Described as "low and melancholy"*: Glover, *The Retreat, York*, 36.
56 *Mathematics and the classics . . . most beneficial literature*: Jones, *Asylums and After.*
56 *delicacies such as oranges and figs*: Ibid.
56 *"I was on the brink of destruction"*: Glover, *The Retreat, York.*
57 *tax on tea from 100 percent to 12 percent*: Ibid.
57 *"What tenderness with strength combined"*: Ibid.
58 *"Darkness and bonds are . . . never applied"*: Connolly, *Treatment of the Insane Without Mechanical Restraints*, 166.
58 *"The old prison-air has departed"*: Ibid., 82.
58 *"Not only to the maniacal patient"*: Ibid., 60.
58 *"Recreation . . . is so necessary"*: A. L. Ashworth, *Stanley Royd Hospital, Wakefield:*

One Hundred and Fifty Years: A History (Wakefield, UK: Stanley Royd Hospital, 1975), 51.

58 *Outside, there was cricket*: S. Rutherford, *The Victorian Asylum* (London: Shire, 2008).

58 *"The convalescent wear an expression"*: Connolly, *Treatment of the Insane Without Mechanical Restraints*, 87.

59 *fed a diet of beef tea and arrowroot*: A. Digby, "Moral Treatment at the Retreat 17-96-1846," in R. Porter, W. F. Bynum, and M. Shepherd, *The Anatomy of Madness*, vol. 2 (Abingdon, UK: Routledge, Kegan, and Paul, 1985), 66.

59 *brewed its own beer on-site*: Ashworth, *Stanley Royd Hospital.*

59 *rhubarb was often given as a natural laxative*: Rutherford, *Victorian Asylum.*

59 *"mournful prophet"*: M. N. Ikuigu and E. A. Ciaravino, *Psychosocial Conceptual Practice Models in Occupational Therapy: Building Adaptive Capability* (London: Elsevier Health Sciences, 2007).

59 *Dorothea Dix helped to spread the values*: According to Dr. James W. Trent Jr., "Dix insisted that hospitals for the insane be spacious, well ventilated, and have beautiful grounds. In such settings, Dix envisioned troubled people regaining their sanity," in J. W. Trent Jr., "Moral Treatment," VCU Libraries Social Welfare History Project, January 24, 2014, https://socialwelfare.library.vcu.edu/issues/moral-treatment-insane/.

59 *After a nervous breakdown in 1836, Dix*: Ikuigu and Ciaravino, *Psychosocial Conceptual Practice Models in Occupational Therapy.*

60 *long been known to spontaneously recover*: Haslam, *Observations on Madness and Melancholy*, 257. "Patients, who are in a furious state, recover in a larger proportion than those who are depressed and melancholic." 62/100 vs. 27/100.

60 *Between 10 to 15 percent of people with depression die by suicide*: X. Gonda et al., "Prediction and Prevention of Suicide in Patients with Unipolar Depression and Anxiety," *Annals of General Psychiatry* 6 (2007): 23.

60 *Chronic depression*: See: M. Ten Have et al., "Recurrence and Chronicity of Major Depressive Disorder and Their Risk Indicators in a Population Cohort," *Acta Psychiatrica Scandinavica* 137, no. 6 (2018): 503–15; J. Verduijn, J. E. Verhoeven, Y. Milaneschi et al., "Reconsidering the Prognosis of Major Depressive Disorder Across Diagnostic Boundaries: Full Recovery Is the Exception Rather Than the Rule," *BMC Medicine* 15, no. 1 (2017): 215.

60 *the term* efficacy: U. P. Gielen, J. M. Fish, and J. G. Draguns, eds., *Handbook of Culture, Therapy, and Healing* (Mahwah, NJ: Lawrence Erlbaum Associates, 2004).

61 *"We have 36 males sleeping without beds"*: E. Shorter, *A History of Psychiatry: From the Era of the Asylum to the Age of Prozac* (Hoboken, NJ: John Wiley & Sons, 1997), 47.

61 *In 1904, a total of 150,000 people*: Ibid., 34.

61 *"The treatment is humane but it necessarily lacks individuality"*: Jones, *Asylums and After.*

61 *"this is just like being in a show"*: Digby, "Moral Treatment at the Retreat 17-96-1846."

61 *a corridor that was over half a kilometer long*: Jones, *Asylums and After.*

61 *a jangle of keys hung from their belts*: Ibid.

61 *"individuals with illnesses of the nerves and the brain"*: Shorter, *History of Psychiatry*.
62 *"[w]e are groping in the dark for what we do not yet know"*: G. H. Savage, *Insanity and Allied Neuroses* (London: Cassell, 1884).

The Talking Cure

63 *"Sigmund Freud did not so much create a revolution"*: G. Makari, *Revolution in Mind: The Creation of Psychoanalysis* (New York: Harper, 2008), 5.
64 *"She spoke an agrammatical jargon"*: H. F. Ellenberger, "The Story of 'Anna O': A Critical Review with New Data," *Journal of the History of the Behavioral Sciences* 8 (1972): 267–79.
64 *"a real one and an evil one"*: Makari, *Revolution in Mind*, 39.
65 *playing with a Newfoundland dog*: Ellenberger, "The Story of 'Anna O,'" 269.
66 *"talking cure"*: Ibid.
66 *Pappenheim would spend the next few years institutionalized*: Makari, *Revolution in Mind*, 39. "Bertha had bounced in and out of sanatoriums between 1883 and 1887."
67 *"the symptoms vanished"*: S. Freud, *Introductory Lectures on Psycho-Analysis* (London: Allen & Unwin, 1968), 236.
67 *"Insight such as this"*: S. Freud, *The Interpretation of Dreams* (New York: Carlton House, 1931), preface.
68 *favorite book from his childhood*: L. Börne, "The Art of Becoming an Original Writer in Three Days" (1823). https://www1.cmc.edu/pages/faculty/LdelaDurantaye/art_of_ignorance_harvard_review.pdf.
68 *"a resolute pursuit of the historical meaning of a symptom"*: S. Freud, *Introductory Lectures on Psycho-Analysis*, 230.
68 *few, if any, depressed patients*: R. Wallenstein, *The Talking Cures: Psychoanalyses and the Psychotherapies* (New Haven, CT: Yale University Press, 1995). "neuropathic degeneracy . . . psychoses, states of confusion and deeply rooted (I might say toxic) depression are . . . not suitable for psychoanalysis; at least not for the method as it has been practiced up to the present."
68 *"who interacts sparsely with others"*: E. Shorter, *How Everyone Became Depressed: The Rise and Fall of the Nervous Breakdown* (New York: Oxford University Press, 2013).
69 *the "Rat Man" . . . the "Wolf Man"*: S. Freud, *Case Histories II*, vol. 9 (New York: Penguin, 1991).
69 *a depletion of sexual energy from too much masturbating*: K. J. Zucker, "Freud's Early Views on Masturbation and the Actual Neuroses," *Journal of the American Academy of Psychoanalysis* 7, no. 1 (1979): 15–32.
69 *only 351 copies were sold*: S. Freud, *Interpretation of Dreams*, 43.
69 *"The attitude adopted by reviewers"*: Ibid., 46.

Love and Hate

74 *The first significant psychoanalytic theory of depression*: U. May-Tolzmann, "Die Entdeckung der bösen Mutter: Ein Beitrag Abrahams zur Theorie der Depres-

sion" ("The Discovery of the Bad Mother: Abraham's Contribution to the Theory of Depression"), *Luzif Amor* 10, no. 20 (1997): 98–131. "it was not until 1911 that the first innovative contributions on the subject of depression appeared. They were not Freud's, rather they originated with Abraham."

74 *proposed by Karl Abraham*: A. Bentinck van Schoonheten, *Karl Abraham: Life and Work, a Biography* (Abingdon, UK: Routledge, 2015). Chapter 8. "although Freud had on several occasions attempted to develop a psychoanalytic theory about depression, nothing had yet come of it."

74 *pull his protégé, Carl Jung, closer*: G. Makari, *Revolution in Mind: The Creation of Psychoanalysis* (New York: Harper, 2008), 374. "Freud himself could not get excited about Abraham." "Freud had first castigated Abraham before coming around to his side."

74 *"the depressive conditions have not found similar consideration"*: E. Shorter, *How Everyone Became Depressed: The Rise and Fall of the Nervous Breakdown* (New York: Oxford University Press, 2013), 114.

74 *when he was five years old, his mother died*: U. May, "Abraham's Discovery of the Bad Mother: A Contribution to the History of the Theory of Depression," *International Journal of Psychoanalysis* 82 (2001): 283–305, p. 289.

75 *"ambivalence"*: Makari, *Revolution in Mind*, 209.

75 *"I cannot love people; I have to hate them"*: "Notes on the Psycho-Analytical Investigation and Treatment of Manic-Depressive Insanity and Allied Conditions" (1911), in K. Abraham, *Selected Papers on Psychoanalysis* (London: Hogarth Press, 1927).

75 *dreams were often full of masochistic desires*: E. Freud, ed., *Letters of Sigmund Freud* (New York: Dover, 1992) (Letter 273A). "Too much violence and criminality was uncovered by the analyses of my melancholic patients. The self-reproaches do indicate repressed hostile feelings."

75 *"A recurring theme in Abraham's work"*: van Schoonheten, *Karl Abraham*, 7.

75 *"His mother was still physically present"*: Ibid.

76 *Segantini had died from an infection in 1899*: May, "Abraham's Discovery of the Bad Mother," 289.

76 *"He spoke little, but his silence was justified"*: van Schoonheten, *Karl Abraham*. Chapter 13. Theodore Reik, sent to Abraham from Vienna.

76 *their first meeting in December 1907*: Ibid. Chapter 5. The first meetings between Freud and staff from the Burghölzli clinic.

76 *nearly five hundred letters*: H. Abraham and E. Freud, eds., *A Psycho-Analytical Dialogue: The Letters of Sigmund Freud and Karl Abraham* (London: Hogarth Press, 1965), Foreword; 492 letters.

77 *"I was actually in Berlin for 24 hours"*: Ibid. (Letter 49F).

77 *"Now, I must beg you, dear Professor"*: Ibid. (Letter 50A).

77 *"That is what interests me particularly"*: Ibid. (Letter 68A).

77 *"What gave me most pleasure"*: Ibid. (Letter 70A).

77 *"This was a partnership that paid rich scientific dividends"*: Ibid., Foreword.

78 *opposite a fire station with a sandpit*: van Schoonheten, *Karl Abraham*. Chapter 9.

78 *"something which had never happened before"*: Abraham, "Notes on the Psycho-Analytical Investigation and Treatment of Manic-Depressive Insanity and Allied Conditions."

79 *"I am happier and more care-free"*: Ibid.
79 *"Naturally . . . a definite opinion as regards a cure"*: Ibid.
79 *"Berlin is a difficult but important soil"*: Abraham and Freud, eds., *Psycho-Analytical Dialogue*, 47F.
80 *"the fundamental features of the Freudian trend"*: H. Decker, *The Making of DSM-III: A Diagnostic Manual's Conquest of American Psychiatry* (New York: Oxford University Press, 2013).
80 *"blackest clique of Munich"*: Specifically chapter 7, "The Munich Hospital Managed by Emil Kraepelin," in H. Hippius et al., *The University Department of Psychiatry in Munich: From Kraepelin and His Predecessors to Molecular Psychiatry* (New York: Springer, 2008).
80 *"Psychiatry has given names to various compulsions"*: S. Freud, *Introductory Lectures on Psycho-Analysis* (London: Allen & Unwin, 1968), 221.
80 *"He has to content himself with a diagnosis"*: Ibid., 213.
80 *"a condemnation instead of an explanation"*: Ibid., 221.
81 *Jacobson, a German émigré*: E. Kronold, "Edith Jacobson (1897–1978)," *Psychiatric Quarterly* (1980).
81 *"between the devil and the deep blue sea"*: E. Jacobson, "Transference Problems in the Psycho-Analytic Treatment of Severely Depressive States," *Journal of the American Psychoanalytic Association* 2, no. 4 (1954): 595–606.
81 *"Words and magic were in the beginning one and the same"*: S. Freud, *Introductory Lectures on Psycho-Analysis*, 13.

A First Sketch

82 *horseshoe-shaped building on a tree-lined street*: H. Hippius et al., *The University Department of Psychiatry in Munich: From Kraepelin and His Predecessors to Molecular Psychiatry* (New York: Springer, 2008), fig. 6.4.
82 *experimental psychology, chemistry, serology . . . genealogy*: R. Blashfield, *The Classification of Psychopathology: Neo-Kraepelinian and Quantitative Approaches* (New York: Springer, 1984).
82 *the storage of all of his* Zählkarte: Hippius et al., *University Department of Psychiatry in Munich*. "A large archive for case histories."
82 Psychiatry *textbook, named it "Alzheimer's disease"*: Ibid. Chapter 7, "The Munich Hospital Managed by Emil Kraepelin."
82 *blood sugar, measures of metabolism, gastric juices, nitrogen levels*: E. Mapother, "Discussion on Manic Depressive Psychosis," *British Medical Journal* 2 (1926): 872–79.
82 *"Whoever has listened to the lessons"*: R. Passione, "Italian Psychiatry in an International Context: Ugo Cerletti and the Case of Electroshock," *History of Psychiatry* 15 (2004): 83–104.
83 *"a kind of thought compulsion"*: A. Lewis, "Melancholia: A Historical Review," *Journal of Mental Science* 80 (1934): 1–42, p. 24.
83 *"What we have formulated here is only a first sketch"*: Blashfield, *Classification of Psychopathology*.
83 *"sharp clinical boundaries had become ever more blurred"*: E. J. Engstrom and K.

Kendler, "Emil Kraepelin: Icon and Reality," *History of Psychiatry* 172, no. 12 (2015): 1190–96, p. 1194.

84 "DSM-III *represented a massive 'turning of the page'*": E. Shorter, "The History of Nosology and the Rise of the Diagnostic and Statistical Manual of Mental Disorders," *Dialogues in Clinical Neuroscience* 17, no. 1 (2015): 59–67, p. 59.

84 *a bucolic setting of lakes and trees*: H. Abraham and E. Freud, eds., *A Psycho-Analytical Dialogue: The Letters of Sigmund Freud and Karl Abraham* (London: Hogarth Press, 1965) (Letter 274A). "Allenstein is very prettily situated in a region with forests and lakes only about 40 kilometers west of the Masurian lakes."

84 *"[I] have found confirmation of the solution of melancholia"*: Ibid. (Letter 272F).

84 *"the genius"*: K. Abraham, "A Short Study of the Development of the Libido, Viewed in Light of Mental Disorders" (abridged), in R. V. Frankiel, ed., *Essential Papers on Object Loss* (New York: New York University Press, 1994).

84 *found something that he had not*: Ibid., 81. "I noticed that I felt a quite unaccustomed difficulty in following his train of thought. I was aware of an inclination to reject the idea."

85 *"retroflected hostility"*: D. B. Smith, "The Doctor Is In," *The American Scholar*, September 1, 2009, https://theamericanscholar.org/the-doctor-is-in/.

85 *"worthless, incapable of any achievement"*: S. Freud, *Mourning and Melancholia*, 14th ed. (London: Vintage, 1917).

85 *he waited a month to reply*: Abraham and Freud, eds., *Psycho-Analytical Dialogue*. (Letter 272F) (from Freud): 18.02.1915. (Letter 273A) (from Abraham): 31.03.1915: "I have long postponed commenting on your outline of a theory of melancholia—and not only because I have no real peace for work."

86 *"I should like to remind you"*: Ibid. (Letter 273A).

86 *"to whom we owe the most important"*: A. Bentinck van Schoonheten, *Karl Abraham: Life and Work, a Biography* (New York: Routledge, 2018), chapter 14, "Creation of a Theory About Early Childhood."

87 *"the earliest sense of satisfaction at the mother's breast"*: Ibid., chapter 8.

87 *"Abraham's [ideas] about a pre-Oedipal"*: Ibid.

88 *"[I]t was Melanie Klein who first elaborated"*: M. Mendelson, *Psychoanalytic Concepts of Depression* (Pasadena, CA: Spectrum, 1974).

88 *depression that turned his hair gray and his body thin*: Van Schoonheten, *Karl Abraham*, chapter 16, "The Final War Years in Allenstein": "Abraham's symptoms dated back to the year in which his father and two aunts died. That was [the] same year in which first Gerd and then Hedwig became so ill. Abraham's depression began after a series of deaths and illnesses in the family, just like his mother's depression forty-seven years earlier."

88 *Grecian stone from Freud's personal collection*: Ibid., chapter 11.

88 *"Abraham will in future be recognized"*: Abraham and Freud, eds., *Psycho-Analytical Dialogue*, Foreword.

89 *fish bone lodged in his throat*: Ibid., per the editors' note to Letter 482A: "This apparent bronchitis was the first manifestation of Abraham's fatal illness. It in fact started with an injury to the pharynx from a fish-bone and was followed by septic broncho-pneumonia, lung abscess, and terminal subphrenic abscess. The

illness took the typical course of septicaemia, prior to the introduction of antibiotics, with swinging temperatures, remissions, and euphoria."

89 *deck chair at the Hotel Victoria*: Ibid. (Letter 488A).

89 *read his favorite philosopher*: Ibid. (Letter 490A).

89 *"a man continually and unfailingly at work"*: Ibid. (Letter 484E).

89 *"I have no substitute for him"*: Ibid. (Letter 501F).

89 *"The loss is quite irreplaceable"*: van Schoonheten, *Karl Abraham*. Chapter 16. "The final war years in Allenstein."

90 *October 1926 . . . lost a father figure*: "Hippius et al., *University Department of Psychiatry in Munich*. Chapter 7, "The Munich Hospital Managed by Emil Kraepelin." 07.10.1926.

90 "Dein Name mag vergehen / Bleibt nur dein Werk bestehen": Research trip to Bergfriedhof Heidelberg by author, October 2018.

Part Two. "The Biological Approach Seems to Be Working"

Fighting Fire with Fire

95 *He offered £30,000*: P. Aldridge, "The Foundation of the Maudsley Hospital," in G. Berrios and H. Freeman, eds., *150 Years of British Psychiatry, 1841–1991* (London: Gaskell/Royal College of Psychiatrists, 1991), 87.

95 *psychiatry was changing from German to English*: R. Hayward, "International Relations in Psychiatry," in V. Roelcke, P. J. Weindling, and L. Westwood, eds., *International Relations in Psychiatry* (Woodbridge, UK: Boydell & Brewer, 2010).

95 *Portland stone and red bricks*: Aldridge, "The Foundation of the Maudsley Hospital," 87.

96 *during the Battle of Loos*: E. Jones, "Aubrey Lewis, Edward Mapother, and the Maudsley," *Medical History Supplement* 22 (2003): 3–38.

96 *previously been used to describe the complete opposite conditions*: E. Shorter, *How Everyone Became Depressed: The Rise and Fall of the Nervous Breakdown* (New York: Oxford University Press, 2013).

96 *"could see it was beautiful but . . . could not feel it"*: Quoting Dr. T. A. Ross, in E. Mapother, "Discussion on Manic Depressive Psychosis," *British Medical Journal* 2 (1926): 872–79, p. 878.

97 *"infinitely in degree"*: Ibid.

97 *the duration, severity, and potential recovery*: Ibid. "It is obvious that in many cases we are dealing with a merely quantitative deviation, that the disturbance consists in a response similar in kind to that which like circumstances might provide in us, but morbidly prolonged or disproportionate or disastrous in degree."

97 *"nor susceptibility to psychotherapy will serve"*: Ibid., 872.

97 *"massage is a useful substitute for exercise"*: Ibid., 876.

97 *"analysis of any kind, is merely harmful"*: Ibid.

97 *"helpless when faced with a depressed patient"*: H. E. Lehman and N. Kline, "Clinical Discoveries with Antidepressant Drugs," in M. J. Parnham and J Bruinvels,

eds., *Discoveries in Pharmacology, Volume 1: Psycho- and Neuro-Pharmacology* (Elsevier Science, 1983). "At the time I entered psychiatry, just before ECT was introduced into the Americas, a clinician was almost helpless when faced with a depressed patient."

97 *A range of treatments . . . had been tried and failed*: Ibid.

97 *opium*: L. Kalinowsky, "Biological Treatments Preceding Pharmacotherapy," in F. J. Ayd and B. Blackwell, eds., *Discoveries in Biological Psychiatry* (New York: Lippincott, 1970). "unconvincing results."

97 *the removal of teeth*: A. C. Bested, A. C. Logan, and E. M. Selhub, "Intestinal Microbiota, Probiotics, and Mental Health: From Metchnikoff to Modern Advances: Part I—Autointoxication Revisited," *Gut Pathogens* 5 (2013): 5.

98 *"Even in the more neurotic types of depression"*: Shorter, *How Everyone Became Depressed.* Chapter 6, "A Different Kind of Nervous Breakdown."

98 *"a large preponderance of women"*: "Discussion on the Diagnosis and Treatment of Milder Forms of the Manic-Depressive Psychosis," Section of Psychiatry and Section of Neurology, *Proceedings of the Royal Society of Medicine*, February 11, 1930.

98 *"could not yet find any relief for his suffering"*: E. Jones and S. Rahman, "Framing Mental Illness, 1923–1939: The Maudsley Hospital and Its Patients," *Social History of Medicine* 21 (2008): 1–18.

98 *a series of familiar stages*: G. Davis, "The Most Deadly Disease of Asylumdom: General Paralysis of the Insane and Scottish Psychiatry, c. 1840–1940," *Journal of the Royal College of Physicians of Edinburgh* 42 (2012): 266–73.

99 *"the case is without hope"*: Quoting Julius Mickle, author of the first English textbook on neurosyphilis, in E. M. Brown, "Why Wagner-Jauregg Won the Nobel Prize for Discovering Malaria Therapy for General Paresis of the Insane," *History of Psychiatry* 11 (2000): 371–82, p. 373.

99 *60 percent achieved remissions*: Ibid., 379.

99 *over 72 percent*: H. C. Solomon, "Value of Treatment in General Paresis," *Boston Medical & Surgical Journal* 188, no. 17 (1923): 635–39.

99 *tertian malaria*: J. Wagner-Jauregg, "The History of the Malaria Treatment of General Paralysis" (1946), *The American Journal of Psychiatry* 151 (1994): 231–35.

99 *patients did die from this treatment*: W. Yorke, "The Malaria Treatment of General Paresis," *Nature* 114 (1924): 615–16.

99 *Although antibiotics such as penicillin were soon introduced*: L. Kalinowsky, "The Discoveries of Somatic Treatments in Psychiatry: Facts and Myths," *Comprehensive Psychiatry* 21 (1980): 428–35, p. 428.

99 *Wagner-Jauregg's legacy*: N. Endler and E. Persad, *Electroconvulsive Therapy: The Myths and the Realities* (Hans Huber, 1988). "Therapeutic treatments did not commence until Jauregg."

99 *"a minimum of highly trained staff"*: E. Valenstein, *Great and Desperate Cures* (New York: Basic Books, 1986), 34.

99 *"was preferable to confessing helplessness"*: Ibid., 44.

Unfixing Thoughts

100 *she'd throw her own feces*: F. Freeman and J. W. Watts, "Psychosurgery During 1936–1946," *Archives of Neurology and Psychiatry* 58, no. 4 (1947): 417–25.

100 *a completely different chimp*: C. F. Jacobsen, "Studies of Cerebral Function in Primates. I. The Functions of the Frontal Association Areas in Monkeys," *Comparative Psychology Monographs* 13, no. 3 (1936): 1–60.

100 *and remain unfazed*: W. Freeman and J. W. Watts, *Psychosurgery in the Treatment of Mental Disorders and Intractable Pain* (Hoboken, NJ: Blackwell Scientific Publications, 1950).

100 *"happiness cult"*: J. F. Fulton, "Physiological Basis of Frontal Lobotomy," *Acta Medica Scandinavica* 128 (1947): 617–25, p. 621. "It was as if the animal had joined the 'happiness cult of the Elder Micheaux.'"

101 *"brain localization" theory*: G. Berrios and H. Freeman, eds., *150 Years of British Psychiatry, 1841–1991* (London: Gaskell/Royal College of Psychiatrists, 1991), 191.

101 *smeared with greasy bits of brain and blood*: J. M. Harlow, "Recovery from the Passage of an Iron Bar Through the Head," *History of Psychiatry* 4 (1993): 274–81.

101 *coherent soon after the accident*: A. Garcia-Molina, "Phineas Gage and the Enigma of the Prefrontal Cortex," *Neurologia* 27, no. 6 (2010): 370–75. "The most singular circumstance connected with this melancholy affair is, that he was alive at two o'clock this afternoon, and in full possession of his reason, and free from pain."

101 *"it conflicts with his desires"*: K. O'Driscoll and J. P. Leach, "'No Longer Gage': An Iron Bar Through the Head. Early Observations of Personality Change After Injury to the Prefrontal Cortex," *British Medical Journal* 317 (1998): 1673–74.

101 *they refused to offer him another job*: Harlow, "Recovery from the Passage of an Iron Bar Through the Head," 274–81.

101 *"Patient A"*: R. Brickner, "Bilateral Frontal Lobectomy: Follow-up Report of a Case," *Archives of Neurology and Psychiatry* 41, no. 3 (1939): 580–85.

102 *sixty-year-old*: J. El-Hai, *The Lobotomist: A Maverick Medical Genius and His Tragic Quest to Rid the World of Mental Illness* (Hoboken, NJ: John Wiley & Sons, 2005), 95.

102 *shiny toupee covering his balding pate*: E. Valenstein, *Great and Desperate Cures* (New York: Basic Books, 1986), 73, fig. 4.4. "Moniz, who was as vain as he was ambitious, was concealing his baldness with a toupee."

102 *Moniz asked Jacobsen*: A. Kucharski, "History of Frontal Lobotomy in the United States, 1935–55," *Neurosurgery* 14, no. 6 (1984): 765–72.

102 *reached back five hundred years in Avança alone*: A. J. Tierney, "Egas Moniz and the Origins of Psychosurgery: A Review Commemorating the 50th Anniversary of Moniz's Nobel Prize," *Journal of the History of the Neurosciences* 9 (2000): 22–36.

102 *he met the Nobel laureate Ramón y Cajal*: Valenstein, *Great and Desperate Cures*, 67.

102 *the first Portuguese scientist to receive the honor*: Ibid., 63. "Portugal's only recipient of the Nobel Prize."

103 *the first photo of cerebral angiography*: Tierney, "Egas Moniz and the Origins of Psychosurgery," 22–36.

103 *over one hundred articles and two books on angiography*: Ibid.

103 *a person who other scientists could look to for guidance*: G. Duarte and A. Goulão, "Egas Moniz, the Pioneer of Cerebral Angiography," *Interventional Neuroradiology* 3 (1997): 107–11.

103 *His photos were plastered over an entire wall*: Valenstein, *Great and Desperate Cures*, 77.

103 *"so quickly and dramatically translated into a therapeutic procedure"*: J. Pressman, *Last Resort: Psychosurgery and the Limits of Medicine* (Cambridge, UK: Cambridge University Press, 1998). Chapter 2, "Sufficient Promise."

103 *November 12, 1935*: S. Yong Tan and A. Yip, "Antonio Egas Moniz (1874–1955), Lobotomy Pioneer and Nobel laureate," *Singapore Medical Journal* 55, no. 4 (2014): 175–76.

103 *severe depression and paranoia*: Valenstein, *Great and Desperate Cures*, 103.

103 *gout that had painfully erupted when he was just twenty-four years old*: Ibid., 65.

104 *"frontal barrier"*: Z. Kotowicz, "Gottlieb Burckhardt and Egas Moniz—Two Beginnings in Psychosurgery," *Gesnerus* 62 (2005): 77–101, p. 81.

104 *"clinical cure"*: Valenstein, *Great and Desperate Cures*, 104.

104 *Nine of the patients (seven women and two men) were diagnosed with depression*: G. E. Berrios, "The Origins of Psychosurgery: Shaw, Burckhardt, and Moniz," *History of Psychiatry* 8 (1997): 61–81, p. 75, table 2.

104 *After the seventh patient*: Valenstein, *Great and Desperate Cures*, 107. "The eighth patient was the first case where the leucotome was used."

104 *"core operation"*: Ibid., 106.

104 *"be made with an organic orientation"*: Ibid., 81.

105 *"fixed connections"*: G. E. Berrios and H. Freeman, eds., *150 Years of British Psychiatry, 1841–1991* (London: Gaskell/Royal College of Psychiatrists, 1991), 188.

105 *"keeping it in constant activity"*: Berrios, "The Origins of Psychosurgery," 74.

105 *"and force thoughts along different paths"*: Ibid.

105 *a cantankerous, poo-flinging alter ego*: Jacobsen, "Studies of Cerebral Function in Primates. I.," 1–60.

106 *"useful to his argument and ignored the rest"*: Valenstein, *Great and Desperate Cures*, 91.

106 *"it rests on pure cerebral mythology"*: Ibid., 99.

106 *in the spring of 1936*: Ibid., 141.

"The Brain Has Ceased to Be Sacred"

107 *"Freeman's lectures substituted for entertainment"*: J. El-Hai, *The Lobotomist: A Maverick Medical Genius and His Tragic Quest to Rid the World of Mental Illness* (Hoboken, NJ: John Wiley & Sons, 2005), 92.

107 *draw on the blackboard with both hands*: E. Valenstein, *Great and Desperate Cures* (New York: Basic Books, 1986), 132.

107 *was appointed as the head of his own department*: Ibid.

107 *became a secretary of the American Medical Association*: Ibid., 135.

108 *she vaulted over a few rows of seats to greet him*: El-Hai, *The Lobotomist*, 90.

108 *"the most beautiful pair of hands"*: See chapter 2, "Sufficient Promise," in J. Pressman, *Last Resort: Psychosurgery and the Limits of Medicine* (Cambridge, UK: Cambridge University Press, 1998).

108 *the same French supplier that Egas Moniz used*: In a letter from Moniz to Freeman: "included the address of the French instrument maker who produced his leucotome," in El-Hai, *The Lobotomist*, 106.

108 *Following her first pregnancy in her twenties and the death of the child*: Ibid., 9.

108 *"no longer cared"*: Valenstein, *Great and Desperate Cures*, 142.

109 *"the patient evinced previous to her operation are relieved"*: Ibid., 143.

109 *"[T]he involutional melancholic would be a thin, elderly man or woman"*: E. Shorter, *How Everyone Became Depressed: The Rise and Fall of the Nervous Breakdown* (New York: Oxford University Press, 2013). Chapter 6, "A Different Kind of Nervous Breakdown."

109 *lived day-to-day, according to a strict routine*: W. Freeman, J. W. Watts, and T. Hunt, *Psychosurgery: Intelligence, Emotion, and Social Behavior Following Prefrontal Lobotomy for Mental Disorders* (Springfield, IL: Charles C. Thomas, 1942). Chapter 19, "Affective Reaction Types."

109 *"a pronounced and rigid ethical code"*: D. Henderson and R. D. Gillespie, *A Text-Book of Psychiatry* (New York: Oxford University Press, 1956), 279.

110 *"mind is occupied with the 'might-have-beens,'"*: Ibid., 278.

110 *"They are perfectly wretched about it"*: E. Kraepelin, *Clinical Psychiatry: A Textbook for Students and Physicians*, abstracted and adapted from the sixth German edition of Kraepelin's *Lehrbuch der psychiatrie* (New York: Macmillan, 1904). Chapter, "Involutional Psychoses."

110 *"All wickedness is due to them"*: Ibid.

110 *the treatment of choice for involutional melancholics*: S. Wohlfahrt, "Psychiatric Views on the Problem of Leucotomy," *Acta Psychiatrica Scandinavica* 22 (1947): 348–67.

110 *"They all responded remarkably to prefrontal lobotomy"*: A. Myerson and P. G. Myerson, "Prefrontal Lobotomy in the Chronic Depressive States of Old Age," *New England Journal of Medicine* 237, no. 14 (1947): 511–12.

111 *"[T]he patients have become more placid"*: Valenstein, *Great and Desperate Cures*, 143.

111 *"some sparkle, some flavor of the personality"*: W. Freeman and J. W. Watts, "Prefrontal Lobotomy in the Treatment of Mental Disorders," *Southern Medical Journal* 30 (1937): 23–31.

111 *"to have little capacity for any emotional experiences—pleasurable or otherwise"*: J. L. Hoffman, "Clinical Observations Concerning Schizophrenic Patients Treated by Prefrontal Leukotomy," *New England Journal of Medicine* 241, no. 6 (1949): 233–36, p. 234.

111 *"My husband may be better, but he is not the same"*: D. Furtado et al., "Personality Changes After Lobotomy," *Psychiatrie und Neurologie* 117, no. 2 (1949): 65–76, p. 67.

111 *"her soul is in some way lost"*: V. W. Swayze, "Frontal Leucotomy and Related Psychosurgical Procedures in the Era Before Antipsychotics (1935–1954): A Historical Review," *American Journal of Psychiatry* 152, no. 4 (1995): 505–15.

112 *"precision method"*: Freeman, Watts, and Hunt, *Psychosurgery*. Chapter 3, "Precision Method."

112 *Their Washington-based manufacturer engraved both their names*: El-Hai, *The Lobotomist*, 144.

112 *Watts orientated himself with a series of landmarks*: Freeman, Watts, and Hunt, *Psychosurgery*.

112 *disconnected the frontal lobes from the emotional center of the brain*: W. Freeman and J. W. Watts, "Psychosurgery During 1936–1946," *Archives of Neurology and Psychiatry* 58, no. 4 (1947): 417–25.

112 *the less emotional oomph the thalamus provided*: Myerson and Myerson, "Prefrontal Lobotomy in the Chronic Depressive States of Old Age," 511.

112 *they performed a "minimal" or "standard"*: W. Freeman and J. W. Watts, "Prefrontal Lobotomy; Survey of 331 Cases," *American Journal of Medical Science* 211 (1946): 1–8.

112 *"depressions disappear as if by magic"*: C. Rylander, "Psychological Tests and Personality Analyses Before and After Frontal Lobotomy," *Association for Research in Nervous and Mental Disease* 27, no. 1 (1948): 691–705.

113 *"This is not an operation, this is a mutilation"*: Valenstein, *Great and Desperate Cures*, 146.

113 *"the consciousness of the self"*: W. Freeman et al., "Neurosurgical Treatment of Certain Abnormal Mental States. Panel Discussion at Cleveland Session," *Journal of the American Medical Association* 117, no. 7 (1941): 517–27, p. 517.

113 *"The past and future seem telescoped for them into the present"*: W. Freeman and J. W. Watts, *Psychosurgery: On the Treatment of Mental Disorders and Intractable Pain* (Hoboken, NJ: Blackwell Scientific Publications, 1950). Chapter 25, "The Functions of the Frontal Lobes."

113 *"and became happy, cheerful and slightly elevated"*: Freeman et al., "Neurosurgical Treatment of Certain Abnormal Mental States," 519.

114 *"the contentment with existence as it is"*: Ibid.

114 *$351,000 every year*: Ibid., 520.

114 *increase to 10 or even 20 percent*: Swayze, "Frontal Leucotomy and Related Psychosurgical Procedures in the Era Before Antipsychotics (1935–1954)," 505–15.

114 *"I have not seen a single case that has not responded favorably"*: Freeman et al., "Neurosurgical Treatment of Certain Abnormal Mental States," 520.

115 *"once one cuts, there is no return"*: Ibid., 525.

115 *journalists were largely positive*: G. J. Diefenbach et al., "Portrayal of Lobotomy in the Popular Press: 1935–1960," *Journal of the History of the Neurosciences* 8, no. 1 (1999): 60–69.

115 *"The brain has ceased to be sacred"*: W. Kaempffert, "Turning the Mind Inside Out," *Saturday Evening Post*, May 24, 1941.

116 *"executed a valid operation for mental disorder"*: Freeman, Watts, and Hunt, *Psychosurgery*.

116 *In 1939, Moniz had been shot four times*: Ibid., introduction.

117 *"The mood was one of enormous depression"*: Ibid. Chapter 19, "Affective Reaction Types."

117 *"the patient left the operating table in excellent condition"*: Ibid.

117 *"They tell me that I look ten years younger"*: Ibid.

119 *"neurological disturbance [that] had overtaken her"*: K. Clifford Larson, *Rosemary: The Hidden Kennedy Daughter* (Boston: Mariner Books, 2016), 157.
119 *Joe Kennedy Sr. spoke to a Boston-based doctor*: Ibid.
119 *a location not far from Rosemary's school*: El-Hai, *The Lobotomist*, 173.
119 *Rosemary suffered from "agitated depression"*: Larson, *Rosemary*, 162.
119 *with only Novocain, a local anesthetic, for comfort*: El-Hai, *The Lobotomist*, 145.
119 *"a grinding sound that is as distressing"*: Ibid., 146.
119 *"We made an estimate on how far to cut"*: R. Kessler, *The Sins of the Father* (New York: Warner Books, 1996), 243.
119 *"Kennedy Cottage"*: L. Leamer, *The Kennedy Women: The Saga of an American Family* (New York: Villard, 1994), 412.
119 *"she left nursing altogether, haunted for the rest of her life"*: Larson, *Rosemary*, 170.
120 *"he cuts but does not know what he sees"*: El-Hai, *The Lobotomist*, 217.
120 *"We claimed ours was the best"*: Ibid., 219.
120 *performing over three thousand operations during his career*: W. McKissock, "Discussion of Psychosurgery," *Proceedings of the Royal Society of Medicine* 52 (1959): 203–10.
120 *"This is not a time-consuming operation"*: W. McKissock, "Rostral Leucotomy," *The Lancet* 2, no. 6673 (1951): 91–94.
120 *"the ideal operation for use in crowded mental hospitals"*: A. Kucharski, "History of Frontal Lobotomy in the United States, 1935–55," *Neurosurgery* 14, no. 6 (1984): 765–72.
121 *driving over eleven thousand miles*: Valenstein, *Great and Desperate Cures*, 229.
121 *"Some of the 'black eyes' are beauties"*: Freeman and Watts, *Psychosurgery*. Chapter 3, "Operative Technique—Precision Method."
121 *"the cat who walks by himself"*: Pressman, *Last Resort*. Chapter 2, "Sufficient Promise."
121 *"Why not use a shotgun? It would be quicker!"*: Valenstein, *Great and Desperate Cures*, 205.
121 *From its first use in 1946*: It was first used in January 1946 on Ellen Ionesco. El-Hai, *The Lobotomist*, 187.
121 *"any surgical procedure involving cutting of brain tissue is a major surgical operation"*: Freeman and Watts, *Psychosurgery*. Chapter 3, "Operative Technique—Precision Method."
121 *"carefully selected cases it can yield much of value"*: S. Wohlfahrt, "Psychiatric Views on the Problem of Leucotomy," *Acta Psychiatrica Scandinavica* 22 (1947): 348–67, p. 358.
122 *"the effect of the operation has been sensational"*: Ibid., 355.
122 *taking the "sting" out of mental disorders*: Freeman, Watts, and Hunt, *Psychosurgery*, preface.
122 *"Patients were more comfortable and families gratified"*: M. Greenblatt and H. C. Solomon, *Frontal Lobes and Schizophrenia* (New York: Springer, 1953), 29.
123 *"the therapeutic value of leucotomy in certain psychoses"*: https://www.nobelprize.org.
123 *"violates the principles of humanity"*: Diefenbach et al., "Portrayal of Lobotomy in the Popular Press: 1935–1960."
123 *only a few hundred operations had been performed since the end of 1944*: B. Zajicek,

"Banning Soviet Lobotomy: Psychiatry, Ethics, and Professional Politics During Late Stalinism," *Bulletin of the History of Medicine* 91, no. 1 (2017): 22–61.

123 *signed a decree on December 9, 1950*: Ibid., 59.

123 *"are serious enough to give us considerable concern"*: Diefenbach et al., "Portrayal of Lobotomy in the Popular Press: 1935–1960," 65.

The Most Powerful Reaction

124 *"Electrical slaughtering"*: U. Cerletti, "Electroshock Therapy," in A. M. Sackler et al., *The Great Physiodynamic Therapies in Psychiatry: An Historical Reappraisal* (New York: Hoeber, 1956).

124 *the humane part of slaughter*: Ibid. "Used at the suggestion of the 'Society for the Prevention of Cruelty to Animals.'"

124 *"It occurred to me that the hogs of the slaughterhouse"*: Ibid.

124 *Only when he left the electricity running for over a minute . . . was the electric shock fatal*: E. Shorter and D. Healy, *Shock Therapy: A History of Electroconvulsive Treatment in Mental Illness* (New Brunswick, NJ: Rutgers University Press, 2007), 36.

125 *"sufficient to ensure a complete convulsive seizure"*: Ibid.

125 *three-story, ocher-colored concrete building*: R. L. Palmer, ed., *Electroconvulsive Therapy: An Appraisal* (New York: Oxford University Press, 1981), 7.

125 *collected by the municipal dogcatcher*: N. Endler, "The Origins of Electroconvulsive Therapy (ECT)," *Convulsive Therapy* 4, no. 1 (1988): 5–23.

125 *"like a falcon high in the sky"*: F. Accornero, "An Eyewitness Account of the Discovery of Electroshock," *Convulsive Therapy* 4, no. 1 (1988): 40–49.

125 *Neither Cerletti nor Bini saw anything that might prevent this procedure*: U. Cerletti and L. Bini, "Le alterazioni istopatologiche del sistema nervoso in seguito all," *E. S. Rivisata Sperimentale di Freniatria* 64 (1940): 311. "The macroscopic findings of the brain and bone marrow have never revealed significant pathological data."

125 *"this most powerful reaction of the nervous system"*: Cerletti, "Electroshock Therapy," 95.

126 *"one species of spasm however occasioned seldom fails"*: M. Fink, *Convulsive Therapy: Theory and Practice* (New York: Raven Press, 1979), 5.

126 *"A complete cure followed"*: Sackler et al., *Great Physiodynamic Therapies in Psychiatry.*

126 *"When the patient awakes from the epileptic attack his reason will return"*: Palmer, ed., *Electroconvulsive Therapy.*

126 *"an opposite, a paralyzing, effect of the glia system"*: L.J. Meduna (1956) "The Convulsive Treatment," in Sackler, et al., *Great Physiodynamic Therapies in Psychiatry.*

126 *"biological antagonism"*: Fink, *Convulsive Therapy:* chapter 2, "History of Convulsive Therapies."

126 *could one disease treat another?*: Meduna, "The Convulsive Treatment." "The beneficial effect of the epileptic attack upon the schizophrenic process reminds me of the beneficial effect of high temperature upon the course of paresis."

126 *Strychnine, thebaine, nikethamide, caffeine, absinthe*: M. Fink, "Meduna and the

Origins of Convulsive Therapy," *American Journal of Psychiatry* 141 (1984): 1034–41, p. 1035.

127 *"impending doom"*: R. A. Dunne and D. M. McLoughlin, "Electroconvulsive Therapy and Therapeutic Neuromodulation," *Core Psychiatry* (2012): 617–27.

127 *grand mal seizure fit progressed through its familiar stages*: G. W. T. H. Fleming and F. L. Golla, "Electric-Convulsion Therapy of Schizophrenia," *The Lancet* 234 (1939): 1353–55.

127 *urine, feces, and, for men, semen are sometimes ejected*: Cerletti, "Electroshock Therapy."

127 *Frothing at the mouth is not unusual*: H. Shepley and J. S. McGregor, "Electrically Induced Convulsions in Treatment of Mental Disorders," *British Medical Journal* 2 (1939): 1269–71.

127 *ten were rid of the hallucinations*: Fink, *Convulsive Therapy: Theory and Practice.* Chapter 2, "History of Convulsive Therapies."

127 *"[A] new art and possibility of cure opens up for us"*: Fink, "Meduna and the Origins of Convulsive Therapy," 1035.

127 *training course in convulsive therapy in Vienna*: Palmer, ed., *Electroconvulsive Therapy*, 21.

128 *studying the effects on the hippocampus*: Cerletti, "Electroshock Therapy." "sclerosis in Ammon's horn, specifically the Sommer's sector."

128 *"these experiments would have any practical application"*: Ibid.

128 *"like a tree that does not give fruit"*: Accornero, "An Eyewitness Account of the Discovery of Electroshock," 44.

128 *a shoebox-sized, cream-colored cube of metal*: Research trip by author to Museum of History of Medicine, University of Rome La Sapienza, October 2018.

128 *Used by a laboratory technician who took naps between shifts*: Endler, "The Origins of Electroconvulsive Therapy (ECT)," 5–23.

128 *younger students to poke his head out*: Accornero, "An Eyewitness Account of the Discovery of Electroshock," 45. "Challiol would leave the room periodically to see whether any undesired person was lurking in the corridor."

128 *A former student of Emil Kraepelin and Alois Alzheimer*: R. Passione, "Italian Psychiatry in an International Context: Ugo Cerletti and the Case of Electroshock," *History of Psychiatry* 15 (2004): 83–104.

128 *"the Maestro"*: Accornero, "An Eyewitness Account of the Discovery of Electroshock," 42.

129 *"Cerletti atmosphere"*: Endler, "The Origins of Electroconvulsive Therapy (ECT)," 5–23.

129 *Inspired by the "youthful enthusiasm" of Bini*: Ibid.

129 *He had started counting*: Accornero, "An Eyewitness Account of the Discovery of Electroshock," 47.

129 *his forehead was beaded with sweat*: Ibid.

130 *"We glanced at each other; in our eyes, there was a new shining light"*: Ibid.

130 *The details of the first session . . . have been lost*: A. Aruta, "Shocking Waves at the Museum: The Bini-Electro-Shock Apparatus," *Medical History* 55 (2011): 407–12.

130 *"talking in his usual loose way"*: L. Bini, "Professor Bini's Notes on the First Electro-Shock Experiment," *Convulsive Therapy* 11 (1995): 260–61.

130 *eleven sessions of electroshock therapy*: Endler, "The Origins of Electroconvulsive Therapy (ECT)," 5–23.

130 *discharged from the university's clinic on June 17, 1938*: Cerletti, "Electroshock Therapy."

131 *Electroshock didn't come with these drawbacks*: G. Berrios, "The Scientific Origins of Electroconvulsive Therapy: A Conceptual History," *History of Psychiatry* 8 (1997): 105–19.

131 *"Loss of consciousness instantaneously follows"*: A. Metastasio and D. Dodwell, "A Translation of 'L'Elettroshock' by Cerletti & Bini, with an Introduction," *European Journal of Psychiatry* 27 (2013): 231–39, p. 236.

131 *"they do not remember anything but having slept"*: Ibid., 237.

131 *blow an out-of-tune trumpet*: Endler, "The Origins of Electroconvulsive Therapy (ECT)," 5–23.

131 *"Although physicians were reluctant to earmark the new treatment"*: "Insanity Treated by Electric Shock," *New York Times*, July 6, 1940.

132 *"particularly in depressive episodes"*: U. Cerletti, "L'Elettroshock," *Rivista Sperimentale di Freniatri* (1940), 64. Translation of original manuscript during author's reporting trip to Rome, October 2018.

132 *In some cases, only four sessions*: Fink, *Convulsive Therapy*. Chapter 2, "History of Convulsive Therapies."

132 *"A series of complete remissions was obtained"*: Cerletti, "L'Elettroshock," 64.

132 *"white as a sheet"*: L. Tondo, "Interview with Lothar B. Kalinowsky (1899–1992)," *Clinical Neuropsychiatry* 8 (1990): 303–18, p. 305.

132 *high-pitched yelp as the air was forced from the patient's lungs*: Ibid., 127. According to Frank Ayd, "not really a scream of pain, but as the air was inspired."

133 *one thousand sessions during his time in Rome*: L. Kalinowsky and S. E. Barrera, "Electric Convulsion Therapy in Mental Disorders," *Psychiatric Quarterly* 14 (1940): 719–30, p. 727.

133 *"depressions of various depths and duration clear up"*: L. B. Kalinowsky, "Present Status of Electric Shock Therapy," *Bulletin of the New York Academy of Medicine* 25 (1949): 541–53.

133 *leaving behind their home, their possessions, and even their two children*: R. Abrams, "Interview with Lothar Kalinowsky," *Convulsive Therapy* 4 (1988): 25–39, p. 31.

133 *a blueprint of Bini's electroshock machine*: Tondo, "Interview with Lothar B. Kalinowsky (1899–1992)," 305.

133 *he showed Bini's designs to directors*: Endler, "The Origins of Electroconvulsive Therapy (ECT)," 5–23.

133 *Aubrey Lewis . . . met Kalinowsky with stern indifference*: L. Rzesnitzek, "'A Berlin Psychiatrist with an American Passport': Lothar Kalinowsky, Electroconvulsive Therapy and International Exchange in the Mid-Twentieth Century," *History of Psychiatry* 26, no. 4 (2015): 433–51, p. 439.

133 *attracted psychiatrists from all over the United States*: Shorter and Healy, *Shock Therapy*, 79.

134 *the Johnny Appleseed of electroshock therapy*: E. Shorter, *A History of Psychiatry: From the Era of the Asylum to the Age of Prozac* (Hoboken, NJ: John Wiley & Sons, 1997), 221.

134 *A small fraction of this actually reached the brain*: L. Kalinowsky, N. Bigelow, and

P. Brikates, "Electric Shock Therapy in State Hospital Practice," *Psychiatric Quarterly* 15 (1941): 450–59, p. 451.

134 *"replaced with typical American enthusiasm"*: Endler, "The Origins of Electroconvulsive Therapy (ECT)," 5–23.

134 *near to 100 percent effectiveness*: A. E. Bennett, "An Evaluation of the 'Shock' Therapies." Presented before the Missouri Psychiatric Association, Kansas City, Missouri, November 15, 1944. "almost a specific treatment."

134 *"effectively terminates depressive psychoses"*: Ibid., 470.

134 *"Patients with severe depressive disorders . . . came back to life"*: Quoting Louis Linn, in Shorter and Healy, *Shock Therapy*, 79.

135 *electroshock therapy was rarely as effective in milder depressions*: "Electroconvulsive Therapy—Consensus Conference," *Journal of the American Medical Association*, 254 (1985): 2103–108. "ECT is not effective for patients with milder depressions."

135 *great debate between biological psychiatrists ensued*: M. Mendelson, *Psychoanalytic Concepts of Depression* (New York: Spectrum, 1974).

136 *"the better the treatment"*: L. Kalinowsky, "Experience with Electric Convulsive Therapy in Various Types of Psychiatric Patients," *Bulletin of the New York Academy of Medicine* 20, no. 9 (1944): 485–94, p. 488.

136 *Psychoanalysts were never in doubt*: Mendelson, *Psychoanalytic Concepts of Depression.*

136 *"until extensive trials of psychotherapy were exhausted"*: Shorter and Healy, *Shock Therapy*, 92.

136 *"when their mothers get depressed, they send them to me"*: Ibid., 85.

Legacy

137 *"they are content with burning my books"*: B. R. Hergenhahn, *An Introduction to the History of Psychology* (Cengage Learning, 1980), 509.

137 *the city in which he had lived and worked for seventy-nine years*: E. Jones, *The Life and Work of Sigmund Freud* (New York: Penguin, 1967), 643.

137 *friends and family at Victoria Station*: Ibid. "Freud's eldest children, Mathilde and Martin, and of course my wife and myself, were waiting, and the reunion was a moving scene."

137 *he had always admired England from afar*: Ibid., 50. "land of his dreams."

137 *he hoped that he wouldn't live to see its inevitable domination in Europe*: Ibid., 633. "There is probably no holding up the Nazi invasion. . . . My only hope is that I shall not live to see it."

138 *the squamous cell carcinoma, which had first been diagnosed in 1923*: W. L. Adeyemo, "Sigmund Freud: Smoking Habit, Oral Cancer, and Euthanasia," *Nigerian Journal of Medicine* 13, no. 2 (2004): 189–95.

138 *"a little island of pain floating on a sea of indifference"*: Jones, *Life and Work of Sigmund Freud*, 653.

138 *flooded with three centigrams of morphine*: P. Gay, *Freud: A Life of Our Time* (London: Papermac, 1995), 651.

138 *four of his five sisters who were murdered by the Nazis*: E. Freud, ed., *Letters of Sigmund Freud* (New York: Dover, 1992), 455.

138 *"no gross distortion or falsification"*: American Psychiatric Association, *Diagnostic and Statistical Manual of Mental Disorders (DSM-I)* (Washington, D.C.: American Psychiatric Publishing, 1952).

138 *"The relatives have often suffered from the same form of the disease"*: E. Kraepelin, *Manic Depressive Insanity and Paranoia* (Edinburgh: E. & S. Livingstone, 1921).

139 *"empirical heredity prognosis"*: G. Kösters et al., "Ernst Rüdin's Unpublished 1922–1925 Study 'Inheritance of Manic-Depressive Insanity': Genetic Research Findings Subordinated to Eugenic Ideology," *PLoS Genetics* 11, no. 11 (2017): 1–14.

139 *"[Rüdin's] demands for negative eugenic measures"*: Ibid., 8.

140 *"only the healthy beget children"*: Ibid., 22.

140 *"nothing but applied biology"*: Ibid., 31.

140 *410,000 people . . . were surgically sterilized*: R. J. Lifton, *Nazi Doctors: Medical Killing and the Psychology of Genocide* (New York: Basic Books, 2000), 25.

140 *buried in mass graves by Polish prisoners (who were then shot)*: Ibid., 78.

140 *seventy thousand people were murdered*: G. Gazdag, G. S. Ungvari, and H. Czech, "Mass Killing Under the Guise of ECT: The Darkest Chapter in the History of Biological Psychiatry," *History of Psychiatry* 28, no. 4 (2017): 482–88.

140 *"one of Upper Austria's most beautiful and most significant Renaissance castles"*: S. Loistl and F. Schwanninger, "Vestiges and Witnesses: Archaeological Finds from the Nazi Euthanasia Institution of Hartheim as Objects of Research and Education," *International Journal of Historical Archaeology* 22, no. 3 (2017): 614–38, p. 614.

140 *over eighteen thousand people were killed*: Ibid., 616.

141 *buried in an ecumenical ceremony*: Research trip to Hartheim Castle by the author, October 2018.

141 *"life unworthy of life"*: Lifton, *Nazi Doctors*, 46.

141 *"decentralized euthanasia," or "wild euthanasia"*: H. Czech, "Nazi Medical Crimes at the Psychiatric Hospital Gugging," Documentation Centre of Austrian Resistance (DÖW), 2016, 18.

141 *Emil Gelny . . . killed three hundred mentally ill patients*: L. Rzesnitzek and S. Lang, "'Electroshock Therapy' in the Third Reich," *Medical History* 61 (2017): 66–88, p. 86.

141 *He had bought an Elkra II*: Ibid.

141 *electrodes that could be attached to a person's wrists and ankles*: Gazdag, Ungvari, and Czech, "Mass Killing Under the Guise of ECT," 482–88.

141 *They were dead in less than ten minutes*: Ibid., 485.

142 *a means of punishing unruly or loud patients*: M. G. Jacoby, "Abuse of E.C.T.," *British Medical Journal* 1 (1958): 282. "Her noisiness had caused many complaints from people living near the hospital and she had been given E.C.T. repeatedly to quieten her. . . . It is obvious that E.C.T. was wrongly used in this case."

142 *"so regressed that he is bedridden, does not know his name"*: Ibid.

142 *"The recovery of the severe, long-standing, agitated depressions"*: L. H. Smith et al., "Electroshock Treatment in the Psychoses," *American Journal of Insanity* 98 (1942): 558–61.

142 *80 and 90 percent*: A. E. Bennett, "An Evaluation of the 'Shock' Therapies."

Presented before the Missouri Psychiatric Association, Kansas City, Missouri, November 15, 1944; Smith et al., "Electroshock Treatment in the Psychoses," 558–61.

142 *"I would not have lasted out in psychiatry"*: F. Post, "Then and Now," *British Journal of Psychiatry* 133 (1978): 83–86.

142 *read reports of ten thousand sessions*: L. Kalinowsky and S. E. Barrera, "Electric Convulsion Therapy in Mental Disorders," *Psychiatric Quarterly* 14 (1940): 719–30, p. 727.

143 *"shock method seems relatively harmless"*: Ibid.

143 *"relatively speaking, a mild approach"*: A. Myerson, "Further Experience with Electric Shock Therapy in Mental Disease," *New England Journal of Medicine* 227, no. 11 (1942): 403–7.

Cerletti's Monster

144 *"some neurotics may be harmed"*: L. B. Kalinowsky, "Present Status of Electric Shock Therapy," *Bulletin of the New York Academy of Medicine* 25 (1949): 541–53, p. 552.

144 *"demand more rapid and drastic treatment"*: F. A. Mettler, ed., *Selective Partial Ablation of the Frontal Cortex. A Correlative Study of Its Effects on Human Psychotic Subjects by Columbia-Greystone Associates* (New York: Paul B. Hoeber, 1949).

144 *families of war veterans were willing to try anything*: E. Valenstein, *Great and Desperate Cures* (New York: Basic Books, 1986), 229. "the relatives of veterans with psychiatric illnesses constituted a pressure group for effective treatment."

145 *a "last resort"*: Ibid., 198. "Whenever electroshock proved ineffective, lobotomy was considered as a possible treatment."

145 *"proper surgical technique and its unnecessary damage eliminated?"*: Mettler, ed., *Selective Partial Ablation of the Frontal Cortex.*

145 *A survey from forty-four neurosurgical centers in the UK*: B. M. Baraclough and N. A. Mitchell-Heggs, "Use of Neurosurgery for Psychological Disorder in British Isles During 1974–6," *British Medical Journal* 2 (1978): 1591–93.

145 *"the burgeoning world of the new psychosurgery"*: J. El-Hai, *The Lobotomist: A Maverick Medical Genius and His Tragic Quest to Rid the World of Mental Illness* (Hoboken, NJ: John Wiley & Sons, 2005), 311.

146 *Fractures in . . . 23 percent of patients*: J. R. Lingley and L. L. Robbins, "Fractures Following Electroshock Therapy," *Radiology* 48 (1947): 124–28.

146 *for chemically induced convulsions . . . 43 percent*: E. Shorter and D. Healy, *Shock Therapy: A History of Electroconvulsive Treatment in Mental Illness* (New Brunswick, NJ: Rutgers University Press, 2007), 65.

146 *"If shock treatment is to survive"*: A. E. Bennett, "Preventing Traumatic Complications in Convulsive Shock Therapy by Curare," *Journal of the American Medical Association* 114, no. 4 (1940): 322–24, p. 322.

146 *"shoulder broken rather than suffer the depression"*: S. Katzenelbogen, A. K. Baur, and A. R. M. Coyne, "Electric Shock Therapy: Clinical, Biochemical, and Morphologic Studies," *Archives of Neurology and Psychiatry* 52, no. 4 (1944): 323–26.

146 *Kalinowsky found no evidence of damage*: L. Kalinowsky, N. Bigelow, and P. Brikates, "Electric Shock Therapy in State Hospital Practice," *Psychiatric Quarterly* 15 (1941): 450–59.

147 *military service could be forgotten*: Interview by the author with Poul Videbech, January 24, 2019.

147 *simply forgot that they were ill*: L. H. Smith et al., "Electroshock Treatment in the Psychoses," *American Journal of Insanity* 98 (1942): 558–61.

147 *0.06 percent of patients died . . . "negligible"*: W. S. Maclay, "Death Due to Treatment," *Proceedings of the Royal Society of Medicine* 46 (1952): 13–20.

147 *The most common cause of death was heart failure*: B. L. Pacella, "Sequelae and Complications of Convulsive Shock Therapy," *Bulletin of the New York Academy of Medicine* 20, no. 11 (1944): 575–87.

147 *"other than the electrical treatment administered to the deceased"*: Maclay, "Death Due to Treatment," 13–20.

147 *"an even greater responsibility to bear"*: Ibid., 20.

147 *electroshock therapy became a staple of psychiatry*: Shorter and Healy, *Shock Therapy*, 51.

148 *"Our evidence, however, argues against this"*: K. J. Tillotson and W. Sulzbach, "A Comparative Study and Evaluation of Electric Shock Therapy in Depressive States," *American Journal of Psychiatry* 101, no. 4 (1945): 455–59.

148 *80 percent vs. 50 percent*: Ibid.

148 *A similar study*: A. E. Bennett and C. B. Wilbur, "Convulsive Shock Therapy in Involutional States After Complete Failure with Previous Estrogenic Treatment," *American Journal of Medical Science* 208 (1944): 170–76, p. 170.

148 *"efficient intellectual as well as emotional adaptivity"*: Ibid.

148 *"[I]n private practice . . . progress cannot be made"*: R. Passione, "Italian Psychiatry in an International Context: Ugo Cerletti and the Case of Electroshock," *History of Psychiatry* 15 (2004): 83–104.

149 *"get away from the use of electroshock"*: U. Cerletti, "Electroshock Therapy," in A. M. Sackler, ed., *The Great Physiodynamic Therapies in Psychiatry: An Historical Reappraisal* (New York: Paul B. Hoeber, 1956), 106.

149 *"I had somehow betrayed these patients"*: Ibid.

149 *"The future lies in the domain of biochemists"*: Ibid., 115.

149 *at the University of Turin . . . Cerletti had failed chemistry*: Passione, "Italian Psychiatry in an International Context: Ugo Cerletti and the Case of Electroshock," 83–104.

149 acroaganines: Cerletti, "Electroshock Therapy."

149 *"highly vitalizing substances with defensive properties"*: Ibid.

149 *gloopy and sickly yellow*: Author trip to the Museum of History of Medicine, University of Rome La Sapienza, October 2018.

150 *Fear of imminent death, punishment from a parent-like figure*: E. Miller, "Psychological Theories of ECT: A Review," *British Journal of Psychiatry* 113 (1967): 301–11.

150 *"breaking through of the patient's autism"*: E. V. Weigert, "Psychoanalytic Notes on Sleep and Convulsion Treatment in Functional Psychoses," *Psychiatry* 3, no. 2 (1940): 189–209.

150 *"[W]hatever doubt may exist about the dynamics of improvement"*: H. Selenski,

"The Selective Use of Electro-Shock Therapy as an Adjuvant to Psychotherapy," *Bulletin of the New York Academy of Medicine* 19 (1943): 245–52.

150 *"no adequate theory for [electroshock therapy] is available"*: Kalinowsky, "Present Status of Electric Shock Therapy," 541–53.

151 *"Risks are therefore justified"*: L. B. Kalinowsky and P. Hoch, *Somatic Treatments in Psychiatry: Pharmacotherapy; Convulsive, Insulin, Surgical, Other Methods* (New York: Grune & Stratton, 1961), 173.

151 *one-tenth of a lethal dose*: Bennett, "Preventing Traumatic Complications in Convulsive Shock Therapy by Curare," 323. "The only contraindication to the use of curare is myasthenia gravis."

151 *Paralyzing the muscles for fifteen to twenty minutes*: A. E. Bennett, "Curare: A Preventative of Traumatic Complications in Convulsive Shock Therapy (Including a Preliminary Report on Synthetic Curare-like Drug)" *Convulsive Therapy* 13, no. 2 (1997): 93–107.

151 *"I had my first and only fatality"*: R. Abrams, "Interview with Lothar Kalinowsky," *Convulsive Therapy* 4 (1988): 25–39, p. 36.

151 *"more dangerous than the complications"*: Kalinowsky, "Present Status of Electric Shock Therapy," 541–53.

152 *cancer more likely and more lethal*: G. D. Batty et al., "Psychological Distress in Relation to Site Specific Cancer Mortality: Pooling of Unpublished Data from 16 Prospective Cohort Studies," *British Medical Journal* 356 (2017): 1–11.

152 *diabetes*: X. Zhang et al., "Depressive Symptoms and Mortality Among Persons With and Without Diabetes," *American Journal of Epidemiology* 161, no. 7 (2005): 652–60.

152 *die from heart complications*: J. Barth, "Depression as a Risk Factor for Mortality in Patients with Coronary Heart Disease: A Meta-Analysis," *Psychosomatic Medicine* 66 (2004): 802–13.

152 *chemist working just a stone's throw*: D. A. Cozanitis, "Daniel Bovet, Nobelist: Muscle Relaxants in Anaesthesia. The Role Played by Two Neglected Protagonists," *Wiener Medizinische Wochenschrift* 166 (2016): 487–99.

152 *"curare of short action"*: Ibid.

153 *"electroplexy"*: S. Smith, "The Use of Electroplexy (E.C.T.) in Psychiatric Syndromes Complicating Pregnancy," *Journal of Mental Science* 102 (1956): 796–800.

153 *"central stimulation with patterned response (CSPR)"*: H. Spiro, "The Stigma of Electroconvulsive Therapy: A Workshop," in P. J. Fink and A. Tansman, eds., *Stigma and Mental Illness* (Washington, D.C.: American Psychiatric Press, 1992). Chapter 17.

153 *"its very existence had been somehow erased"*: Shorter and Healy, *Shock Therapy*, 161.

The Psychic Energizers

154 *It seemed to buzz with energy*: M. Chessin, E. R. Kramer, and C. C. Scott, "Modifications of the Pharmacology of Reserpine and Serotonin by Iproniazid," *Journal of Pharmacology and Experimental Therapeutics* 119 (1956): 453–60.

154 *"twinkle in his eye"*: F. Ayd, "Discovery of Antidepressants," in D. Healy, *The Psychopharmacologists* (London: Chapman & Hall, 1996).

154 *was visiting the laboratories of Charles Scott*: N. S. Kline, "MAOIs: An Unfinished Picaresque Tale," in F. J. Ayd and B. Blackwell, eds., *Discoveries in Biological Psychiatry* (New York: Lippincott, 1970).

154 *sarpagandha . . . used for a variety of ills*: J. Monachino, "*Rauvolfia serpentina*—Its History, Botany, and Medical Use," *Economic Botany* 8 (1954): 349–65.

154 *"attain states of introspection and meditation"*: M. Meyers, *Happy Accidents: Serendipity in Modern Medical Breakthroughs* (New York: Arcade, 2007), 274.

155 *Even the hospital's glazier*: D. Healy, *The Antidepressant Era* (Cambridge, MA: Harvard University Press, 1997), 55.

155 *didn't come with the lethargic side effects*: N. Kline, "Use of *Rauwolfia serpentina Benth*," in "Neuropsychiatric Conditions," *Annals of the New York Academy of Sciences* 59, no. 1 (1954): 107–32.

155 *"I could sleep through the night"*: Author interview with Thomas Ban, March 20, 2018.

155 *"one of the most exciting periods"*: M. Platt, *Storming the Gates of Bedlam: How Dr. Nathan Kline Transformed the Treatment of Mental Illness* (London: Depew, 2012), 31.

155 *Barsa, a stocky, balding, and conservative colleague*: Ayd, "Discovery of Antidepressants."

156 *"compound that could swing it up"*: H. E. Lehmann and N. S. Kline, "Clinical Discoveries with Antidepressant Drugs," in M. J. Parnham and J. Bruinvels, eds., *Discoveries in Pharmacology, Volume 1: Psycho- and Neuro-Pharmacology* (Elsevier Science, 1983), 209–21.

156 *blooms in white flowers across India*: Monachino, "*Rauvolfia serpentina*—Its History, Botany, and Medical Use," 349–65.

156 *bought these redundant reserves at low cost*: M. Sandler, "Monoamine Oxidase Inhibitors in Depression: History and Mythology," *Journal of Psychopharmacology* 4, no. 3 (1990): 136–39, p. 136.

156 *In 1951 . . . isoniazid*: J. D. Aronson, S. L. Ehrlich, and W. Flagg, "Effects of Isonicotinic Acid Derivatives on Tubercle Bacilli," *Proceedings of the Society for Experimental Biology and Medicine* 80, no. 2 (1952), 259–62. Isoniazid from E. R. Squib and Sons, New Brunswick.

156 *Herbert Fox . . . created iproniazid*: T. Ban, "The Role of Serendipity in Drug Discovery," *Dialogues in Clinical Neuroscience* 8 (2006): 335–44, p. 342.

157 *It killed* Mycobacterium tuberculosis: H. Fox, "The Chemical Approach to the Control of Tuberculosis," *Science* 116 (1952): 129–34.

157 *Guinea pigs, rabbits, rhesus monkeys*: Ibid.

157 *"WONDER DRUG FIGHTS TB"*: "Medicine: TB—and Hope," *Time*, March 3, 1952.

157 *fifteen-minute radio broadcast*: F. Ryan, *Tuberculosis: The Greatest Story Never Told* (Worcestershire, UK: Swift, 1992), 360.

157 *a four-million-dollar complex*: G. A. Norman et al., "New York City Farm Colony—Seaview Hospital," New York City Landmarks Commission, Designation Report, 1985, 18.

157 *"the largest and finest hospital ever built"*: Ibid.

157 *"coughing up their lives"*: F. López-Muñoz and C. Alamo, "Monoaminergic Neurotransmission: The History of the Discovery of Antidepressants from 1950s Until Today," *Current Pharmaceutical Design* 15 (2009): 1563–86, p. 1566.

157 *"dancing in the halls"*: E. Bullmore, *The Inflamed Mind: A Radical New Approach to Depression* (London: Short Books, 2018), 93, fig. 7: *Life* magazine, 1952.

157 *"No bed cases remain"*: G. Crane, "The Psychiatric Side-Effects of Iproniazid," *American Journal of Psychiatry* 112 (1956): 494–501, p. 494.

158 *While reserpine reduced the levels*: P. A. Shore et al., "Role of Brain Serotonin in Reserpine Action," *Annals of the New York Academy of Sciences* 66, no. 3 (1957): 609–15.

158 *iproniazid increased them*: P. A. Shore et al., "On the Physiologic Significance of Monoamine Oxidase in the Brain," *Science* 126 (1957): 1063–64.

158 *Betty Twarog . . . first found evidence*: B. M. Twarog and I. H. Page, "Serotonin Content of Some Mammalian Tissues and Urine and a Method for Its Determination," *American Journal of Physiology* 175 (1953): 157–61.

158 *Independently, researchers in Edinburgh*: A. R. Green, "Gaddum and LSD," *British Journal of Pharmacology* 154 (2008): 1590.

158 *studied in the blood serum*: I. H. Page, "The Discovery of Serotonin," *Perspectives in Biology and Medicine* 20 (1976): 1–8.

158 *and the gut*: P. M. Whitaker-Azmitia, "The Discovery of Serotonin and Its Role in Neuroscience," *Neuropsychopharmacology* 21, no. 5 (1999): 2S–8S. Vittorio Erspamer and enterochromaffin cells of the gut.

158 *90 percent of serotonin*: Based on studies in mice: T. Hata et al., "Regulation of Gut Luminal Serotonin by Commensal Microbiota in Mice," *PLoS ONE* 12, no. 7 (2017): e0180745.

158 *"quiet, withdrawn, and unresponsive"*: H. P. Loomer, J. C. Saunders, and N. S. Kline, "Evaluation of Iproniazid as a Psychic Energizer," *Psychiatric Research Reports* 8 (1957): 129–41, p. 136.

158 *"she just wasn't there"*: Ibid.

159 *70 percent*: Ibid.

159 *drove his black Thunderbird soft-top*: M. Platt, *Storming the Gates of Bedlam: How Dr. Nathan Kline Transformed the Treatment of Mental Illness* (London: Depew, 2012), 93.

159 *"metastasis from the Museum of Modern Art"*: Author research trip to Nathan Kline Institute, November 2018.

159 *five nurses and three psychiatrists*: Ibid.

159 *Over two-thirds of his patients*: N. S. Kline, *Depression: Its Diagnosis and Treatment* (Basel, Switzerland: Karger, 1969).

159 *"[s]uffer is often too mild a term"*: Ibid.

159 *"You see, Doctor"*: N. S. Kline et al., "Iproniazid in Depressed and Regressed Patients," in M. Rinkel and C. B. Denber, eds., *Proceedings of the Symposium on Chemical Concepts of Psychosis Held at the Second International Congress of Psychiatry in Zurich, Switzerland, September 1 to 7, 1957* (London: Peter Owen, 1957).

159 *thirty "responded with complete remission"*: Ibid.

159 *"There is a revolution in psychiatry"*: Platt, *Storming the Gates of Bedlam*, 37.

159 *MAO inhibitor . . . is certainly more accurate*: E. A. Zeller and J. Barksy, "In

Vivo Inhibition of Liver and Brain Monoamine Oxidase by 1-Isonicotinoyl-2-Isopropyl Hydrazine," *Proceedings of the Society for Experimental Biology and Medicine* 81, no. 2 (1952): 459–61. Review: T. A. Ban, "Pharmacotherapy of Depression: A Historical Analysis," *Journal of Neural Transmission* 108 (2001): 707–716.

160 *Our brains aren't just electrical organs*: E. S. Valenstein, *The War of the Soups and the Sparks: The Discovery of Neurotransmitters and the Dispute Over How Nerves Communicate* (New York: Columbia University Press, 2005).

161 *trained by a leading psychoanalyst*: Platt, *Storming the Gates of Bedlam.*

161 *"One supposed mode of development"*: N. S. Kline, *From Sad to Glad: Kline on Depression* (New York: Ballantine, 1974), 50.

162 *There was no place for iproniazid in this technique*: López-Muñoz and Alamo, "Monoaminergic Neurotransmission," 1563–86. "the pharmacological treatment of depressive symptoms . . . was viewed by a large part of the psychiatric community as a real error, since it would prevent patients from discovering the 'true' roots of their internal conflicts."

162 *"I was making a fool of myself"*: Kline, *From Sad to Glad*, 67.

162 *"No one in their right mind"*: Heinz Lehmann in"Drug for Treating Schizophrenia Identified," PBS, 1998.

162 *"had flair, flamboyance, was politically amoral"*: "Boss 'Nucky' Johnson Is Dead at 85—Unconscious 25 Hours Before 'Time Took Him,'" *Atlantic City Press*, December 10, 1968.

163 *"We have whisky, wine, women"*: "Enoch L. Johnson, Ex-Boss in Jersey. Prohibition-Era Ruler of Atlantic City, 85, Dies," *New York Times*, December 10, 1968.

163 *"I thought everyone lived that way"*: Platt, *Storming the Gates of Bedlam*, 59.

163 *inviting the company's president to lunch*: López-Muñoz and Alamo, "Monoaminergic Neurotransmission," 1567; Sandler, "Monoamine Oxidase Inhibitors in Depression: History and Mythology," 136–39.

163 *known for his womanizing*: Platt, *Storming the Gates of Bedlam*, 129. "It was generally believed by those who knew him that he had frequent brief liaisons with other women throughout his marriage."

163 *intentionally antagonized his peers*: Ayd, "Discovery of Antidepressants."

164 *"Depression: You Can Conquer It Without Analysis!"*: Kline, *From Sad to Glad.*

The Shoes that Prozac Would Fill

165 *Originally created as a potential decongestant*: N. Rasmussen, "Making the First Anti-Depressant: Amphetamine in American Medicine, 1929–1950," *Journal of the History of Medicine and Allied Sciences* 61, no. 3 (2006): 288–323.

165 *"everybody showed an increased tendency to talk"*: S. A. Peoples and E. Guttmann, "Hypertension Produced with Benzedrine: Its Psychological Accompaniments," *The Lancet* 227 (1936): 1107–9, p. 1109.

165 *"exulted being"*: Ibid.

165 *"energy and self-confidence"*: G. Guttmann, "The Effect of Benzedrine on Depressive States," *Journal of Mental States* 82, no. 340 (1936): 618–25.

165 *"As soon as I started taking the benzedrine"*: D. W. Wilbur, A. R. MacClean, and E. V. Allen, "Clinical Observations on the Effect of Benzedrine Sulfate," *Journal of the American Medical Association* 109, no. 8 (1937): 549–54, p. 552.

166 *"Benzedrine is a stimulant"*: "Discussion on Benzedrine: Uses and Abuses," Section of Medicine with the Section of Therapeutics and Pharmacology, *Proceedings of the Royal Society of Medicine* 32 (1938): 385–98, p. 391.

166 *British armies allocated Benzedrine . . . the Nazis preferred Pervitin*: N. Rasmussen, "America's First Amphetamine Epidemic 1929–1971: A Quantitative and Qualitative Retrospective with Implications for the Present," *American Journal of Public Health* 98, no. 6 (2008): 974–85.

166 *reports of students fainting and dying*: "Pep-Pill Poisoning," *Time*, May 10, 1937.

166 *"a therapeutic weapon capable of alleviating depression"*: Rasmussen, "America's First Amphetamine Epidemic 1929–1971," 975.

167 *"is not in any sense curative and its effects are not permanent"*: A. Myerson, "Effect of Benzedrine Sulfate on Mood and Fatigue in Normal and in Neurotic Persons," *Archives of Neuropsychiatry* 36, no. 4 (1936): 816–22, p. 822.

167 *"'mutually corrective'"*: D. Legge and H. Steinberg, "Actions of a Mixture of Amphetamine and a Barbiturate in Man," *British Journal of Pharmacology* 18 (1962): 490–500.

167 *eight hundred metric tons of amphetamine pills every year*: Rasmussen, "America's First Amphetamine Epidemic 1929–1971," 977.

167 *"the shoes that Prozac would ultimately have to fill"*: N. Rasmussen, "Making the First Anti-Depressant: Amphetamine in American Medicine, 1929–1950," *Journal of the History of Medicine and Allied Sciences* 61, no. 3 (2006): 288–323.

169 *"fatalities may occur"*: E. P. Kraai and S. A. Seifert, "Citalopram Overdose: A Fatal Case," *Journal of Medical Toxicology* 11 (2014): 232–36.

170 *7.2 percent of the U.S. adult population*: Y. Luo et al., "National Prescription Patterns of Antidepressants in the Treatment of Adults with Major Depression in the US Between 1996 and 2015: A Population Representative Survey Based Analysis," *Frontiers in Psychiatry* 11, no. 35 (2020).

170 *16.6 percent of the UK population*: J. Marsden et al., "Medicines Associated with Dependence or Withdrawal: A Mixed-Methods Public Health Review and National Database Study in England," *Lancet Psychiatry* 6, no. 11 (2019): 935–50.

170 *In Australia, these figures stand at 15 percent*: M. Whitley and M. Raven, "1 in 8 (Over 3 Million) Australians Are on Antidepressants—Why Is the Lucky Country So Miserable?," table 1, PsychWatch Australia, updated August 25, 2019, https://www.psychwatchaustralia.com/post/1-in-8-over-3-million-australians-are-on-antidepressants-why-is-the-lucky-country-so-miserable.

170 *only worth a patient's while if they are severely depressed*: "The NICE threshold for clinical significance was met for intake HDRS scores of 25 or greater," in B. Arrol et al., "Antidepressants Versus Placebo for Depression in Primary Care," *Cochrane Database of Systematic Review*, 3 (2009). And: J. C. Fournier et al., "Antidepressant Drug Effects and Depression Severity: A Patient-Level Meta-Analysis," *Journal of the American Medical Association* 303, no. 1 (2010): 47–53.

170 *round after round of contradictory reviews*: J. C. Jakobsen, C. Gluud, and I. Kirsch,

"Should Antidepressants Be Used for Major Depressive Disorder?" *BMJ Evidence-Based Medicine* 25, no. 4 (2020): 130.

171 *Reviewing studies on twenty-one different antidepressants*: A. Cipriani et al., "Comparative Efficacy and Acceptability of 21 Antidepressant Drugs for the Acute Treatment of Adults with Major Depressive Disorder: A Systematic Review and Network Meta-Analysis," *The Lancet* 391 (2018): 1357–66.

171 *"our results are more readily generalizable"*: G. Lewis et al., "The Clinical Effectiveness of Sertraline in Primary Care and the Role of Depression Severity and Duration (PANDA): A Pragmatic, Double-Blind, Placebo-Controlled Randomized Trial," *Lancet Psychiatry* 6 (2019): 903–14.

G22355

173 *"developed into a small cyclone"*: M. Rinkel and C. B. Denber, eds., *Proceedings of the Symposium on Chemical Concepts of Psychosis Held at the Second International Congress of Psychiatry in Zurich, Switzerland, September 1 to 7, 1957* (London: Peter Owen, 1960).

173 *test his latest quip on his daughter*: M. Platt, *Storming the Gates of Bedlam: How Dr. Nathan Kline Transformed the Treatment of Mental Illness* (London: Depew, 2012), 42.

173 the crafty Schizococcus!: Rinkel and Denber, eds., *Proceedings of the Symposium on Chemical Concepts of Psychosis Held at the Second International Congress of Psychiatry in Zurich, Switzerland, September 1 to 7, 1957*.

174 *received two Lasker Awards*: The other is Robert Edward Gross, a pediatric cardiovascular surgeon.

174 *"It is almost a truism"*: Ibid.

174 *Taciturn, sober, and with a whisper-soft voice*: Frank Ayd is quoted as saying, "Kuhn is a very tall man, slender, very soft spoken, very cultured, very dignified and very erudite," in E. Shorter, *Before Prozac: The Troubled History of Mood Disorders in Psychiatry* (New York: Oxford University Press, 2009), 60.

174 *Kline didn't even know who Kuhn was*: "Antidepressant drug therapy . . . was still such a pioneer endeavor, however, that those who were involved often didn't know of each other's existence until discoveries converged along common paths," in N. S. Kline, *From Sad to Glad: Kline on Depression* (New York: Ballantine, 1974), 107.

174 *a thick fog blankets the streets*: Author research trip to Münsterlingen, September 2018.

175 *trio of tall buildings*: *150 Jahre Münsterlingen: Das Thurgauische Kantonsspital und die Psychiatrische Klinik 1840-1990*. pg. 115: Spitalanlage Münsterlingen, 1874, Massstab 1:5000.

175 *might choose to holiday*: Paraphrased from author interview with Ralf-Peter Gebhardt at Münsterlingen, September 2018.

175 *Münsterlingen firmly on the map*: E. Shorter, *A History of Psychiatry: From the Era of Asylums to the Age of Prozac* (Hoboken, NJ: John Wiley & Sons, 1997). Switzerland was the center of biological psychiatry and nearby Kreuzlingen was

famous for its psychoanalytic institute (the Bellevue Sanatorium, led by Ludwig Binswanger), but Kuhn and Münsterlingen were little known.

175 *G22355, a direct descendent of chlorpromazine*: H. E. Lehmann, C. H. Cahn, and R. L. de Verteuil, "The Treatment of Depressive Conditions with Imipramine (G22355)," *Canadian Psychiatric Association Journal* 3, no. 4 (1958): 155–64, p. 156. "phenothiazine derivatives."

175 *in place of an atom of sulphur*: Ibid.

175 *Paula J. . . . had "undergone a transformation"*: "The First Patient Treated with Imipramine." Photocopy from medical history #21502 of the "Kantonal Treatment and Care Clinic in Munsterlingen," concerning female patient Paula F. J., born April 30, 1907.

176 *elevate moods in forty of them*: W. A. Brown and M. Rosdolsky, "The Clinical Discovery of Imipramine," *American Journal of Psychiatry* 172 (2015): 426–29.

176 *"Feelings of guilt, delusions"*: Ibid., 427.

176 *"depression . . . gave way to friendly, joyous"*: F. López-Muñoz and C. Alamo, "Monoaminergic Neurotransmission: The History of the Discovery of Antidepressants from 1950s Until Today," *Current Pharmaceutical Design* 15 (2009): 1563–86, p. 1569.

176 psychic, ego, *and* id: N. S. Kline, "MAOIs: An Unfinished Picaresque Tale," in F. J. Ayd and B. Blackwell, eds., *Discoveries in Biological Psychiatry* (New York: Lippincott, 1970).

176 *the room remained quiet*: D. Healy, *The Antidepressant Era* (Cambridge, MA: Harvard University Press, 1997), 52. "The presentation didn't electrify many in the audience. The rest of the meeting remained unaware that they had missed something important."

176 *The dozen or so people who turned up*: López-Muñoz and Alamo, "Monoaminergic Neurotransmission," 1569.

176 *Lehmann . . . didn't make it to Kuhn's talk*: T. A. Ban, "Heinz Lehmann and Psychopharmacology," 1999. Typewritten document at http://inhn.org.

176 *became aware of the Swiss psychiatrist's work as he was flying back*: H. E. Lehmann and N. S. Kline, "Clinical Discoveries with Antidepressant Drugs," in M. J. Parnham and J. Bruinvels, eds., *Discoveries in Pharmacology, Volume 1: Psycho- and Neuro-Pharmacology* (Elsevier Science, 1983), 209–21.

177 *skiing holiday in Quebec . . . Verdun Protestant Hospital*: J. Paris, "Heinz Lehmann: A Pioneer of Modern Psychiatry," *Canadian Journal of Psychiatry* 44 (1999): 441–42.

177 *(60 percent) either fully recovered*: Lehmann, Cahn, and de Verteuil, "The Treatment of Depressive Conditions with Imipramine (G22355)," 163.

177 *"The effect is striking"*: R. Kuhn, "The Treatment of Depressive States with G22355 (Imipramine Hydrochloride)," *American Journal of Psychiatry* 115, no. 5 (1958): 459–64.

178 *They weren't "fatigued"*: Ibid.

178 *different chemical pathway from MAO inhibitors*: T. A. Ban, "Pharmacotherapy of Depression: A Historical Analysis," *Journal of Neural Transmission* 108 (2001): 707–16.

178 *"The pharmacological uniqueness of imipramine is of special significance"*: F. Freyhan,

"The Modern Treatment of Depressive Disorders," *American Journal of Psychiatry* 116 (1960): 1057–64, p. 1061.

178 *"much less spectacular than . . . electroconvulsive treatment"*: Lehmann, Cahn, and de Verteuil, "The Treatment of Depressive Conditions with Imipramine (G22355)," 161.

179 *"the most effective treatment for depression"*: I. M. McDonald et al., "A Controlled Trial Comparison of Amitriptyline and ECT in the Treatment of Depression," *American Journal of Psychiatry* 122, no. 12 (1966): 1427–31.

179 *two large-scale studies*: M. Greenblatt, G. H. Grosser, and H. Wechsler, "Differential Response of Hospitalized Depressed Patients to Somatic Therapy," *American Journal of Psychiatry* 120, no. 10 (1964): 935–43; Medical Research Council Clinical Psychiatry Committee, "Clinical Trial of the Treatment of Depressive Illness," *British Medical Journal* 1 (1965): 881–86.

179 *"psychotic depression" . . . higher rate of suicide*: R. Gournellis et al., "Psychotic (Delusional) Depression and Completed Suicide: A Systematic Review and Meta-Analysis," *Annals of General Psychiatry* 17 (2018): 39.

179 *"presence of 22 patients with delusions"*: Clinical Research Centre, Division of Psychiatry, "The Northwick Park ECT Trial: Predictors of Response to Real and Simulated ECT," *British Journal of Psychiatry* 144 (1984): 227–37, p. 235.

179 *"[A] clear and significant advantage for ECT over medication"*: N. A. Payne and J. Prudic, "Electroconvulsive Therapy Part I: A Perspective on the Evolution of Current Practice of ECT," *Journal of Psychiatric Practice* 15, no. 5 (2009): 346–68.

180 *"[O]ther candies"*: N. S. Kline, *Depression: Its Diagnosis and Treatment* (Basel, Switzerland: Karger, 1969).

180 *first reported case of an overdose*: N. P. Lancaster and A. R. Foster, "Suicidal Attempt by Imipramine Overdose," *British Medical Journal* 2, no. 5164 (1959): 1458.

181 *response rate of 75 percent . . . as low as 25 percent*: E. D. West and P. J. Dally, "Effects of Iproniazid in Depressive Syndromes," *British Medical Journal* 1, no. 5136 (1959): 1491–94.

181 *By 1959, bipolar disorder*: M. Mendelson, *Psychoanalytic Concepts of Depression* (New York: Spectrum, 1974), 25. Leonard (1957).

182 *"difficult to treat by any method"*: W. Sargant, "Some Newer Drugs in the Treatment of Depression and Their Relation to Other Somatic Treatments," *Psychosomatics* 1 (1960): 14–17, p. 16.

182 *within five to eight days*: J. T. Davidson et al., "Atypical Depression," *Archives of General Psychiatry* 39 (1982): 527–34, p. 532.

182 *Relapse . . . defining feature of MAO inhibitors*: Sargant, "Some Newer Drugs in the Treatment of Depression and Their Relation to Other Somatic Treatments," 16.

182 *imipramine might take four or more weeks*: Davidson et al., "Atypical Depression," 532.

182 *ECT . . . avoided for such patients*: Ibid., 533.

183 *"such improvements may be of greatest importance"*: West and Dally, "Effects of Iproniazid in Depressive Syndromes," 1491–94.

183 *one patient in every five hundred . . . became jaundiced*: Sargant, "Some Newer Drugs in the Treatment of Depression and Their Relation to Other Somatic Treatments," 15.

183 *"some who had ski accidents—and some who got jaundice"*: Kline, *From Sad to Glad*, 103.

183 *iproniazid was withdrawn*: López-Muñoz and Alamo, "Monoaminergic Neurotransmission," 1572.

183 *"The whole iproniazid story was one of these 'acres of diamonds' affairs"*: Lehmann and Kline, "Clinical Discoveries with Antidepressant Drugs," 209–21.

184 *Parnate was considered the safest*: R. M. Atkinson and K. S. Ditman, "Tranylcypromine: A Review," *Clinical Pharmacology and Therapeutics* 6 (1965): 631–55, p. 632. "The fact that tranylcypromine was not a hydrazine derivative . . . added to enthusiasm for it as an agent that might show far less toxicity than the other antidepressants."

The Mysterious Case of the Lethal Headaches

185 *a twenty-seven-year-old man walked into the Maudsley*: J. L. McClure, "Reactions Associated with Tranylcypromine," *The Lancet* 1, no. 1351 (1962).

185 *"devastating headache"*: B. Blackwell, "Hypertensive Crisis Due to Monoamine-Oxidase Inhibitors," *The Lancet* 282 (1963): 849–51.

185 *Blackwell listened to the reports and quickly lost interest*: B. Blackwell, *Bits and Pieces of a Psychiatrist's Life: A Shrunken Life* (Bloomington, IN: Xlibris, 2012).

186 *spent more time at the local pubs and playing rugby*: Ibid.

186 *"I was transformed by contact with patients"*: Ibid.

186 *rows of pendant lights . . . smell of coffee*: Photo of cafeteria in the late 1950s (when it opened) at Bethlem Museum of the Mind. Author research trip, December 2017.

186 *"Was she taking Parnate?"*: Blackwell, *Bits and Pieces of a Psychiatrist's Life.*

187 *a short letter in* The Lancet: B. Blackwell, "Tranylcypromine," *The Lancet* 281, no. 7273 (1963): 167–68.

187 *a pharmacist working in Nottingham*: B. Blackwell, "The Process of Discovery," in F. J. Ayd and B. Blackwell, eds., *Discoveries in Biological Psychiatry* (New York: Lippincott, 1970).

187 *"No effects are caused by butter"*: "Adumbration: A History Lesson," *International Network for the History of Neuropsychopharmacology*, December 18, 2014.

188 *He knew she was vegetarian*: Blackwell, "The Process of Discovery."

188 *twenty milligrams of Parnate . . . sat down to a breakfast of cheese*: Ibid.

188 *two imbalance theories . . . one for norepinephrine and one for serotonin*: F. López-Muñoz and C. Alamo, "Monoaminergic Neurotransmission: The History of the Discovery of Antidepressants from 1950s Until Today," *Current Pharmaceutical Design* 15 (2009): 1563–86, p. 1575.

188 *depressive-like condition . . . treatment with electroconvulsive therapy*: R. W. P. Achor, N. O. Hanson, and R. W. Gifford, "Hypertension Treated with *Rauwolfia serpentina* (Whole Root) and with Reserpine," *Journal of the American Medical Association* 159 (1955): 841–45.

188 *cast doubt on this association*: A. A. Baumeister, M. F. Hawkins, and S. M. Uzelac, "The Myth of Reserpine-Induced Depression," *Journal of the History of the Neurosciences* 12, no. 2 (2003): 207–20.

189 *"a reductionist oversimplification of a very complex biological state"*: J. J. Schildkraut, "The Catecholamine Hypothesis of Affective Disorders: A Review of Supporting Evidence," *American Journal of Psychiatry* 122, no. 5 (1965): 509–22, p. 517.

189 *Although they offered no improvement in efficacy*: T. A. Ban, "Pharmacotherapy of Depression: A Historical Analysis," *Journal of Neural Transmission* 108 (2001): 7070–716. "clinical expectations from the SSRIs were not fulfilled."

189 *studies in rat brains in 1978*: A. Carlsson and M. Lindqvist, "Effects of Antidepressant Agents on the Synthesis of Brain Monoamines," *Journal of Neural Transmission* 43 (1978): 73–91.

189 *eventual ban in 1983*: J. Fagius et al., "Guillain-Barré Syndrome Following Zimeldine Treatment," *Journal of Neurology, Neurosurgery, and Psychiatry* 48 (1985): 65–69.

189 *twenty-five-fold increased risk of Guillain-Barré syndrome*: Ibid.

189 *it would take fifteen years for this drug to be released*: López-Muñoz and Alamo, "Monoaminergic Neurotransmission," 1576. From creation in 1972 to release in Belgium in 1987.

192 *another can lead them into complete remission*: A. J. Rush et al., "Bupropion-SR, Sertraline, or Venlafaxine-XR After Failure of SSRIs for Depression," *New England Journal of Medicine* 354, no. 12 (2006): 1231–42.

193 *Blackwell was moonlighting as a medical officer in an ambulance*: Blackwell, *Bits and Pieces of a Psychiatrist's Life.*

194 *She had developed a terribly painful headache*: Blackwell, "The Process of Discovery."

194 *a colleague rushed through the corridor*: Ibid. "Coincidence continued to play a part. While walking down a Maudsley corridor one night, I was overtaken by the duty resident en route to a ward where two patients were simultaneously complaining of a headache."

194 *Eight had definitely eaten cheese*: Blackwell, "Hypertensive Crisis Due to Monoamine-Oxidase Inhibitors," 849–51.

194 *"the idea that a common dietary substance might kill someone"*: Blackwell, "The Process of Discovery."

194 *"unscientific and premature"*: Blackwell, *Bits and Pieces of a Psychiatrist's Life.*

194 *a molecule called tyramine*: B. Blackwell and L. A. Mabbitt, "Tyramine in Cheese Related to Hypertensive Crises After Monoamine-Oxidase Inhibition," *The Lancet* 285 (1965): 938–40.

195 *Kline read the worrying reports of the cheese reaction*: Papers in Kline's personal files, passed to him by his assistant. Author research trip to Nathan Kline Institute, Orangeburg, October 2018.

195 *Parnate was withdrawn*: R. M. Atkinson and K. S. Ditman, "Tranylcypromine: A Review," *Clinical Pharmacology and Therapeutics* 6 (1965): 631–55. "Feb. 24, 1964." Parnate was remarketed with "restrictive prescribing information" in mid-August 1964.

195 *five hundred people suffered from the devastating headaches*: Ibid., 640.

195 *Forty people died*: Ibid.

195 *yeast extract*: B. Blackwell, "Effects of Yeast Extract After Monoamine-Oxidase Inhibition," *The Lancet* 285 (1965): 1166.

196 *growing list of restricted substances*: E. Marley and B. Blackwell, "Interactions of

Monoamine Oxidase Inhibitors, Amines, and Foodstuffs," *Advances in Pharmacology* 8 (1971): 185–239.

196 *"The people who really benefit"*: Author interview with Ted Dinan in Cork, Ireland, November 2018.

196 *Several clinical trials in the 1980s and '90s*: F. M. Quitkin et al., "Phenelzine Versus Imipramine in the Treatment of Probable Atypical Depression: Defining Syndrome Boundaries of Selective MAOI Responders," *American Journal of Psychiatry* 145 (1988): 306–11. Review: M. E. Thase, M. H. Trivedi, and A. J. Rush, "MAOIs in the Contemporary Treatment of Depression," *Neuropsychopharmacology* 12, no. 3 (1995): 185–219.

196 *"[And] they respond . . . like magic"*: Author interview with Ted Dinan in Cork, November 2018.

196 *eighteen leaders in biological psychiatry gathered*: F. J. Ayd and B. Blackwell, eds., *Discoveries in Biological Psychiatry* (New York: Lippincott, 1970).

196 *"[N]o drug in history was so widely used"*: N. S. Kline, "MAOIs: An Unfinished Picaresque Tale," ibid.

197 *Given to four hundred thousand people in the U.S.*: López-Muñoz and Alamo, "Monoaminergic Neurotransmission: The History of the Discovery of Antidepressants from 1950s Until Today," 1567.

197 *Kuhn told his audience, his trials started in January 1956*: R. Kuhn, "The Imipramine Story," in Ayd and Blackwell, eds., *Discoveries in Biological Psychiatry*.

197 *featured on the front cover of* Fortune *magazine*: D. Healy, *Antidepressant Era* (Cambridge, MA: Harvard University Press, 1997), 68.

197 *Kline was one of three psychiatrists*: Along with Henry Brill and Frank Ayd.

197 *"The amount involved . . . in the 1950s was vast"*: Healy, *Antidepressant Era*, 66.

197 *"An Unfinished Picaresque Tale"*: Kline, "MAOIs: An Unfinished Picaresque Tale."

197 *looked it up in his* Oxford English Dictionary: B. Blackwell, *Pioneers and Controversies in Psychopharmacology* (e-Book) (International Network for the History of Neuropsychopharmacology, 2018). Chapter 8, "Nathan ('Nate') Kline and the Monoamine Oxidase Inhibitors."

198 *"a wheeler-dealer's arrogance"*: J. Goldberg, *Anatomy of a Scientific Discovery* (New York: Bantam, 1989).

198 *his host was "courageous"*: Ibid.

199 *thanks largely to . . . Frances Kelsey*: B. Fintel, A. T. Samaras, and E. Carias, "The Thalidomide Tragedy: Lessons for Drug Safety and Regulation," Helix, July 28, 2009.

199 *"liked the limelight too much"*: Goldberg, *Anatomy of a Scientific Discovery*, 137.

199 *"resemblance to Charlton Heston's Moses"*: Ibid., 121.

199 *"a nonsense study"*: Ibid., 137.

199 *"slow agony"*: Ibid., 202.

200 *"any investigational new drug"*: "Psychiatrist Barred from Administering Experimental Drug," *New York Times*, May 25, 1982.

200 *retired from a thirty-year career*: "Nathan Kline, Developer of Antidepressants, Dies," *New York Times*, February 14, 1983.

200 *"With this unfortunate matter behind me"*: "Psychiatrist Barred from Administering Experimental Drug."

200 *boxed away in a disused room next to a photocopier*: Author research trip to Nathan Kline Institute, November 2018.
200 *"Literally hundreds of thousands of people"*: Ibid.

Part Three. Getting Therapy

In Your Dreams, Freud

205 *"He will* not *die!"*: M. E. Weishaar, *Aaron T. Beck (Key Figures in Counseling & Psychotherapy Series)* (Thousand Oaks, CA: SAGE Publications, 1993), 9.
205 *90 percent mortality rate*: Ibid.
205 *"if I got into a hole I could dig myself out"*: Ibid., 10.
205 *He was told that a Jewish boy*: Ibid., 12.
206 *"deep, dark invisible forces"*: Ibid., 15.
206 *"I thought it was nonsense"*: Ibid., 14.
206 *"had answers for everything"*: Ibid., 15.
206 *What he found was, in truth, quite banal*: D. B. Smith, "The Doctor Is In," *American Scholar*, September 1, 2009.
207 *Beck started playing a rigged card game*: "A Psychiatrist Who Wouldn't Take No for an Answer," *New York Times*, August 11, 1981.
207 *"A brilliant academician questioned his basic intelligence"*: A. T. Beck, "Thinking and Depression I: Idiosyncratic Content and Cognitive Distortions," *Archives of General Psychiatry* 9 (1963): 324–33, p. 326.
208 *"the depressed patient was prone to magnify any failures"*: Ibid., 326.
208 *"Intrinsic to this type of thinking"*: Ibid., 328.
208 *pigeons pecking at levers and rats pressing pedals*: W. N. Dember, "Motivation and the Cognitive Revolution," *American Psychologist* 29, no. 3 (1974): 161–68.
209 *Medium-sized "mongrel dogs"*: M. E. P. Seligman and S. F. Maier, "Failure to Escape Traumatic Shock," *Journal of Experimental Psychology* 74, no. 1 (1967): 1–9.
209 *"severe, pulsating" . . . shocks didn't abate for fifty seconds*: M. E. P. Seligman, "Learned Helplessness," *Annual Reviews of Medicine* 23 (1972): 407–12, p. 408.
209 *"The initial problem seems to be one of 'getting going'"*: Ibid., 410.
209 *lower levels of norepinephrine in their brain*: Ibid., 408.
210 *"positively reinforced behaviors that are missing"*: C. B. Ferster, "A Functional Analysis of Depression," *American Psychologist* 28, no. 10 (1973): 857–70.
210 *behavior therapy was successful in over 80 percent*: J. Wolpe, "Behavior Therapy Versus Psychoanalysis: Therapeutic and Social Implications," *American Psychologist* 36 (1981): 159–64, p. 161.
210 *"The cognitivists reject the conditioning theory"*: J. Wolpe, "The Derailment of Behavior Therapy: A Tale of Conceptual Misdirection," *Journal of Behavior Therapy & Experimental Psychiatry* 20, no. 1 (1989): 3–15.
210 *"Cognitive therapy is a subclass of behavior therapy"*: J. Wolpe, "Behavior Therapy and Its Malcontents—II. Multimodal Eclecticism, Cognitive Exclusivism and 'Exposure' Empiricism," *Journal of Behavior Therapy & Experimental Psychiatry* 7 (1976): 109–16, p. 114.

211 *"change the 'minds and hearts of men and women'"*: B. F. Skinner, "Why I Am Not a Cognitive Psychologist," *Behaviorism* 5, no. 2 (1977): 1–10.
211 *"We felt very much on our own"*: Weishaar, *Aaron T. Beck*, 23.
211 *"excitement of being the underdog"*: Ibid., 27.
212 *"[B]ehavioural techniques are used to promote cognitive change"*: Ibid., 90.
212 *cured his mother's depression the day he was born*: Ibid., 9.
212 *Beck Depression Inventory*: A. T. Beck et al., "An Inventory for Measuring Depression," *Archives of General Psychiatry* 4 (1961): 53–63.
213 *trained himself to highlight his own cognitive distortions*: Smith, "The Doctor Is In."
213 *playing tennis, napping, and going to the cinema*: R. Rosner, "The 'Splendid Isolation' of Aaron Beck," *Isis* 105, no. 4 (2014): 734–58, p. 751.
214 *"the explosion after 1962 of spiral-bound notebooks"*: Ibid., 755.
214 *"She was my reality tester"*: "Scientist at Work: Aaron T. Beck; Pragmatist Embodies His No-Nonsense Therapy," *New York Times*, January 11, 2000.
214 *a bright red one for special occasions*: Ibid.
214 *friendly demeanor and his bright, pale blue eyes*: Ibid.
214 *run-down offices with tattered chairs*: Smith, "The Doctor Is In."
214 *"cognitive therapy is more effective"*: "A Psychiatrist Who Wouldn't Take No for an Answer," *New York Times*, August 11, 1981.

More Than One Psychotherapy

215 *an expert on tricyclic antidepressants*: G. L. Klerman and J. O. Cole, "Clinical Pharmacology of Imipramine and Related Antidepressant Compounds," *Pharmacological Reviews* 17, no. 2 (1965): 101–41.
215 *"inverse correlation between recovery and psychotherapy"*: H. J. Eysenck, "The Effects of Psychotherapy: An Evaluation," *Journal of Consulting Psychology* 16 (1952): 319–24.
216 *framed photograph of Aaron Beck*: Author interview with Myrna Weissman at the New York State Psychiatric Institute, November 2018.
216 *"We were friends"*: Ibid.
216 *thinking that he was God*: Ibid.
217 *"as much the hallmark of the [human] species"*: J. Bowlby, "The Nature of the Child's Tie to Its Mother," *International Journal of Psycho-Analysis* 39 (1958): 350–73.
217 Attachment and Loss: J. Bowlby, *Attachment and Loss. Volume 1: Attachment* (New York: Basic Books, 1969).
217 *"departures from the social field"*: E. S. Paykel et al., "Life Events and Depression: A Controlled Study," *Archives of General Psychiatry* 21 (1969): 753–60.
217 *"loss and disappointment rather than change"*: G. W. Brown and T. Harris, *Social Origins of Depression: A Study of Psychiatric Disorder in Women* (London: Tavistock, 1978), 105.
218 *"find out what is going on in their life"*: Author interview with Myrna Weissman, November 2018.
218 *sick role*: M. M. Weissman and J. C. Markowitz, "Interpersonal Psychotherapy: Current Status," *Archives of General Psychiatry* 51 (1994): 599–606.
219 *79 percent vs. imipramine's 24 percent*: A. J. Rush et al., "Comparative Efficacy of

Cognitive Therapy and Pharmacotherapy in the Treatment of Depressed Outpatients," *Cognitive Therapy & Research* 1 (1977): 17–37.

219 *"sense of mastery"*: "New Theories of Depression Hold Promise of Simpler Remedy," *New York Times*, June 2, 1981.

220 *Beck complained that this wasn't sufficient*: M. E. Weishaar, *Aaron T. Beck (Key Figures in Counseling & Psychotherapy Series)* (Thousand Oaks, CA: SAGE Publications, 1993), 38. "Beck replied with dismay."

220 *"Some of the beauty . . . got lost in the translation"*: Ibid., 39.

220 *compared the therapists . . . to . . . junior doctors*: Ibid., 38.

220 *"The immediate goal of the training programs"*: I. Elkin et al., "NIMH Treatment of Depression Collaborative Research Program: Background and Research Plan," *Archives of General Psychiatry* 42 (1985): 305–16, p. 308.

221 *"'this won't [turn] out that well,'"*: Weishaar, *Aaron T. Beck*, 39.

221 *"[T]he most intensive training program is that for CB"*: Elkin et al., "NIMH Treatment of Depression Collaborative Research Program," 309.

221 *the number of patients who relapsed . . . lowest for . . . cognitive therapy*: M. T. Shea et al., "Course of Depressive Symptoms Over Follow-up. Findings from the National Institute of Mental Health Treatment of Depression Collaborative Research Program," *Archives of General Psychiatry* 49 (1992): 782–87, p. 786. "Patients recovering in CBT had the lowest rates of receiving treatment for depression during the follow-up, and particularly at the 1-year point, they had a low rate of relapse when defined as MDD or treatment (14% compared with 50% for imipramine . . . 43% for IPT, and 31% for placebo)."

222 *"that's cognitive therapy's major contribution"*: Weishaar, *Aaron T. Beck*, 41.

222 *"strongest evidence is for CBT"*: Author interview with Helen Christensen, August 2018.

224 *a study of four hundred people living in northeast England*: S. Ali et al., "How Durable Is the Effect of Low Intensity CBT for Depression and Anxiety? Remission and Relapse in a Longitudinal Cohort Study," *Behavior Research and Therapy* 94 (2017): 1–8.

224 *benefits of twelve to eighteen sessions of CBT are maintained for at least four years*: N. J. Wiles, "Long-Term Effectiveness and Cost-Effectiveness of Cognitive Behavioral Therapy as an Adjunct to Pharmacotherapy for Treatment-Resistant Depression in Primary Care: Follow-up of the CoBalT Randomized Controlled Trial," *Lancet Psychiatry* 3 (2016): 137–44.

226 *"so little on improving its* quality*"*: R. Layard and D. M. Clark, *Thrive: The Power of Evidence-Based Therapies* (New York: Penguin, 2015), 89.

"If Mom Ain't Happy Ain't Nobody Happy"

227 *"There shouldn't be one psychotherapy"*: Author interview with Myrna Weissman at the New York State Psychiatric Institute, November 2018.

227 *"I did not fit the historical picture of a Yalie"*: M. M. Weissman, "Depression," *Annals of Epidemiology* 19, no. 4 (2009): 264–67.

228 *In her first survey*: M. M. Weissman and J. K. Myers, "Affective Disorders in a US Urban Community," *Archives of General Psychiatry* 35 (1978): 1304–11.

229 *depression was predominantly a disorder of menopausal women*: Author interview with Myrna Weissman, November 2018. Also: Weissman, "Depression," 265.

229 *Weissman sent the handwritten drafts*: Weissman, "Depression," 265.

229 *Epidemiologic Catchment Area (ECA) survey*: D. A. Reiger et al., "The NIMH Epidemiologic Catchment Area Program," *Archives of General Psychiatry* 41 (1984): 934–41.

230 *On average, nearly 5 percent of American people*: M. M. Weissman et al., "Affective Disorders in Five United States Communities," *Psychological Medicine* 18 (1988): 141–53.

230 *lower than in New Zealand and France*: M. M. Weissman et al., "Cross-National Epidemiology of Major Depression and Bipolar Disorder," *Journal of the American Medical Association* 276, no. 4 (1996): 293–94.

230 *two to four times higher in people who had been divorced*: Ibid.

230 *"astronomical in terms of lost productivity alone"*: R. M. Glass and D. X. Freedman, "Psychiatry," *Journal of the American Medical Association* 254, no. 16 (1985): 2280–83, p. 2282.

231 *$185 billion per year*: Ibid.

231 *interpersonal therapy provided a boost*: A. DiMascio et al., "Differential Symptom Reduction by Drugs and Psychotherapy in Acute Depression," *Archives of General Psychiatry* 36 (1979): 1450–56.

231 *"combination of drugs and psychotherapy has an additive effect"*: "Yale Researchers," *New York Times,* August 6, 1978.

231 *roughly half of the patients*: I. Elkin et al., "National Institute of Mental Health Treatment of Depression Collaborative Research Program: General Effectiveness of Treatments," *Archives of General Psychiatry* 46 (1989): 971–83, p. 977, fig. 2. Also: P. Cuijpers et al., "Interpersonal Psychotherapy for Depression: A Meta-Analysis," *American Journal of Psychiatry* 168, no. 6 (2011): 581–92, p. 8. "There is no indication that IPT is superior to CBT, and the two seem equally effective overall."

231 *useful for depression in mothers*: H. A. Swartz et al., "Brief Interpersonal Psychotherapy for Depressed Mothers Whose Children Are Receiving Psychiatric Treatment," *American Journal of Psychiatry* 165, no. 9 (2008): 1155–62.

232 *"The world of medicine has lost"*: M. B. Keller, "In Memoriam: Gerald L. Klerman, MD, 1928–1992," *Journal of Clinical Psychopharmacology* 12, no. 6 (1992): 379–81.

232 *"Gerry always brought a kind of cleansing clarity"*: Ibid., 381.

232 *"I put [interpersonal therapy] aside"*: Author interview with Myrna Weissman, November 2018.

233 *"Depression [can be] associated with abnormal grief reactions"*: G. L. Klerman et al., *Interpersonal Psychotherapy of Depression* (New York: Basic Books, 1984), 96.

233 *"That's not rocket science"*: Author interview with Myrna Weissman, November 2018.

233 *At the ten-year mark*: M. M. Weissman et al., "Offspring of Depressed Parents. 10 Years Later," *Archives of General Psychiatry* 54 (1997): 932–40.

233 *three times as high*: M. M. Weissman et al., "A 30-Year Study of 3 Generations at High Risk and Low Risk for Depression," *JAMA Psychiatry* 73, no. 9 (2016): 970–77.

234 *"you can modify that, by treating them"*: Author interview with Myrna Weissman, November 2018.

234 *The crucial element was remission*: M. M. Weissman et al., "Remissions in Maternal Depression and Child Psychopathology: A Star*D-Child Report," *Journal of the American Medical Association* 295, no. 12 (2006): 1389–398.

234 *"If mom ain't happy ain't nobody happy"*: Author interview with Myrna Weissman, November 2018.

234 *"We do not know from this study"*: Author interview with Judy Garber, September 2019.

235 *"So we did another study"*: Author interview with Myrna Weissman, November 2018.

235 *Published in the* Journal of American Psychiatry *in 2015*: M. M. Weissman et al., "Treatment of Maternal Depression in a Medication Clinical Trial and Its Effect on Children," *American Journal of Psychiatry* 172, no. 5 (2015): 450–59.

235 *"if the mother got better, with treatment, the kids got better"*: Author interview with Myrna Weissman, November 2018.

235 *replicated Weissman's STAR*D results*: Swartz, "Brief Interpersonal Psychotherapy for Depressed Mothers Whose Children Are Receiving Psychiatric Treatment," 1155–62.

235 *"You don't know who's going to get depressed"*: Author interview with Helen Christensen, August 2018.

236 *clinical trials in Chile*: R. Araya et al., "School Intervention to Improve Mental Health of Students in Santiago, Chile: A Randomized Clinical Trial," *JAMA Pediatrics* 167, no. 11 (2013): 1004–10.

236 *and the UK*: P. Stallard et al., "Classroom Based Cognitive Behavioural Therapy in Reducing Symptoms of Depression in High Risk Adolescents: Pragmatic Randomised Controlled Trial," *British Medical Journal* 345 (2012): e6058.

236 *"Universal depression prevention remains a challenge"*: R. Whittaker et al., "MEMO: An mHealth Intervention to Prevent the Onset of Depression in Adolescents: A Double-Blind, Randomized, Placebo-Controlled Trial," *Journal of Child Psychology and Psychiatry* 58, no. 9 (2017): 1014–22.

236 *Judy Garber and her colleagues*: J. Garber et al., "Prevention of Depression in At-Risk Adolescents: A Randomized Controlled Trial," *Journal of the American Medical Association* 301, no. 21 (2009): 2215–24.

236 *"What we found . . . a moderator effect"*: Author interview with Judy Garber, September 2019.

"Happier Than We Europeans"

237 *"They were hardly eating, hardly moving"*: Author interview with Melanie Abas, February 2018.

237 *"That gave me a lot of early confidence"*: Ibid.

238 *twenty thousand Ndebele citizens*: T. M. Mashingaidze, "The 1987 Zimbabwe National Unity Accord and Its Aftermath: A Case of Peace Without Reconciliation," in C. Hendricks and L. Lushaba, eds., *From National Liberation to Democratic Renaissance in Southern Africa* (Dakar: Codesria, 2005), 82–92.

238 *"the early rain that washes away the chaff"*: H. Cameron, "British Policy Towards Zimbabwe During Matabeleland Massacre: Licence to Kill," *The Conversation*, September 17, 2017.

238 *"worst drought in living memory"*: B. Maphosa, "Lessons from the 1992 Drought in Zimbabwe: The Quest for Alternative Food Policies," *Nordic Journal of African Studies* 3, no. 1 (1994): 53–58.

238 *eighty-seven children died before the age of five*: "Zimbabwe: Key Demographic Indicators," UNICEF for Every Child, https://data.unicef.org/country/zwe/. 86.6 per 1,000 live births.

238 *eleven times higher than that of the UK*: "United Kingdom: Key Demographic Indicators," UNICEF for Every Child, https://data.unicef.org/country/gbr/. 7.4 per 1,000 live births.

238 *HIV . . . infect a quarter of the population by the mid-1990s*: O. Mugurungi, S. Gregson, A. D. McNaghten, S. Dube, and N. C. Grassly, "HIV in Zimbabwe 1985–2003: Measurement, Trends, and Impact," in M. Caraël and J. R. Glynn, eds., *HIV, Resurgent Infections and Population Change in Africa (International Studies in Population)*, vol. 6 (Berlin: Springer, 2007).

238 *one in every four thousand patients (0.001 percent)*: J. Broadhead and M. Abas, "Depressive Illness—Zimbabwe," *Tropical Doctor* 24 (1994): 27–30, p. 27.

238 *"In rural clinics"*: Ibid.

238 *9 percent of women in Camberwell*: M. Abas and J. Broadhead, "Depression and Anxiety Among Women in an Urban Setting in Zimbabwe," *Psychological Medicine* 27 (1997): 59–71.

239 *"[N]othing is so gentle as man in his primitive state"*: J. J. Rousseau, *Discourse on the Origin and Foundations of Inequality Among Men*, I. Johnstone, translation (1754).

239 *"happier than we Europeans"*: A. Page, "'They Are All Dead': For Indigenous People, Cook's Voyage of 'Discovery' was a Ghostly Visitation," *The Conversation*, April 28, 2020.

240 *"toward the decay of the species"*: Rousseau, *Discourse on the Origin and Foundations of Inequality Among Men*.

240 *"Scramble for Africa"*: T. Pakenham, *The Scramble for Africa* (Harbor, ME: Abacus, 2015).

240 *entire villages were stolen*: H. Schofield, "Human Zoos: When Real People Were Exhibits," BBC News, December 27, 2011.

241 *"classical depressive syndromes are seldom seen"*: J. Carothers, *The African Mind in Health and Disease: A Study in Ethnopsychiatry*, World Health Organization Monograph Series, no. 17 (Geneva, Switzerland: World Health Organization, 1953).

241 *"'without them life would be wonderful'"*: D. Ben-Tovim, "Mental Health and Primary Health Care in Botswana," in M. Rutter, ed., *Development Psychiatry* (Washington, D.C.: American Psychiatric Press, 1987).

241 *Portuguese explorers . . . in the sixteenth century*: M. Hall and R. Stefoff, *Great Zimbabwe* (New York: Oxford University Press, 2006), 10.

241 *trade networks with China, India, and the Middle East*: G. Pwiti, "Trade and Economies in Southern Africa: The Archaeological Evidence," *Zambezia* 18, no. 2 (1991): 119–29.

242 *ten and eighteen thousand people lived at Great Zimbabwe*: S. Chirikure et al.,

"What Was the Population of Great Zimbabwe (CE 1000–1800)?" *PLoS ONE* 12, no. 6 (2017): e0178335.

242 *the work of the mythical Queen of Sheba*: D. Randall-MacIver, "The Rhodesia Ruins: Their Probable Origin and Significance," *Geographical Journal* 27, no. 4 (1906): 325–36.

242 *Erythaean people . . . travelers from southern India*: S. T. Carroll, "Solomonic Legend: The Muslims and the Great Zimbabwe," *International Journal of African Historical Studies* 21, no. 2 (1988): 233–47.

242 *Arabs from the Middle East*: J. E. Mullan, *The Arab Builders of Zimbabwe* (Salisbury, Rhodesia, 1969).

242 *"Censorship of guidebooks"*: P. Garlake, *The Ruins of Zimbabwe* (Zambia: Historical Association of Zambia, 1974).

243 *"has but few gifts for work"*: Carothers, *African Mind in Health and Disease.*

243 *"the condition as a whole is relatively rare"*: Ibid.

243 *"glorified pseudo-scientific novels"*: R. H. Prince, "John Carothers and African Colonial Psychiatry," *Transcultural Psychiatric Research Review* 33 (1996): 226–40, p. 231.

243 *"a force-field that radiated power and presence"*: "Remembering Thomas Adeoye Lambo and the Mysteries of the African Mind," *The News* (Nigeria), January 9, 2018.

243 *"that they can no longer be seriously presented"*: Prince, "John Carothers and African Colonial Psychiatry," 231.

243 *rates of depression soared as decolonization spread*: R. H. Prince, "The Changing Picture of Depressive Syndromes in Africa: Is It Fact or Diagnostic Fashion?" *Canadian Journal of African Studies* 1, no. 2 (1967): 177–92.

243 *Studies in two Ugandan villages*: J. Orley and J. Wing, "Psychiatric Disorders in Two African Villages," *Archives of General Psychiatry* 36 (1979): 513–20.

244 *Among the Yoruba tribe . . . four times as high*: A. H. Leighton et al., *Psychiatric Disorder Among the Yoruba* (New York: Cornell University Press, 1963).

244 *"For one trained in one's own culture"*: V. Rao, "Depressive Illness in India," *Indian Journal of Psychiatry* 26, no. 4 (1984): 301–11.

Kufungisisa

245 *two hundred households in Glen Norah*: M. Abas and J. Broadhead, "Depression and Anxiety Among Women in an Urban Setting in Zimbabwe," *Psychological Medicine* 27 (1997): 59–71.

245 *"thinking too much," was the most common descriptor*: M. Abas et al., "Defeating Depression in the Developing World: A Zimbabwean Model: One Country's Response to the Challenge," *British Journal of Psychiatry* 164 (1994): 293–96.

245 *"pretty classical depression"*: Author interview with Melanie Abas, February 2018.

246 *"one in every five people"*: Abas and Broadhead, "Depression and Anxiety Among Women in an Urban Setting in Zimbabwe," 59–71.

246 *Adopting the methods of George Brown*: G. W. Brown and T. Harris, *Social Origins of Depression: A Study of Psychiatric Disorder in Women* (London: Tavistock, 1978).

246 *strong pattern emerge from her surveys*: J. C. Broadhead and M. Abas, "Life Events,

Difficulties, and Depression Among Women in an Urban Setting in Zimbabwe," *Psychological Medicine* 28 (1998): 29–38.

246 *"in Zimbabwe, there were a lot more of these events"*: Author interview with Melanie Abas, February 2018.

246 *twice as likely to be depressed*: D. Chibanda et al., "Validation of Screening Tools for Depression and Anxiety Disorders in a Primary Care Population with High HIV Prevalence in Zimbabwe," *Journal of Affective Disorders* 198 (2016): 50–55.

246 *poverty: it is both a trigger*: C. Lund, "Poverty and Mental Disorders: Breaking the Cycle in Low-Income and Middle-Income Countries," *The Lancet* 378 (2011): 1502–14.

246 *in the 1990s, over one hundred million people*: Q. Liu et al., "Changes in the Global Burden of Depression from 1990 to 2017: Findings from the Global Burden of Disease Study," *Journal of Psychiatric Research* 126 (2020): 134–40.

246 *90 percent*: D. Chisholm et al., "Scaling-Up Treatment of Depression and Anxiety: A Global Return on Investment Analysis," *Lancet Psychiatry* 3 (2016): 415–24, table 1.

247 *"we are all developing countries"*: D. Mohammadi, "Shekhar Saxena," *The Lancet Psychiatry* 4, no. 5 (2017): 359.

247 *"[T]ry and share their sadness"*: Abas et al., "Defeating Depression in the Developing World," 294.

247 *ten psychiatrists serving a country of ten million*: Ibid., 296.

247 *"Thinking too much makes people ill"*: Ibid., 295.

248 *psychiatrists . . . dropped to just one or two*: Author interview with Dixon Chibanda in Harare, April 2018.

249 *disability-adjusted life year*: The World Bank, *World Development Report 1993: Investing in Health* (New York: Oxford University Press, 1993).

249 *"one lost year of 'healthy' life"*: "Metrics: Disability-Adjusted Life Year (DALY)," World Health Organization.

249 *over 8 percent of all healthy years lost*: R. Desjarlais et al., *World Mental Health: Problems and Priorities in Low-Income Countries* (New York: Oxford University Press, 1995), 4.

249 *"Yet despite the importance of these problems"*: Ibid.

249 *"I watched us being knocked off the agenda"*: Author interview with Arthur Kleinman at Harvard University, May 2018.

250 *"That's the only way to go"*: Presentation by Vikram Patel at the Harvard Center for Global Health Delivery in Dubai.

251 *"Will it happen in the twenty-first century? Absolutely"*: Author interview with Arthur Kleinman at Harvard University, May 2018.

251 *they asked their relatives if they could help*: Author interview with Helena Verdeli at the Harvard Center for Global Health Delivery in Dubai.

252 *"The clinical outcomes that paraprofessionals achieve"*: J. A. Durlak, "Comparative Effectiveness of Paraprofessional and Professional Helpers," *Psychological Bulletin* 86, no. 1 (1979): 80–92.

252 *nearly a quarter of the people . . . were infected with HIV*: P. Bolton et al., "Group Interpersonal Psychotherapy for Depression in Rural Uganda: A Randomized Controlled Trial," *Journal of the American Medical Association* 289, no. 23 (2003): 3117–24.

253 *21 percent of people*: H. Verdeli et al., "Adapting Group Interpersonal Psychotherapy for a Developing Country: Experience in Rural Uganda," *World Psychiatry* 2, no. 2 (2003): 114–20, p. 114.
253 *A previous study had found a similar rate*: Ibid.
253 *inability to help people with depression-like illnesses*: Ibid., 115.
253 Yo'kwekyawa *and* okwekubazida: Bolton, "Group Interpersonal Psychotherapy for Depression in Rural Uganda," 3119.
253 *a few nuances of their own*: Verdeli, "Adapting Group Interpersonal Psychotherapy for a Developing Country," 115.
253 *"deeply embedded in one's social world"*: Author interview with Vikram Patel, March 2018.
253 *In China* . . . neurasthenia: A. Kleinman, "Depression, Somatization, and the 'New Cross-Cultural Psychiatry,'" *Social Science and Medicine* 11 (1977): 3–10.
253 *socially acceptable for Chinese citizens*: S. Lee, "Diagnosis Postponed: Shenjing Shuairuo and the Transformation of Psychiatry in Post-Mao China," *Culture, Medicine, and Psychiatry* 23 (1999): 349–80.
254 *In India . . . depression is felt as "tension"*: T. Roberts et al., "'Is There a Medicine for These Tensions?' Barriers to Treatment-Seeking for Depressive Symptoms in Rural India: A Qualitative Study," *Social Science and Medicine* 246 (2020): 112741.
254 *"We never use the word 'depression'"*: Author interview with Vikram Patel, March 2018.
254 *The most common descriptor . . . "thinking too much"*: K. Kidia et al., "'I Was Thinking Too Much': Experiences of HIV-Positive Adults with Common Mental Disorders and Poor Adherence to Antiretroviral Therapy in Zimbabwe," *Tropical Medicine and International Health* 20, no. 7 (2015): 903–13.
254 *"people tend to see themselves as part of a family"*: Verdeli, "Adapting Group Interpersonal Psychotherapy for a Developing Country," 115.
255 *"The dead are living among us"*: Ibid., 117.
255 *reduction in symptoms was three times as great*: Bolton, "Group Interpersonal Psychotherapy for Depression in Rural Uganda," 3117–24.
256 *similar projects in . . . Santiago, Chile*: R. Araya et al., "Treating Depression in Primary Care in Low-Income Women in Santiago, Chile: A Randomised Controlled Trial," *The Lancet* 361 (2003): 995–1000.
256 *and Goa, India*: V. Patel et al., "Efficacy and Cost-Effectiveness of Drug and Psychological Treatments for Common Mental Disorders in General Health Care in Goa, India: A Randomised, Controlled Trial," *The Lancet* 361 (2003): 33–39.

Care by the Community

258 *"I've lost quite a number of patients through suicide"*: Author interview with Dixon Chibanda in Harare, April 2018.
258 *"If you come to Harare and don't visit Mbare"*: Author chat with waiter at Alechi Portuguese Grill, Harare, April 2018.
259 *Operation Murambatsvina, or "Clear Out the Rubbish"*: A. Tibaijuka, "Report of the Fact-Finding Mission to Zimbabwe to Assess the Scope and Impact of Op-

eration Murambatsvina by the UN Special Envoy on Human Settlements Issues in Zimbabwe," United Nations, 2005.

259 *seven hundred thousand people across the country*: Ibid.

259 *a large majority . . . met the clinical threshold for depression*: Author interview with Dixon Chibanda in Harare, April 2018.

259 *"no one knew what we could do"*: Ibid.

260 *fourteen elderly women*: "Dixon Chibanda: Grandmothers Help to Scale up Mental Health Care," *Bulletin of the World Health Organization*, June 2018.

260 *"They are the custodians of health"*: Author interview with Nigel James at Mbare clinic, April 2018.

260 *"How could this possibly work"*: Author interview with Dixon Chibanda in Harare, April 2018.

260 *translate medical terms into everyday words*: M. Abas et al., "'Opening Up the Mind': Problem-Solving Therapy Delivered by Female Lay-Health Workers to Improve Access to Evidence-Based Care for Depression and Other Common Mental Disorders Through the Friendship Bench Project in Zimbabwe," *International Journal of Mental Health Systems* 10 (2016): 39.

260 *built by local craftsmen*: D. Chibanda et al., "Problem-Solving Therapy for Depression and Common Mental Disorders in Zimbabwe: Piloting a Task-Shifting Primary Mental Health Care Intervention in a Population with High Prevalence of People Living with HIV," *BMC Public Health* 11 (2011): 828, p. 4.

261 *Chigaro Chekupanamazano*: Ibid. And: Abas et al., "'Opening Up the Mind,'" 39.

261 *"Scientific knowledge tends to travel"*: R. Araya et al., "Cost-Effectiveness of a Primary Care Treatment Program for Depression in Low-Income Women in Santiago, Chile," *American Journal of Psychiatry* 163 (2006): 1379–87, p. 1385.

261 *"You don't know me"*: Author interview with Melanie Abas, February 2018.

262 *in 320 patients, there was a significant reduction in depressive symptoms*: Chibanda et al., "Problem-Solving Therapy for Depression and Common Mental Disorders in Zimbabwe," 6, fig. 2.

262 *In October 2011, the first study . . . was published*: Ibid.

263 *"We have an epidemic of that"*: Author interview with Dixon Chibanda in Harare, April 2018.

263 *"'No longer are we fighting about money'"*: Author interview with the grandmothers at Mbare clinic, April 2018. Interpreter: Nigel James.

264 *serodiscordant relationships*: O. Eyawo et al., "HIV Status in Discordant Couples in Sub-Saharan Africa: A Systematic Review and Meta-Analysis," *Lancet Infectious Disease* 10, no. 11 (2010): 770–77. "Data from the first demographic and health surveys (DHS) to include results from HIV tests suggest that in at least two-thirds of couples in whom at least one of the partners is HIV positive, only one person is infected."

264 *she had to kill herself and her two children*: Tanya Ngwenya speaking at a Circle Kubatana Tose, "Holding Hands Together," meeting in Harare, April 2018. Interpreter: Ropalloyd Dzapasi.

265 *the multicenter clinical trial*: D. Chibanda et al., "Effect of a Primary Care–Based Psychological Intervention on Symptoms of Common Mental Disorders in Zimbabwe: A Randomized Clinical Trial," *Journal of the American Medical Association* 316, no. 24 (2016): 2618–26.

265 *Patel's SSQ-14*: V. Patel et al., "The Shona Symptom Questionnaire: The Development of an Indigenous Measure of Common Mental Disorders in Harare," *Acta Psychiatrica Scandinavica* 95 (1997): 469–75.

265 *"They are superstars"*: Author interview with Tarisai Bere in Harare, April 2018.

265 *"The Friendship Bench has become imaginary"*: Author interview with four grandmothers at Mbare clinic, April 2018. Interpreter: Nigel James.

"I Live and Breathe Peer"

266 *Skipper discovered the extracurricular activities*: Author interview with Helen Skipper in East Harlem, New York, April 2018.

266 *more likely to be addicted to crack*: J. J. Palomar et al., "Powder Cocaine and Crack Use in the United States: An Examination of Risk for Arrest and Socioeconomic Disparities in Use," *Drug and Alcohol Dependence* 149 (2015): 108–16.

266 *"The subcultural behaviors associated with crack use"*: E. Dunlap, A. Golub, and B. D. Johnson, "The Severely-Distressed African American Family in the Crack Era: Empowerment Is Not Enough," *Journal of Sociology and Social Welfare* 33, no. 1 (2006): 115–39, p. 121.

267 *"We never wanted to see a doctor"*: Author interview with Helen Skipper in East Harlem, New York, April 2018.

268 *"If you're working that area"*: Ibid.

268 *"You're not going to do me like that"*: Ibid.

268 *"I had a good therapist"*: Ibid.

268 *"I've been in every situation"*: Ibid.

269 *only peer supervisor in the DOHMH*: Ibid.

269 *bright orange plastic and looked like giant Lego bricks*: Author reporting trip to New York, April 2018.

270 *highest rate of lethal overdoses in the country*: A. Yensi, "The Bronx Continues to See Highest Number of Opioid Overdose Deaths," Spectrum News NY1, November 23, 2019.

270 *"This is darkness"*: Author interview with Bethsaida George-Rodriguez in New York, April 2018.

270 *An estimated one in ten people*: J. D. Livingstone, "Contact Between Police and People with Mental Disorders: A Review of Rates," *Psychiatric Services* 67 (2016): 850–57, p. 852.

270 *from 558,922 in 1955 to 159,405 in 1977*: B. Pepper, M. C. Kirshner, and H. Ryglewicz, "The Young Adult Chronic Patient: Overview of a Population," *Hospital and Community Psychiatry* 32 (1981): 463–69, p. 463.

270 *"only 14 percent were receiving mental health treatment"*: H. R. Lamb and R. Grant, "Mentally Ill Women in a County Jail," *Archives for General Psychiatry* 40 (1983): 363–68.

271 *"stop it before it gets hold o' you"*: Author interview with Helen Skipper in East Harlem, New York, April 2018.

271 *project had reached (over forty thousand)*: "Friendship Benches. Program Overview: July–December," printed document.

Part Four. The Universe Within

It Feels Like Spring

277 *it reminded her of a cooked piece of spaghetti*: Author interview with Helen Mayberg in New York, October 2018.

277 *thousands of successful operations*: C. Hamani, J. Neimat, and A. M. Lozano, "Deep Brain Stimulation for the Treatment of Parkinson's Disease," in P. Riederer et al., eds., *Parkinson's Disease and Related Disorders* (Vienna: Springer, 2006). "Approximately 30,000 patients."

277 *buzz of five milliamps*: Author interview with Helen Mayberg in New York, October 2018. Also: H. S. Mayberg et al., "Deep Brain Stimulation for Treatment-Resistant Depression," *Neuron* 45 (2005): 651–60. Range: 0.0–9.0 volts, average of 4.0 volts.

278 *"What are you doing?"*: Author interview with Helen Mayberg in New York, October 2018.

278 *even ECT had failed*: Mayberg et al., "Deep Brain Stimulation for Treatment-Resistant Depression," 653, table 1.

278 *"These people are morbidly ill"*: Author interview with Helen Mayberg, February 2018.

278 *"their language is just ridiculous"*: Author interview with Helen Mayberg in New York, October 2018.

279 *Parkinson's, Huntington's, stroke*: H. S. Mayberg, "Frontal Lobe Dysfunction in Secondary Depression," *Journal of Neuropsychiatry and Clinical Neurosciences* 6 (1994): 428–42.

279 *"I like controlling as many variables"*: Author interview with Helen Mayberg in New York, October 2018.

280 *40 percent of patients*: Mayberg, "Frontal Lobe Dysfunction in Secondary Depression," 428–42.

280 *"[T]he tendency to insanity"*: Ibid., 434.

280 *"not as an acceptable or normal part of growing old"*: Author interview with Charles Reynolds, August 2018.

280 *Depression came with a significant reduction in activity*: Based on the amount of glucose consumption in these areas, "activity" was a measure of meta-bolism.

280 *prelimbic frontal and temporal cortex*: Mayberg, "Frontal Lobe Dysfunction in Secondary Depression," 428–42.

280 *brains of patients with primary depression*: Ibid.

280 *literal markers of "the blues"*: A. Lozano, "Parkinson's, Depression, and the Switch That Might Turn Them Off," TEDxCaltech, January 18, 2013.

281 *most famous studies*: H. S. Mayberg et al., "Reciprocal Limbic-Cortical Function and Negative Mood: Converging PET Findings in Depression and Normal Sadness," *American Journal of Psychiatry* 156, no. 5 (1999): 675–82.

282 *"I talk too much"*: Author interview with Helen Mayberg in New York, October 2018.

283 *area 25 was quiet—hypoactive*: Mayberg et al., "Reciprocal Limbic-Cortical Function and Negative Mood," 678, fig. 1.

284 *"psychotherapy-responsive" form of depression*: B. W. Dunlop and H. S. Mayberg, "Neuroimaging-Based Biomarkers for Treatment Selection in Major-Depressive Disorder," *Dialogues in Clinical Neuroscience* 16 (2014): 479–90.

284 *one of the safest procedures in all of medicine*: M. Fink, *Convulsive Therapy: Theory and Practice* (New York: Raven Press, 1979). Chapter 4, "Risks of Convulsive Therapy." 0.03% death rate.

Rebirth

285 *"ECT," Alice says*: Author visit to ECT clinic in Brooklyn, New York: October 2018.

286 *cellular damage could occur*: G. Holmberg, "The Factor of Hypoxemia in Electroshock Therapy," *American Journal of Psychiatry* 110, no. 2 (1953): 115–18.

286 *"It's my job to keep the patient breathing"*: Author discussion with anaesthesiologist in Brooklyn, New York, October 2018.

286 *neurotransmitters and growth-inducing molecules*: A. Jorgensen et al., "Regional Brain Volumes, Diffusivity, and Metabolite Changes After Electroconvulsive Therapy for Severe Depression," *Acta Psychiatrica Scandinavica* 133 (2016): 154–64. Also: M. J. A. Van Den Bossche et al., "Hippocampal Volume Change Following ECT Is Mediated by rs699947 in the Promotor Region of VEGF," *Translational Psychiatry* 9 (2019): 191.

286 *prune the brain's delicate wiring*: J. L. Warner-Schmidt and R. S. Duman, "Hippocampal Neurogenesis: Opposing Effects of Stress and Antidepressant Treatment," *Hippocampus* 16 (2006): 239–49.

286 *Parts of the brain start to noticeably shrink*: P. Videbech and B. Ravnkilde, "Hippocampal Volume and Depression: A Meta-Analysis of MRI Studies," *American Journal of Psychiatry* 161 (2004): 1957–66.

287 *"I've always been a bit of a difficult case"*: Author interview, January 2019.

287 *performed thirty-five thousand sessions*: Author interview with Charlie Kellner in Brooklyn, New York, October 2018.

287 *literally wrote the book*: C. H. Kellner, *Handbook of ECT: A Guide to Electroconvulsive Therapy for Practitioners* (Cambridge, UK: Cambridge University Press, 2018).

287 *first session was in 1978*: Author interview with Charlie Kellner on the Upper East Side, New York City, October 2018.

287 *some didn't have an automated pulse*: "ECT in Britain: A Shameful State of Affairs," *The Lancet* 2 (1981): 1207–8.

287 *"If ECT is ever legislated against"*: Ibid.

288 *"For the right kind of illness"*: Author interview with Charlie Kellner on the Upper East Side, New York, October 2018.

288 *psychosis, mania, catatonia, and even a form of self-harming autism*: Ibid.

288 *nearly a fifth*: M. M. Ohayon and A. F. Schatzberg, "Prevalence of Depressive Episodes with Psychotic Features in the General Population," *American Journal of Psychiatry* 159 (2002): 1855–61.

288 *half of all cases*: R. Gournellis and L. Lykouras, "Psychotic (Delusional) Major

Depression in the Elderly: A Review," *Current Psychiatry Reviews* 2 (2006): 235–44.

288 *"I felt like I had some sort of life"*: Author interview, January 2019.

289 *"[They] are very real"*: Ibid.

289 *varies from $300 to $1,000*: E. L. Ross, K. Zivin, and D. F. Maixner, "Cost-Effectiveness of Electroconvulsive Therapy vs. Pharmacotherapy/Psychotherapy for Treatment-Resistant Depression in the United States," *JAMA Psychiatry* 75, no. 7 (2018): 713–22.

289 *most effective methods to prevent suicides*: D. Avery and G. Winokur, "Suicide, Attempted Suicide, and Relapse Rates in Depression: Occurrence After ECT and Antidepressant Therapy," *Archives of General Psychiatry* 35, no. 6 (1978): 749–53. And: J. Prudic and H. A. Sackheim, "Electroconvulsive Therapy and Suicide Risk," *Journal of Clinical Psychiatry* 60, no. 2 (1999): 104–10.

289 *"People are allowed to remain sick for years"*: Author interview with Charlie Kellner on the Upper East Side, New York, October 2018.

290 *harder to treat with each successive episode*: S. M. Monroe and K. L. Harkness, "Recurrence in Major Depression: A Conceptual Analysis," *Psychological Review* 118, no. 4 (2011): 655–74.

290 *82 percent respond to ECT*: D. Kroessler, "Relative Efficacy Rates for Therapies of Delusional Depression," *Convulsive Therapy* 1, no. 3 (1985): 173–82. Similar findings that ECT is more effective than medications alone or in combination: Parker et al., "Psychotic (Delusional) Depression," 17–24.

290 *ECT became cost-effective after two failed medications*: Ross, Zivin, and Maixner, "Cost-effectiveness of Electroconvulsive Therapy vs. Pharmacotherapy/Psychotherapy for Treatment-Resistant Depression in the United States," 713–22.

290 *"Two failed medication trials, not twenty-two"*: Author interview with Charlie Kellner on the Upper East Side, New York, October 2018.

290 *70 to 95 percent of patients*: A. J. Flint and N. Gagnon, "Effective Use of Electroconvulsive Therapy in Late-Life Depression," *Canadian Journal of Psychiatry* 47 (2002): 734–41.

290 *effective treatments in the whole of medicine*: T. Bolwig, "Electroconvulsive Therapy Reappraised," *Acta Psychiatrica Scandinavica* 129 (2014): 415–16.

290 *is estimated to be 0.2 to 0.4 per 10,000 patients*: Ibid. Also: M. Fink, "Convulsive and Drug Therapies of Depression," *Annual Review of Medicine* 32 (1981): 405–12.

291 *"I asked them to do it"*: M. Manning, *Undercurrents: A Therapist's Reckoning with Her Own Depression* (San Francisco: Harper, 1995).

291 *"Feeling this good is truly amazing"*: K. Dukakis and L. Tye, *Shock: The Healing Power of Electroconvulsive Therapy* (New York: Avery, 2014), 120.

291 *"All the heat and fear purged itself"*: C. H. Kellner, "Electroconvulsive Therapy (ECT) in Literature: Sylvia Plath's *Bell Jar*," *Progress in Brain Research* 206 (2013): 214–28, p. 225.

291 *"If you consider yourself a 'victim,'"*: "ECT and Me: A Peak Inside the Enraging Anti-ECT 'Movement' and Why I Will Fight It Forever," *Upside Down and Inside Out*, November 14, 2018.

291 *"So WHEN can we give ours?"*: Ibid.

292 *nearly banned its use entirely*: C. Buccelli et al., "Electroconvulsive Therapy in Italy: Will Public Controversies Ever Stop?" *Journal of ECT* 32 (2016): 207–11.

292 *neither Cerletti nor Bini spoke out*: Ibid., 207.

292 *"[T]he unconscionable misuse"*: Ibid.

292 *rarely used in Nazi Germany*: L. Rzesnitzek and S. Lang, "'Electroshock Therapy' in the Third Reich," *Medical History* 61 (2017): 66–88, p. 81. "the notion of a quick and extensive introduction of electroshock therapy in the Third Reich seems to be nothing more than a myth; a 'cliché.'"

292 *"make emotionally disturbed people fit for work again"*: Ibid., 85.

292 *"psychiatry [was] becoming more and more superfluous"*: Ibid., 78.

292 *rarely performed in Germany*: N. Loh et al., "Accessibility, Standards, and Challenges of Electroconvulsive Therapy in Western Industrialized Countries: A German Example," *World Journal of Biological Psychiatry* 14, no. 6 (2013): 432–40.

293 *highest utilization rates of ECT*: D. Bjørnshauge, S. Hjerrild, and P. Videbech, "Electroconvulsive Therapy Practice in the Kingdom of Denmark: A Nationwide Register- and Questionnaire-Based Study," *Journal of ECT* 35, no. 4 (2019): 258–63.

293 *"Tiny little Denmark"*: Author interview with Charlie Kellner on the Upper East Side, New York, October 2018.

293 *"Well, that's bullshit"*: Author interview with Martin Jørgensen in Copenhagen, January 2019.

293 *significant decrease in dementia*: M. Osler et al., "Electroconvulsive Therapy and Risk of Dementia in Patients with Affective Disorders: A Cohort Study," *Lancet Psychiatry* 5, no. 4 (2018): 348–56.

293 *"[We know] there is an increased risk of dementia"*: Author interview with Poul Videbech in Glostrup, January 2019.

294 *she found very early signs of neural regeneration*: G. Ende et al., "The Hippocampus in Patients Treated with Electroconvulsive Therapy: A Proton Magnetic Resonance Spectroscopic Imaging Study," *Archives of General Psychiatry* 57, no. 10 (2000): 937–43.

294 *a similar effect occurred in the frontal cortex*: K. Gbyl et al., "Cortical Thickness Following Electroconvulsive Therapy in Patients with Depression: A Longitudinal MRI Study," *Acta Psychiatrica Scandinavica* 140, no. 3 (2019): 205–16.

294 *through a brain scanner, animal studies, or blood tests*: Author interview with Poul Videbech in Glostrup, January 2019.

295 *"at least we know he has a brain"*: Ibid.

295 *rates of ECT are starting to increase*: N. Davis and P. Duncan, "Electroconvulsive Therapy on the Rise Again in England," *The Guardian*, April 17, 2017.

295 *"Psychiatry has to come back into the medical fold"*: Author interview with Charlie Kellner on the Upper East Side, New York, October 2018.

296 *"it was just so significant for me"*: Author interview, January 2019.

The Epitome of Hopelessness

297 *In her lectures*: H. Mayberg, "Tuning Depressions Circuits Using Deep Brain Stimulation," Stockholm Psychiatry Lecture, April 19, 2012.

297 *"In depression"*: W. Styron, *Darkness Visible: A Memoir of Madness* (New York: Random House, 1991).

297 *other laboratories using different methods*: P. E. Holtzheimer, M. E. Kelley, R. E. Gross et al., "Subcallosal Cingulate Deep Brain Stimulation for Treatment-Resistant Unipolar and Bipolar Depression," *Archives of General Psychiatry* 69, no. 2 (2012): 150–58. And: D. Puigdemont, R. Perez-Egea, M. J. Portella et al., "Deep Brain Stimulation of the Subcallosal Cingulate Gyrus: Further Evidence in Treatment-Resistant Major Depression," *International Journal of Neuropsychopharmacology* 15, no. 1 (2012): 121–33; D. A. Pizzagalli et al., "Pretreatment Rostral Anterior Cingulate Cortex Theta Activity in Relation to Symptom Improvement in Depression: A Randomized Clinical Trial," *JAMA Psychiatry* 75, no. 6 (2018): 547–54.

298 *a few basic decisions*: Author interview with Helen Mayberg in New York, October 2018.

299 *50 percent improvement*: A. L. Crowell et al., "Long-Term Outcomes of Subcallosal Cingulate Deep Brain Stimulation for Treatment-Resistant Depression," *American Journal of Psychiatry* 176, no. 11 (2019): 949–56.

Mind on Fire

301 *an old barn converted into an office*: Author visit to Brian Leonard's home, February 2019.

301 *In 1999, Leonard retired*: "Lifetime Achievement Award 2008: Professor Brian Leonard," British Association for Psychopharmacology, https://www.bap.org.uk/pdfs/biogs/awards2008_BrianLeonard.pdf.

301 *opened its doors to students in 1849*: "Our History & Heritage," NUI Galway, https://www.nuigalway.ie/about-us/who-we-are/our-history.html.

301 *Leonard was a prominent voice*: Author interview with Michael Berk, June 2018.

302 *Patients who don't respond to conventional antidepressants*: T. Eller et al., "Pro-Inflammatory Cytokines and Treatment Response to Escitalopram in Major Depressive Disorder," *Progress in Neuro-Psychopharmacology & Biological Psychiatry* 32 (2008): 445–50. And: S. Lanquillon et al., "Cytokine Production and Treatment Response in Major Depressive Disorder," *Neuropsychopharmacology* 22 (2000): 370–79.

302 *In 2013, a randomized controlled trial*: C. L. Raison et al., "A Randomized Controlled Trial of the Tumor Necrosis Factor Antagonist Infliximab in Treatment-Resistant Depression: The Role of Baseline Inflammatory Biomarkers," *JAMA Psychiatry* 70, no. 1 (2013): 31–41.

302 *predicted the onset of depressive symptoms twelve years later*: D. Gimeno et al., "Associations of C-Reactive Protein and Interleukin-6 with Cognitive Symptoms of Depression: 12-year Follow-up of the Whitehall II Study," *Psychological Medicine* 39 (2009): 413–23.

302 *"Now everybody believes it"*: Author interview with Michael Berk, June 2018.
302 *"the hut"*: Author interview with Brian Leonard in Galway, February 2019.
302 *forty PhD students*: "Lifetime Achievement Award 2008: Professor Brian Leonard."
303 *"The lack of scientific insight"*: R. S. Smith, "The Macrophage Theory of Depression," *Medical Hypotheses* 35 (1991): 298–306.
303 *"put the immune system back in the body"*: J. L. Marx, "The Immune System 'Belongs in the Body,'" *Science* 227 (1985): 1190–92.
303 *"it is difficult to say definitively what is happening"*: Ibid.
304 *several labs based in the U.S. and the Netherlands*: Z. Kronfol and J. D. House, "Lymphocyte Mitogenesis, Immunoglobulin and Competent Levels in Depressed Patients and Normal Controls," *Acta Psychiatrica Scandinavica* 80 (1989): 142–47.
304 *Cai Song, had found similar patterns*: C. Song, T. Dinan, and B. E. Leonard, "Changes in Immunoglobulin, Complement and Acute Phase Protein Levels in the Depressed Patients and Normal Controls," *Journal of Affective Disorders* 30 (1994): 283–88.
304 *A core group of inflammatory markers*: L. Capuron and A. H. Miller, "Immune System to Brain Signaling: Neuropsychopharmacological Implications," *Pharmacology & Therapeutics* 130, no. 2 (2011): 226–38.
304 *"That's not what we're talking about"*: Author interview with Brian Leonard in Galway, February 2019.
305 *Even a clogged artery*: A. Malhotra, R. F. Redberg, and P. Meier, "Saturated Fat Does Not Clog the Arteries: Coronary Heart Disease Is a Chronic Inflammatory Condition, the Risk of Which Can Be Effectively Reduced from Healthy Lifestyle Interventions," *British Journal of Sports Medicine* 51 (2017): 1111–12.
305 *"Patients ignored their eating"*: A. Niiranen et al., "Behavioral Assessment of Patients Treated with Alpha-Interferon," *Acta Psychiatrica Scandinavica* 78, no. 5 (1988): 622–26.
305 *"A patient only needs to exhibit five"*: R. S. Smith, "The Macrophage Theory of Depression," *Medical Hypotheses* 35 (1991): 298–306, p. 300.
306 *"There was no psychological context"*: Author interview with Andrew Miller, March 2020.
306 *symptoms disappeared within two weeks*: Niiranen et al., "Behavioral Assessment of Patients Treated with Alpha-Interferon," 624.
306 *sickness behavior*: B. L. Hart, "Biological Basis of the Behavior of Sick Animals," *Neuroscience & Biobehavioral Reviews* 12 (1988): 123–37.
307 *"The annual rate of suicide"*: T. R. Norman and B. E. Leonard, "Fast-Acting Antidepressants: Can the Need Be Met?" *CNS Drugs* 2, no. 2 (1994): 120–31.
307 *Beta-blockers and anti-cortisol*: J. F. Cryan, C. McGrath, B. E. Leonard, and T. R. Norman, "Combining Pindolol and Paroxetine in an Animal Model of Chronic Antidepressant Action—Can Early Onset of Action Be Detected?" *European Journal of Pharmacology* 352, no. 1 (1998): 23–28.
307 *no breakthroughs*: Author interview with John Cryan, September 2017.
307 *"I sold my soul for a while"*: Ibid.
308 *$985 million to take a drug*: O. J. Wouters, M. McKee, and J. Luyten, "Estimated Research and Development Investment Needed to Bring a New Medicine

to Market, 2009–2018," *Journal of the American Medical Association* 323, no. 9 (2020): 844–53.

308 *Connie Sánchez . . . who developed escitalopram*: C. Sánchez et al., "Escitalopram, the S-(+)-Enantiomer of Citalopram, Is a Selective Serotonin Reuptake Inhibitor with Potent Effects in Animal Models Predictive of Antidepressant and Anxiolytic Activities," *Psychopharmacology* 167 (2003): 353–62.

308 *focusing his attention on . . . the gut*: O. Mahony et al., "Early Life Stress Alters Behavior, Immunity, and Microbiota: Implications for Irritable Bowel Syndrome and Psychiatric Illnesses," *Biological Psychiatry* 65, no. 3 (2008): 263–67.

308 *University College Cork had set up a microbiome unit*: APC Microbiome Ireland, https://apc.ucc.ie/about-2/introduction/.

308 *raise their own germ-free mice*: P. Yi and L.-J. Li, "The Germfree Murine Animal: An Important Animal Model for Research on the Relationship Between Gut Microbiota and the Host," *Veterinary Microbiology* 157 (2012): 1–7.

309 *95 percent of tryptophan usage*: I. Cervenka, L. Z. Agudelo, and J. L. Ruas, "Kynurenines: Tryptophan's Metabolites in Exercise, Inflammation, and Mental Health," *Science* 357 (2017): 369, p. 2.

310 *kynurenine is broken down into quinolinic acid*: Ibid.

310 *neurotoxic*: Ibid.

310 *"neurodegeneration hypothesis of depression"*: A. M. Myint and Y. K. Kim, "Cytokine-Serotonin Interaction Through IDO: A Neurodegeneration Hypothesis of Depression," *Medical Hypotheses* 61 (2003): 519–25.

310 *"It's very complicated"*: Author interview with Brian Leonard in Galway, February 2019.

310 *"He was the one who pinpointed this"*: I. P. Lapin, "Interaction of Kynurenine and Its Metabolites with Tryptamine, Serotonin and Its Precursors and Oxotremorine," *Psychopharmacology* (Berlin) 26 (1972): 237–47.

310 *piano . . . rubbed shoulders with Russian intelligentsia*: G. Oxenkrug and P. Riederer, "Early Serotonin and Kynurenine Research: Slava Lapin (1930–2012)," *Journal of Neural Transmission* 119 (2012): 1465–66.

310 *"enlivened conferences too numerous to count"*: "Lifetime Achievement Award 2008: Professor Brian Leonard."

311 *"The body doesn't work in bits"*: Author interview with Brian Leonard in Galway, February 2019.

"For Life"

312 *made famous by Nobel laureate Joshua Lederberg*: J. Lederberg and A. T. McCray, "'Ome Sweet 'Omics—A Genealogical Treasury of Words," *The Scientist*, April 2, 2001.

312 *"from a harmful to an innocuous one"*: E. Metchnikoff and H. S. Williams, "Why Not Live Forever?" *Cosmopolitan* 53 (1912): 436–46.

312 *finished a four-year degree in just two*: L. Vikhanski, *Immunity: How Elie Metchnikoff Changes the Course of Modern Medicine* (Chicago: Chicago Review Press, 2016), 24.

313 *"fields of wheat after a thunderstorm"*: Ibid., 143.

313 *"Metchnikoff's great discoveries"*: A. C. Bested, A. C. Logan, and E. M. Selhub, "Intestinal Microbiota, Probiotics and Mental Health: From Metchnikoff to Modern Advances: Part 1—Autointoxication Revisited," *Gut Pathogens* 5 (2013): 5.

313 *"promotes physical and mental health"*: Ibid.

313 *manufacturers replaced* L. bulgaricus *with* L. acidophilus: Ibid.

313 *"The results . . . are nothing short of amazing"*: Ibid.

314 *Kirsten Tillisch and her colleagues*: K. Tillisch et al. "Consumption of Fermented Milk Product with Probiotic Modulates Brain Activity," *Gastroenterology* 144 (2013): 1394–1401.

314 *"for life"*: L. Desbonnet et al., "The Probiotic *Bifidobacteria infantis*: An Assessment of Potential Antidepressant Properties in the Rat," *Journal of Psychiatric Research* 43 (2008): 164–74, p. 165.

314 *"improve the health of their human host"*: Ibid., 164.

314 psychobiotics: T. G. Dinan, C. Stanton, and J. F. Cryan, "Psychobiotics: A Novel Class of Psychotropic," *Biological Psychiatry* 74 (2013): 720–26.

314 *developed in the late 1970s, the forced swim test*: R. D. Porsolt, M. Le Pichon, and M. Jalfre, "Depression: A New Animal Model Sensitive to Antidepressant Treatments," *Nature* 266 (1977): 730–32.

315 *many SSRIs were first given to mice*: S. Reardon, "Depression Researchers Rethink Popular Mouse Swim Tests," *Nature*, July 18, 2019.

315 B. infantis . . . *90 percent of the microbes*: Desbonnet et al., "The Probiotic *Bifidobacteria infantis*," 164–74.

315 *no change in how long they swam*: Ibid.

315 Lactobacillus rhamnosus . . . *passed*: J. A. Bravo et al., "Ingestion of Lactobacillus Strain Regulates Emotional Behavior and Central GABA Receptor Expression in a Mouse via the Vagus Nerve," *Proceedings of the National Academy of Sciences* 108 (2011): 16050–55.

315 *"every flippin' thing under the sun"*: Author interview with Ted Dinan in Cork, Ireland, November 2018.

316 *"[Y]ou couldn't fit more negative data"*: J. Gallagher, "The Second Genome," BBC Radio 4, April 24, 2018.

316 *"Lost in Translation"*: J. R. Kelly et al., "Lost in Translation? The Potential Psychobiotic *Lactobacillus Rhamnosus* (JB-1) Fails to Modulate Stress or Cognitive Performance in Healthy Male Subjects," *Brain, Behavior, and Immunity* 61 (2017): 50–59.

316 *"we've had loads of negative studies"*: Author interview with Ted Dinan in Cork, Ireland, November 2018.

316 *first randomized controlled trial*: A. Kazemi et al., "Effect of Prebiotic and Probiotic Supplementation on Circulating Pro-Inflammatory Cytokines and Urinary Cortisol Levels in Patients with Major Depressive Disorder: A Double-Blind, Placebo-Controlled Randomized Clinical Trial," *Journal of Functional Foods* 52 (2019): 596–602.

316 *benefits for people with mild*: Q. Xiang et al., "A Meta-Analysis of the Use of Probiotics to Alleviate Depressive Symptoms," *Journal of Affective Disorders* 228 (2017): 13–19.

317 *listened to a podcast and heard Laura Steenbergen*: Author interview with Scot Bay at Mind, Mood & Microbes, January 2019.
317 *"They got better"*: Ibid.
317 *"Are we there yet?"*: Closing Remarks by John Cryan at Mind, Mood & Microbes, January 2019.
317 *microbial "fingerprint"*: M. Valles-Colomer et al., "The Neuroactive Potential of the Human Gut Microbiota in Quality of Life and Depression," *Nature Microbiology* 4 (2019): 623–32.
318 *"it's kind of a loop"*: Author interview with Mireia Valles-Colomer and Sara Vieira-Silva, June 2019.
318 *"more effects from prebiotics than probiotics"*: Ibid.
318 *It requires long-term dietary changes*: N. Voreades, A. Kozil, and T. L. Weir, "Diet and the Development of the Human Intestinal Microbiome," *Frontiers in Microbiology* 5 (2014): 494, p. 4.
319 *Mediterranean-style diet can act as an antidepressant*: F. N. Jacka et al., "A Randomized Controlled Trial of Dietary Improvement for Adults with Major Depression (the 'SMILES' Trial)," *BMC Medicine* 15 (2017): 23.
319 *replicated by independent research groups*: N. Parletta et al., "A Mediterranean-Style Dietary Intervention Supplemented with Fish Oil Improves Diet Quality and Mental Health in People with Depression: A Randomized Controlled Trial (HELFIMED)," *Nutritional Neuroscience* 22, no. 7 (2017): 474–87.
320 *omega-3 oils are actually produced by oceanic algae*: R. J. Winwood, "Algal Oil as a Source of Omega-3 Fatty Acids," in C. Jacobsen et al., eds., *Food Enrichment with Omega-3 Fatty Acids* (Cambridge, UK: Woodhead, 2013).
320 *5.6 percent . . . have a healthy proportion of fruits and vegetables*: Jacka et al., "A Randomized Controlled Trial of Dietary Improvement for Adults with Major Depression (the 'SMILES' Trial)," 12.
320 *19 percent of children*: I. Wilkinson, "Food Poverty: Agony of Hunger the Norm for Many Children in the UK," *The Conversation*, April 30, 2019.
320 *13 percent of adults and 18 percent of children*: "Obesity and Overweight," World Health Organization, April 1, 2020, https://www.who.int/news-room/fact-sheets/detail/obesity-and-overweight.
320 *over 40 percent in 2018*: C. M. Hales et al., "Prevalence of Obesity and Severe Obesity Among Adults: United States, 2017–2018," *NCHS Data Brief* 360 (2020).
320 *twenty-eight times more likely to develop diabetes*: P. G. Kopelman, "Obesity as a Medical Problem," *Nature* 404 (2000): 635–43.
321 *"an epidemic that threatens global well being"*: Ibid., 635.
321 *over fifty genes*: H. Choquet and D. Meyre, "Genetics of Obesity: What Have We Learned?" *Current Genomics* 12, no. 3 (2011): 169–79.
321 *"Traditional foods such as fish"*: N. L. Hawley and S. T. McGarvey, "Obesity and Diabetes in Pacific Islanders: The Current Burden and the Need for Urgent Action," *Current Diabetes Reports* 15 (2015): 29.
321 *genetic predisposition to weight gain*: R. L. Minster et al., "A Thrifty Variant in *CREBRF* Strongly Influences Body Mass Index in Samoans," *Nature Genetics* 48, no. 9 (2016): 1049–54.

322 *"With few exceptions"*: M. Berk et al., "So Depression Is an Inflammatory Disease, but Where Does the Inflammation Come From?" *BMC Medicine* 11, no. 200 (2013).

322 *risk of depression by over 50 percent*: Ibid., 5.

322 *"atypical" symptom profile*: Y. Milaneschi et al., "Depression and Obesity: Evidence of Shared Biological Mechanisms," *Molecular Psychiatry* 24 (2019): 18–33.

323 *children achieving the top grades*: M. Belot, "Healthy School Meals and Educational Outcomes," Institute for Social and Economic Research, January 2009.

323 *"Feed Me Better" petition*: "TV Chef Welcomes £280m Meals Plan," BBC News, March 30, 2005.

323 *"I think he's very important"*: Author interview with Michael Berk, June 2018.

323 *dozens of clinical trials*: J. A. Blumenthal et al., "Exercise and Pharmacotherapy in the Treatment of Major Depressive Disorder," *Psychosomatic Medicine* 69, no. 7 (2007): 587–96.

323 *same efficacy in treating depression as first-line treatments*: A. L. Dunn et al., "Exercise Treatment for Depression: Efficacy and Dose Response," *American Journal of Preventive Medicine* 28, no. 1 (2005): 1–8.

323 *in 2013, a study from Madhukar Trivedi*: C. D. Rethorst et al., "Pro-Inflammatory Cytokines as Predictors of Antidepressant Effects of Exercise in Major Depressive Disorder," *Molecular Psychiatry* 18 (2013): 1119–24.

324 *A large clinical trial from 2014*: R. Uher et al., "An Inflammatory Biomarker as a Differential Predictor of Outcome of Depression Treatment with Escitalopram and Nortriptyline," *American Journal of Psychiatry* 171 (2014): 1278–86.

324 *Three years later, a similar study*: M. K. Jha et al., "Interleukin 17 Selectively Predicts Better Outcomes with Bupropion-SSRI Combination: Novel T Cell Biomarker for Antidepressant Medication Selection," *Brain, Behavior, and Immunity* 66 (2017): 103–10.

325 *intense exercise . . . can activate the kynurenine pathway*: B. Strasser et al., "Effects of Exhaustive Aerobic Exercise on Tryptophan-Kynurenine Metabolism in Trained Athletes," *PLoS ONE* 11, no. 4 (2016): e0153617.

The Beginning

326 *"psychological stress activates our immune response"*: Author interview with Golam Khandaker, March 2020.

326 *longitudinal studies . . . over one hundred thousand people*: D. Gimeno et al., "Associations of C-Reactive Protein and Interleukin-6 with Cognitive Symptoms of Depression: 12-Year Follow-up of the Whitehall II Study," *Psychological Medicine* 39 (2009): 413–23; M. K. Wium-Andersen et al., "Elevated C-Reactive Protein Levels, Psychological Distress, and Depression in 73131 Individuals," *JAMA Psychiatry* 70, no. 2 (2013): 176–84; G. M. Khandaker et al., "Association of Serum Interleukin 6 and C-Reactive Protein in Childhood with Depression and Psychosis in Young Adult Life," *JAMA Psychiatry* 71, no. 10 (2014): 1121–28.

326 *genetic predisposition*: G. M. Khandaker et al., "Association Between a Functional Interleukin-6 Receptor Genetic Variant and Risk of Depression and Psychosis

in a Population-Based Birth Cohort," *Brain, Behavior, and Immunity* 69 (2018): 264–72.

326 *anti-inflammatories . . . decrease in their depressive symptoms*: N. Kappelman et al., "Antidepressant Activity of Anticytokine Treatment," *Molecular Psychiatry* 23 (2018): 335–43.

327 *"What's exciting about this field now"*: Author interview with Golam Khandaker, March 2020.

327 *The Insight Study*: G. M. Khandaker et al., "Protocol for the Insight Study: A Randomized Controlled Trial of Single-Dose Tocilizumab in Patients with Depression and Low-Grade Inflammation," *BMJ Open* 8, no. 9 (2018): e025333.

327 *disrupt brain function*: C. L. Raison and A. H. Miller, "Do Cytokines Really Sing the Blues?" *Cerebrum*, August 2013.

327 *"This is just the beginning"*: Author interview with Golam Khandaker, March 2020.

328 *"this relatively young subspecialty will have staying power"*: A. Halaris and B. E. Leonard, eds., *Inflammation in Psychiatry* (Basel, Switzerland: Karger, 2013), preface.

Surfing in the Brain Scanner

329 *part of their ceremonies since the 1930s*: "The History of a Doctrine," https://santodaime.com/en/doctrine/history/.

329 mirações: *"seeings"*: D. B. de Araujo et al., "Seeing with the Eyes Shut: Neural Basis of Enhanced Imagery Following Ayahuasca Ingestion," *Human Brain Mapping* 33, no. 11 (2011): 2550–60.

329 *DMT . . . the bush* Psychotria viridis: F. Palhano-Fontes et al., "The Psychedelic State Induced by Ayahuasca Modulates the Activity and Connectivity of the Default Mode Network," *PLoS ONE* 10, no. 2 (2015): e0118143, p. 2.

330 *specific type of serotonin receptor*: D. E. Nichols, "Hallucinogens," *Pharmacology & Therapeutics* 101 (2004): 131–81. It's the 5-HT(2a) receptor.

330 Banisteriopsis caapi, *harmine*: Palhano-Fontes et al., "The Psychedelic State Induced by Ayahuasca Modulates the Activity and Connectivity of the Default Mode Network," 2.

330 *four or more hours*: de Araujo et al., "Seeing with the Eyes Shut," 2550–60.

330 *"labor"*: Palhano-Fontes et al., "The Psychedelic State Induced by Ayahuasca Modulates the Activity and Connectivity of the Default Mode Network," 8.

330 *"that was just astonishing"*: Author interview with Draulio de Araujo, November 2017.

330 *legal in religious ceremonies since the late 1980s*: Erowid, "Legal Status of Ayahuasca in Brazil," The Vaults of Erowid, March 2001.

331 *"This is the level of experience"*: Author interview with Draulio de Araujo, November 2017.

331 *ten volunteers who had surfed the psychedelic waves*: de Araujo et al., "Seeing with the Eyes Shut," 2550–60.

331 *"a status of reality to inner experiences"*: Ibid., 2559.

331 *People with alcohol dependency or depression*: Author interview with Draulio de Araujo, November 2017.

"Turn On, Tune In, and Drop Out"

333 *have been used by indigenous cultures*: M. J. Miller et al., "Chemical Evidence for the Use of Multiple Psychotropic Plants in a 1,000-Year-Old Ritual Bundle from South America," *Proceedings of the National Academy of Sciences* 116, no. 23 (2019): 11207–12.

333 *Siberia to the Sahara and down into Mesoamerica*: B. P. Akers et al., "A Prehistoric Mural in Spain Depicting Neurotropic Psilocybe Mushrooms?" *Economic Botany* 65, no. 2 (2011): 121–28.

333 *"Just take a pinch of psychedelic"*: J. H. Tanne, "Humphry Osmond," *British Medical Journal* 20 (2004): 328.

333 *First synthesized in 1938*: A. Hofmann, "The Discovery of LSD and Subsequent Investigations on Naturally Occurring Hallucinogens," in F. J. Ayd and B. Blackwell, eds., *Discoveries in Biological Psychiatry* (New York: Lippincott, 1970).

334 *"LSD makes available"*: B. Eisner, *Remembrances of LSD Therapy Past* (unpublished memoir, 2002), 129.

334 *"universe had collapsed" on her*: Ibid., 16.

334 *equivalent to four years of psychoanalysis*: Ibid., 20.

334 *"I think the function of the therapist"*: Ibid., 55.

334 *"Chopin's first piano concerto is better than anything"*: Ibid., 31.

334 *anxiety to alcoholism and impotence to homosexuality*: R. A. Sandison and J. D. A. Whitelaw, "Further Studies in the Therapeutic Value of Lysergic Acid Diethylamide in Mental Illness," *Journal of Mental Science* 103, no. 431 (1957): 332–43.

334 *"sociopathic personality disturbance"*: J. Drescher, "Out of DSM: Depathologizing Homosexuality," *Behavioral Sciences* 5, no. 4 (2015): 565–75.

334 *Charles Savage . . . gave LSD to fifteen patients*: C. Savage, "Lysergic Acid Diethylamide (LSD-25): A Clinical-Psychological Study," *American Journal of Psychiatry* 108 (1952): 896–900.

335 *"reticence rather than confidence"*: Ibid., 900.

335 *natural gift for making people feel better*: Eisner, *Remembrances of LSD Therapy Past.*

335 *"The one thing I have noticed"*: Ibid., 24.

336 *"raging fires, the void, dragons, a vortex"*: Ibid., 53.

336 *from human to God*: Ibid., 25.

336 *patients were often left without a therapist*: Ibid., 83.

337 *"psychedelics in all their ramifications"*: Ibid., 84.

337 *small audience . . . "I felt like a true pioneer" . . . "fitting end to a magical trip"*: Ibid., 89.

337 *Albert Hofmann . . . isolated psilocybin*: Hofmann, "The Discovery of LSD and Subsequent Investigations on Naturally Occurring Hallucinogens."

337 *"It does get discouraging"*: Eisner, *Remembrances of LSD Therapy Past*, 188.

338 *In total, forty-four replied*: S. J. Novak, "LSD Before Leary: Sidney Cohen's Critique of 1950s Psychedelic Research," *History of Science Society* 88 (1997): 87–110.

338 *"directly due to LSD"*: Ibid., 105.

338 *"untoward events in connection with LSD-25"*: S. Cohen and K. Ditman, "Complications Associated with Lysergic Acid Diethylamide," *Journal of the American Medical Association* 181, no. 2 (1962): 161–82, p. 181.

338 *"not an innocent aid like milk"*: "LSD—A Dangerous Drug," *New England Journal of Medicine* 273, no. 1280 (1965).

339 *article published in* Good Housekeeping: R. Gehman, "Ageless Cary Grant," *Good Housekeeping*, September 1960, p. 64.

339 *helped push its use into the mainstream*: M. Meyers, *Happy Accidents: Serendipity in Modern Medical Breakthroughs* (New York: Arcade, 2007).

339 *"Turn on, tune in, and drop out"*: Novak, "LSD Before Leary," 87–110.

339 *a ten-year-old boy*: Cohen and Ditman, "Complications Associated with Lysergic Acid Diethylamide (LSD-25)," 161–62.

339 *customers from two hundred to seventy*: Novak, "LSD Before Leary," 108.

339 *"LSD therapy should not be seen"*: Ibid. "Charles Savage wrote the chairman of the California State Assembly Committee On Criminal Procedure, 5 Nov. 1963. California State Assembly Committee On Criminal Procedure, 13, 14, 15 Nov. 1963, Narcotics and Dangerous Drugs (Sacramento, 1964), App. IV-c."

340 *"remarkably safe"*: Ibid., 87–110.

340 *"lost to society, lost to themselves"*: "When Bobby Kennedy Defended LSD," MAPS, July 11, 2012, https://maps.org/news/media/3152-when-bobby-kennedy-defended-lsd.

340 *psychotic killer*: "A Slaying Suspect Tells of LSD Spree: Medical Student Charged in Mother-in-Law's Death," *New York Times,* April 12, 1966.

340 *"very helpful in our society if used properly"*: "When Bobby Kennedy Defended LSD."

340 *illegal substance in the U.S. in October 1968*: Public Law 90-639, October 24, 1968.

340 *aid to help concentration camp survivors*: "The LSD Therapy Career of Jan Bastiaans, M.D.," *MAPS* 8, no. 1 (1998): 18–20.

Building a New System

341 *ecstasy was safer than riding a horse*: D. J. Nutt, "Equasy—An Overlooked Addiction with Implications for the Current Debate on Drug Harms," *Journal of Psychopharmacology* 23 (2009): 3.

341 *"It's the worst censorship of research"*: Author interview with David Nutt at Imperial College London, December 2017.

342 *their first study in 2011*: R. L. Carhart-Harris et al., "The Administration of Psilocybin to Healthy, Hallucinogen-Experienced Volunteers in a Mock-Functional Magnetic Resonance Imaging Environment: A Preliminary Investigation of Tolerability," *Journal of Psychopharmacology* 25, no. 11 (2011): 1562–67.

342 *Mayberg center*: Author interview with David Nutt at Imperial College London, December 2017.

342 *"psychedelics usually disrupt that"*: Ibid.

342 *The hospital room at Imperial College*: Author interview with patient, January 2018.

342 *"I cried a lot during the experience"*: Ibid.

343 *"We must open this treatment up"*: Author interview with Amanda Feilding, August 2017.

343 *"establish whether it's a treatment"*: Author interview with James Rucker in London, November 2018.

343 *significantly less effective than escitalopram*: A. Cipriani et al., "Comparative Efficacy and Acceptability of 21 Antidepressant Drugs for the Acute Treatment of Adults with Major Depressive Disorder: A Systematic Review and Network Meta-Analysis," *The Lancet* 391 (2018): 1357–66. Figure 4: Head-to-head comparison between "Esci" and "Fluo" = 1.34 (1.11–1.61), in favor of escitalopram.

344 *"Depression runs through my family"*: Author interview with James Rucker in London, November 2018.

344 *multicenter trial includes 216 patients from eleven cities*: "COMPASS Pathways Granted Patent Covering Use of Its Psilocybin Formulation in Addressing Treatment-Resistant Depression," COMPASS News, January 15, 2020.

344 *"We came at this not from a position"*: M. Henriques, "Two Parents' Fight to Set Up the Largest-Ever Magic Mushroom Trial for Depression Is Nearly Over," *International Business Times*, September 15, 2017.

344 *"There are very strict rules"*: Author interview with James Rucker in London, November 2018.

345 *cover of magazines like* New Scientist: "Psychedelic Medicine: How Mind-Benders Became Mind-Menders," *New Scientist*, November 21, 2017.

345 *Early trials with psilocybin*: R. R. Griffiths et al., "Psilocybin Produces Substantial and Sustained Decreases in Depression and Anxiety in Patients with Life-Threatening Cancer: A Randomized Double-Blind Trial," *Journal of Psychopharmacology* 30, no. 12 (2016): 1181–97.

347 *"Trust, let go, be open"*: Author interview with Ros Watts, October 2017.

347 *"Therapy takes a really long time"*: Ibid.

Seeing with Eyes Shut

349 *first randomized placebo-controlled trial*: F. Palhano-Fontes et al., "Rapid Antidepressant Effects of the Psychedelic Ayahuasca in Treatment-Resistant Depression: A Randomized Placebo-Controlled Trial," *Psychological Medicine* 49 (2018): 655–63.

349 *57 percent*: Ibid., 658.

349 *water, yeast, citric acid, zinc sulphate, and caramel*: Ibid., 656.

349 *a three-story, whitewashed building*: Photo by "Kulinai," Wikimedia Commons, May 2, 2019.

350 *"our patients come from very difficult realities"*: Author interview with Draulio de Araujo, November 2017.

350 *"You know what, Doctor?"*: Ibid.
351 *given several sessions of ECT*: Palhano-Fontes et al., "Rapid Antidepressant Effects of the Psychedelic Ayahuasca in Treatment-Resistant Depression," 658.
351 *"The placebo effect"*: Author interview with James Rucker in London, November 2018.
351 *"Ayahuasca reflected my beliefs back to me"*: J. Evan, "Caves All the Way Down," *Aeon*, July 17, 2018.
352 *"mind wandering"*: F. Palhano-Fontes et al., "The Psychedelic State Induced by Ayahuasca Modulates the Activity and Connectivity of the Default Mode Network," *PLoS ONE* 10, no. 2 (2015): e0118143, p. 2.
352 *DMN . . . intention, memory, and vision*: Ibid.
352 *"a private, continuous and often unnoticed"*: Ibid.
352 *A famous study from 2012*: J. A. Brewer et al., "Meditation Experience Is Associated with Differences in Default Mode Network Activity and Connectivity," *Proceedings of the National Academy of Sciences* 108, no. 50 (2012): 20254–59.
352 *eight brain regions . . . significant drops*: Palhano-Fontes et al., "The Psychedelic State Induced by Ayahuasca Modulates the Activity and Connectivity of the Default Mode Network," e0118143, table 1.
352 *"I have a lot of issues"*: Author interview with Draulio de Araujo, December 2019.
353 *as alpha waves, remained stable*: L. F. Tófoli and D. B. de Araujo, "Treating Addiction: Perspectives from EEG and Imaging Studies on Psychedelics," *International Review of Neurobiology* 129 (2016): 157–85.
353 *researchers in London using DMT*: C. Timmermann et al., "Neural Correlates of the DMT Experience Assessed with Multivariate EEG," *Scientific Reports* 9 (2019): 2045–322.
354 *"That's gotta be powerful"*: Author interview with Draulio de Araujo, December 2019.
354 *approved by the European Commission*: "Esketamine Is Approved in Europe for Treating Resistant Major Depressive Disorder," *British Medical Journal* (2019): 267.
354 *and the U.S. Food and Drug Administration*: "FDA Approves New Nasal Spray Medication for Treatment-Resistant Depression; Available Only at a Certified Doctor's Office or Clinic," U.S. Food and Drug Administration, March 5, 2019.
354 *in just four hours, seven people*: R. M. Berman et al., "Antidepressant Effects of Ketamine in Depressed Patients," *Biological Psychiatry* 47 (2000): 351–54.
355 *"one of the most significant advances"*: R. S. Duman and N. Li, "A Neurotrophic Hypothesis of Depression: Role of Synaptogenesis in the Actions of NMDA Receptor Antagonists," *Philosophical Transactions of the Royal Society B* 367 (2012): 2475–84, p. 2478.
355 *"[Ketamine] led to a loss of time"*: E. Khorramzadeh and A. O. Lofty, "The Use of Ketamine in Psychiatry," *Psychosomatics* 14 (1973): 344–46.
355 *rave parties since the 1990s*: C. J. A. Morgan and H. V. Curran, "Ketamine Use: A Review," *Addiction* 107 (2011): 27–38, p. 28.
355 *irreversible damage to the bladder and kidney*: Ibid., 31.
355 *"Currently, there is limited evidence"*: "Statement on Ketamine to Treat Depression," Royal College of Psychiatrists, February 2017.

356 *Manufactured in 1962*: "Fact File on Ketamine," World Health Organization, March 2016.
356 *American soldiers during the Vietnam War*: Morgan and Curran, "Ketamine Use," 27.
356 *"One thing I have to do as a scientist"*: Author interview with Draulio de Araujo, December 2019.
356 *"We're hardly scratching the surface"*: Author interview with Draulio de Araujo, November 2017.

Epilogue: New Life

360 *one in every ten pregnant women*: L. M. Howard et al., "Non-psychotic Mental Disorders in the Perinatal Period," *The Lancet* 384, no. 9956 (2014): 1775–88.
360 *Once the baby is born, the risks are increased*: Ibid.
360 *half of all new mothers*: J. Fisher et al., "Prevalence and Determinants of Common Perinatal Mental Disorders in Women in Low- and Lower-Middle-Income Countries: A Systematic Review," *Bulletin of the World Health Organization* 90, no. 2 (2012): 139–49.
362 *emerging argument that all mental illnesses are interconnected*: A. Caspi and T. E. Moffitt, "All for One and One for All: Mental Disorders in One Dimension," *American Journal of Psychiatry* 175 (2018): 831–44.
363 *neuroticism*: J. Ormel et al., "Neuroticism and Common Mental Disorders: Meaning and Utility of a Complex Relationship," *Clinical Psychology Review* 33 (2013): 686–97.
364 *"this was going viral, so to speak"*: Author interview with Dixon Chibanda and Vikram Patel in London, May 2018.
364 *"It's really pretty grim"*: Ibid.
365 *"The final story will never really be the final story"*: Author interview with Myrna Weissman, October 2020.
365 *"This is a historic opportunity"*: The Lancet-WPA Commission on Depression, preprint word document. 2020.

Selected Bibliography

Abas, M., and J. Broadhead. "Depression and Anxiety Among Women in an Urban Setting in Zimbabwe," *Psychological Medicine* 27 (January 1997): 59–71.

Abas, M., et al. "Defeating Depression in the Developing World: A Zimbabwean Model: One Country's Response to the Challenge," *British Journal of Psychiatry* 164 (April 1994): 293–96.

Alexander, F., and S. Selesnick. *History of Psychiatry*. New York: Harper & Row, 1966.

Arrol, B., et al. "Cochrane Review of SSRIs and TCAs. Antidepressants Versus Placebo for Depression in Primary Care." *Cochrane Database of Systematic Review*, Issue 3 (2009).

Atkinson, R. M., and K. S. Ditman. "Tranylcypromine: A Review." *Clinical Pharmacology and Therapeutics* 6 (1965): 631–55.

Ban, T. A. "Pharmacotherapy of Depression: A Historical Analysis." *Journal of Neural Transmission* 108 (2001): 707–16.

Beck, A. T. *Depression: Causes and Its Treatment*. Philadelphia: University of Pennsylvania Press, 1970.

Bernfeld, S. "Freud's Studies on Cocaine, 1884–1887." *Journal of the American Psychoanalytic Association* 35 (1954): 445.

Berrios, G. *The History of Mental Symptoms*. Cambridge, UK: Cambridge University Press, 1996.

———. "The Scientific Origins of Electroconvulsive Therapy: A Conceptual History." *History of Psychiatry* viii (1997): 105–19.

Bolton, P., et al. "Group Interpersonal Psychotherapy for Depression in Rural Uganda: A Randomized Controlled Trial." *Journal of the American Medical Association* 289, no. 23 (2003): 3117–24.

Bowlby, J. *Attachment and Loss: Volume 1: Attachment*. New York: Basic Books, 1969.

Broadhead, J. C., and M. Abas. "Depressive Illness—Zimbabwe." *Tropical Doctor* 24 (1994): 27–30.

———. "Life Events, Difficulties and Depression among Women in an Urban Setting in Zimbabwe." *Psychological Medicine* 28 (1998): 29–38.

Brown, G. W., and T. Harris. *Social Origins of Depression: A Study of Psychiatric Disorder in Women*. London: Tavistock, 1978.

Chibanda, D., et al. "Effect of a Primary Care–Based Psychological Intervention on Symptoms of Common Mental Disorders in Zimbabwe: A Randomized

Clinical Trial." *Journal of the American Medical Association* 316, no. 24 (2016): 2618–26.

———. "Problem-Solving Therapy for Depression and Common Mental Disorders in Zimbabwe: Piloting a Task-Shifting Primary Mental Health Care Intervention in a Population with High Prevalence of People Living with HIV." *BMC Public Health* 11 (2011): 828.

Chisholm, D., et al. "Scaling-Up Treatment of Depression and Anxiety: A Global Return on Investment Analysis." *Lancet Psychiatry* 3 (2016): 415–24.

Crowell, A. L., et al. "Long-Term Outcomes of Subcallosal Cingulate Deep Brain Stimulation for Treatment-Resistant Depression." *American Journal of Psychiatry* 176, no. 11 (2019): 949–56.

Davidson, J. T., et al. "Atypical Depression." *Archives of General Psychiatry* 39 (1982): 527–34, p. 532.

de Araujo, D. B., et al. "Seeing with the Eyes Shut: Neural Basis of Enhanced Imagery Following Ayahuasca Ingestion." *Human Brain Mapping* 33, no. 11 (2011): 2550–60.

DiMascio, A., et al. "Differential Symptom Reduction by Drugs and Psychotherapy in Acute Depression." *Archives of General Psychiatry* 36 (1979): 1450–56.

Dinan, T. G., C. Stanton, and J. F. Cryan. "Psychobiotics: A Novel Class of Psychotropic." *Biological Psychiatry* 74 (2013): 720–26.

Dunlop, B. W., and H. S. Mayberg. "Neuroimaging-Based Biomarkers for Treatment Selection in Major-Depressive Disorder." *Dialogues in Clinical Neuroscience* 16 (2014): 479–90.

Engstrom, E. J., and K. Kendler. "Emil Kraepelin: Icon and Reality." *History of Psychiatry* 172, no. 12 (2015): 1190–96.

Freeman, W., and J. W. Watts. *Psychosurgery: On the Treatment of Mental Disorders and Intractable Pain*. Hoboken, NJ: Blackwell Scientific Publications, 1950.

Freud, E., ed. *Letters of Sigmund Freud*. New York: Dover, 1992.

Freud, S. *Cocaine Papers*. New York: Stonehill, 1974.

———. *Introductory Lectures on Psycho-Analysis*. London: Allen & Unwin, 1968.

Garber, J., et al. "Prevention of Depression in At-Risk Adolescents: A Randomized Controlled Trial." *Journal of the American Medical Association* 301, no. 21 (2009): 2215–24.

Gbyl, K., et al. "Cortical Thickness Following Electroconvulsive Therapy in Patients with Depression: A Longitudinal MRI Study." *Acta Psychiatrica Scandinavica* 140, no. 3 (2019): 205–16.

Glover, M. *The Retreat, York: An Early Quaker Experiment in the Treatment of Mental Illness*. York, UK: William Sessions, 1984.

Hippius, H., et al. *The University Department of Psychiatry in Munich: From Kraepelin and His Predecessors to Molecular Psychiatry*. New York: Springer, 2008.

Jones, E. *The Life and Work of Sigmund Freud*. New York: Penguin, 1967.

Jones, K. *Asylums and After: A Revised History of the Mental Health Services: From the Early 18th Century to the 1990s*. New York: Continuum International Publishing Group, 1993.

Khandaker, G., et al. "Association of Serum Interleukin 6 and C-Reactive Protein in Childhood with Depression and Psychosis in Young Adult Life." *JAMA Psychiatry* 71, no. 10 (2014): 1121–28.

Klerman, G. L., et al. *Interpersonal Psychotherapy of Depression*. New York: Basic Books, 1984.

Kline, N. S. *Depression: Its Diagnosis and Treatment*. Basel, Switzerland: Karger, 1969.

———. *From Sad to Glad: Kline on Depression*. New York: Ballantine, 1974.

Kraepelin, E. *Manic Depressive Insanity and Paranoia*. Edinburgh: E. & S. Livingstone, 1921.

———. *Memoirs*. Berlin: Springer-Verlag, 1987.

Layard, R., and D. M. Clark. *Thrive: The Power of Evidence-Based Therapies*. London: Allen Lane, 2014.

Lewis, A. "Melancholia: A Historical Review." *Journal of Mental Science* 80 (1934): 1–42.

Loomer, H. P., J. C. Saunders, and N. S. Kline. "Evaluation of Iproniazid as a Psychic Energizer." *Psychiatric Research Reports* 8 (1957): 129–41.

López-Muñoz, F., and C. Alamo. "Monoaminergic Neurotransmission: The History of the Discovery of Antidepressants from 1950s Until Today." *Current Pharmaceutical Design* 15 (2009): 1563–86.

Makari, G. *Revolution in Mind: The Creation of Psychoanalysis*. New York: Harper, 2008.

Mapother, E. "Discussion on Manic Depressive Psychosis." *British Medical Journal* 2 (1926): 872–79.

Mayberg, H. S. "Frontal Lobe Dysfunction in Secondary Depression." *Journal of Neuropsychiatry and Clinical Neurosciences* 6 (1994): 428–42.

Mayberg, H. S., et al. "Deep Brain Stimulation for Treatment-Resistant Depression." *Neuron* 45 (2005): 651–60.

———. "Reciprocal Limbic-Cortical Function and Negative Mood: Converging PET Findings in Depression and Normal Sadness." *American Journal of Psychiatry* 156, no. 5 (1999): 675–82.

Mendelson, M. *Psychoanalytic Concepts of Depression*. Pasadena, CA: Spectrum, 1974.

Müller, U., P. Fletcher, and H. Steinberg. "The Origin of Pharmacopsychology: Emil Kraepelin's Experiments in Leipzig, Dorpat, and Heidelberg (1882–1892)." *Psychopharmacology* 184 (2006): 131–38.

Palhano-Fontes, F., et al. "Rapid Antidepressant Effects of the Psychedelic Ayahuasca in Treatment-Resistant Depression: A Randomized Placebo-Controlled Trial." *Psychological Medicine* 49 (2018): 655–63.

Palmer, R. L., ed. *Electroconvulsive Therapy: An Appraisal*. New York: Oxford University Press, 1981.

Passione, R. "Italian Psychiatry in an International Context: Ugo Cerletti and the Case of Electroshock." *History of Psychiatry* 15 (2004): 83–104.

Patel, V., et al. "Efficacy and Cost-Effectiveness of Drug and Psychological Treatments for Common Mental Disorders in General Health Care in Goa, India: A Randomised, Controlled Trial." *The Lancet* 361 (2003): 33–39.

Paykel, E. S., et al. "Life Events and Depression: A Controlled Study." *Archives of General Psychiatry* 21 (1969): 753–60.

Platt, M. *Storming the Gates of Bedlam: How Dr. Nathan Kline Transformed the Treatment of Mental Illness*. London: DePew, 2012.

Pletcher, A. "The Discovery of Antidepressants: A Winding Path." *Experientia* 47 (1991): 4–8.

Radden, J. *The Nature of Melancholy: From Aristotle to Kristeva.* New York: Oxford University Press, 2000.

Rasmussen, N. "America's First Amphetamine Epidemic 1929–1971: A Quantitative and Qualitative Retrospective with Implications for the Present." *American Journal of Public Health* 98, no. 6 (2008): 974–85.

———. "Making the First Anti-Depressant: Amphetamine in American Medicine, 1929–1950." *Journal of the History of Medicine and Allied Sciences* 61, no. 3 (2006): 288–323.

Rosner, R. "The 'Splendid Isolation' of Aaron Beck." *Isis* 105, no. 4 (2014): 734–58.

Sackler, A. M., ed. *The Great Physiodynamic Therapies in Psychiatry: An Historical Reappraisal.* New York: P. B. Hoeber, 1956.

Sandler, M. "Monoamine Oxidase Inhibitors in Depression: History and Mythology." *Journal of Psychopharmacology* 4, no. 3 (1990): 136–39.

Shorter, E. *Before Prozac: The Troubled History of Mood Disorders in Psychiatry.* New York: Oxford University Press, 2009.

———. *A History of Psychiatry: From the Era of Asylums to the Age of Prozac.* Hoboken, NJ: John Wiley & Sons, 1997.

———. *How Everyone Became Depressed: The Rise and Fall of the Nervous Breakdown.* New York: Oxford University Press, 2013.

Shorter, E., and D. Healy. *Shock Therapy: A History of Electroconvulsive Treatment in Mental Illness.* New Brunswick, NJ: Rutgers University Press, 2007.

Steinberg, H., and M. C. Angermeyer. "Emil Kraepelin's Years at Dorpat as Professor of Psychiatry in 19th-Century Russia." *History of Psychiatry* 12 (2001): 297–327.

Thase, M. E., M. H. Trivedi, and A. J. Rush. "MAOs in the Contemporary Treatment of Depression." *Neuropsychopharmacology* 12, no. 3 (1995): 185–219.

Valenstein, E. *Great and Desperate Cures.* New York: Basic Books, 1986.

Valles-Colomer, M., et al. "The Neuroactive Potential of the Human Gut Microbiota in Quality of Life and Depression." *Nature Microbiology* 4 (2019): 623–32.

Videbech, P., and B. Ravnkilde. "Hippocampal Volume and Depression: A Meta-Analysis of MRI Studies." *American Journal of Psychiatry* 161 (2004): 1957–66.

Weishaar, M. E. *Aaron T. Beck (Key Figures in Counseling & Psychotherapy Series).* Thousand Oaks, CA: SAGE Publications, 1993.

Weissman, M. M., and J. K. Myers. "Affective Disorders in a U.S. Urban Community." *Archives of General Psychiatry* 35 (1978): 1304–11.

Weissman, M. M., et al. "Affective Disorders in Five United States Communities." *Psychological Medicine* 18 (1988): 141–53.

———. "Cross-National Epidemiology of Major Depression and Bipolar Disorder." *Journal of the American Medical Association* 276, no. 4 (1996): 293–94.

———. "Remissions in Maternal Depression and Child Psychopathology: A Star*D-Child Report." *Journal of the American Medical Association* 295, no. 12 (2006): 1389–1399.

———. "A 30-Year Study of 3 Generations at High Risk and Low Risk for Depression." *JAMA Psychiatry* 73, no. 9 (2016): 970–77.

Acknowledgments

First and foremost, I would like to thank the people who spoke to me about their experiences with depression and its treatment. Most have been anonymized and their identifying details blurred, but their stories are no less real. Each of their experiences is a testament to the struggles faced by people with depression around the world, and how recovery is still possible even in the most hopeless scenarios. This book couldn't have been written without their generosity, kindness, and courage.

I have spoken to many researchers—scientists, psychiatrists, psychologists—and their research and knowledge provides the backbone to this book. Myrna Weissman, Helen Mayberg, Arthur Kleinman, Dixon Chibanda, Vikram Patel, Ted Dinan, John Cryan, Melanie Abas, and Draulio Barros de Araujo are just a few names that have been indispensable throughout this process. (Many researchers have also read this book in various forms of completion; thanks to Brian Leonard, James Rucker, Charles Reynolds, Charlie Kellner, Eric Engstrom, Michael Berk, Carmine Pariante, Chrissie Giles, Siri Carpenter, and Chris Dowrick for their time, comments, and support.) I have also been guided by some welcoming archivists and historians who have given me access to often precious documents and books that were bound before Sigmund Freud was born. I would like to thank Colin Gale from Museum of the Mind, Daniela Finzi at the Freud Museum in Vienna, Ralf Gebhardt and Gerhard Dammann at the psychiatric institute in Münsterlingen, Bill Roberts at the Berkeley Historical Society, and Stuart Moss at the Nathan Kline Institute in New York. A special thanks to the kind woman from Hartheim Castle who drove me back to Linz at a time when all buses had stopped running, then gave me herbal tea and dried apple slices from her parents' orchard.

ACKNOWLEDGMENTS

I wouldn't have been able to make the transition from academia to science writing without the help of some very patient editors. Lina Zeldovich, Ed Lake, Tim De Chant, Michael Marshall, Adrienne Mason, Sally Davies, Richard Fisher, Catherine Brahic, Chrissie Giles, and David Robson all deserve recognition for their teachings. A special mention is required for Siri Carpenter, editor in chief of *The Open Notebook*, who gave me the confidence, freedom, and support to write openly about my experience with depression and suicide as a freelance writer. This book naturally followed our collaboration in spring 2017.

A huge thank you to my editors Sally Howe at Scribner and Robyn Drury at Ebury who read through every draft, answered every email, and offered indispensable suggestions that made this book a much smoother read from start to finish. I would also like to mention Daniel Loedel, former commissioning editor at Scribner, who first saw the potential in my proposal and helped to broaden its scope without losing its central story. I reserve my warmest appreciation and thanks to my agent, Carrie Plitt, an indefatigable ally who helped transform a nebulous set of ideas into a hundred-page proposal fit for publishers. As well as being my agent, she has also been a friend who I can turn to for guidance and counsel.

Support also came from old friends and family over the last few years. Thanks to my brother Rupert for his continued interest and excitement in my work; to my Dad who has read every article I have ever written at least twice (largely because he dropped all science classes before starting his O-levels); to my Mum for her openness when discussing her past; and to my friends Angus and Will who frequently let me sleep on their sofa in South London after I returned from a long day at the British Library and Wellcome Library. Ben, Jim, Ketki, Conal, Cat, and Steve: you have provided refuges for me over the years, both when I was stable and also when I was not. Finally, to Barbara Boyd ("Granny B.") who single-handedly created a welcoming home for so many generations before she passed in 2020.

There is one person I would like to thank most of all: Lucy. She has seen every twist and turn of this book, a personal drama played out in the flats and houses we've called home over the last few years. I depended on her support. She was constant comfort and saw a brighter future even when I couldn't even imagine one. For her love, strength, compassion, and kindness, this book is dedicated to her.

Index

Note: "Freud" used alone refers to Sigmund Freud.

INDEX

INDEX